The Error of Truth

THE ERROR OF TRUTH

How History and Mathematics Came Together to Form Our Character and Shape Our Worldview

STEVEN J. OSTERLIND

Great Clarendon Street, Oxford, OX2 6DP,
United Kingdom

Oxford University Press is a department of the University of Oxford.
It furthers the University's objective of excellence in research, scholarship,
and education by publishing worldwide. Oxford is a registered trade mark of
Oxford University Press in the UK and in certain other countries

© Steven J. Osterlind 2019

The moral rights of the author have been asserted

First Edition published in 2019

Impression: 1

All rights reserved. No part of this publication may be reproduced, stored in
a retrieval system, or transmitted, in any form or by any means, without the
prior permission in writing of Oxford University Press, or as expressly permitted
by law, by licence or under terms agreed with the appropriate reprographics
rights organization. Enquiries concerning reproduction outside the scope of the
above should be sent to the Rights Department, Oxford University Press, at the
address above

You must not circulate this work in any other form
and you must impose this same condition on any acquirer

Published in the United States of America by Oxford University Press
198 Madison Avenue, New York, NY 10016, United States of America

British Library Cataloguing in Publication Data

Data available

Library of Congress Control Number: 2018956453

ISBN 978–0–19–883160–0

DOI: 10.1093/oso/9780198831600.001.0001

Printed and bound by
CPI Group (UK) Ltd, Croydon, CR0 4YY

Links to third party websites are provided by Oxford in good faith and
for information only. Oxford disclaims any responsibility for the materials
contained in any third party website referenced in this work.

I dedicate this book to those who fill my life: my wife, Nancy, and our children Alex, Janey, and Anna, and their families of Trent, Brody, and Billy.

ACKNOWLEDGMENTS

Although my name appears singly on the title page, I could not have written this book without the help of many people. I cannot thank each of them enough for providing valuable information, comments, and advice. Many of them patiently read drafts of chapters and offered extended critiques and reviews. Others offered a quick comment here and there to a question I posed. I am particularly indebted to those who read the manuscript in toto. To all, thank you—especially John Petrocik, Dan Knauss, Richard Robinson, Michel Belsky, and Ze Wang. A special thanks to my editor, Lynne Jensen Lampe, who corrected my composition and offered helpful suggestions during preparation of the work. Thank you, also, to Dan Taber, Oxford University Press editor for mathematics, statistics, and computing, as well as to the other reviewers from OUP. The book is much better because of your help. Any errors, misstatements, and omissions are my own.

Encouragement from family and friends was paramount, including from my wife, Nancy, and our children Alex, Janey, and Anna and their families of Trent, Brody, and Billy. Finally, special appreciation goes to our dear friends Glen and Marilyn Cameron, Ron and Deanna Walkenbach, and Dan and Rhonda Schoenleber. I know, too, that Phil and Sue are always there for me. It is a true blessing to have you in my life. Thank you, sincerely.

CONTENTS

1.	The Remarkable Story	1
2.	The Context	15
3.	Beginning in Observation	23
4.	The Patterns of Large Numbers	43
5.	The Bell Curve Takes Shape	67
6.	Evidence and Probability Data	83
7.	At Least Squares	101
8.	Coming to Everyman	119
9.	Probably a Distribution	143
10.	Average Man	163
11.	Rare Events	181
12.	Regression to the Mean	203
13.	Interrelated and Correlated	227
14.	Discrepancy to Variability	245
15.	Related to Relativity	265
16.	Psychometrics and Psychological Tests	285
17.	The Arts and the Age of the Chip	303
18.	The Sum of It All	317
	Bibliography	325
	Index	337

It is not knowledge, but the act of learning, not possession but the act of getting there, which grants the greatest enjoyment.

Carl Gauss (1808)

CHAPTER 1

The Remarkable Story

This is the story of a remarkable idea that has shaped humankind beyond the scope of any single historical event or clever invention. It recounts the astonishing and unexpected tale of how quantitative thinking was invented and rose to primacy in our lives in the nineteenth and early twentieth centuries, bringing us to an entirely new perspective on what we know about the world and how we know it—even on what we each think about ourselves: our very nature and identity. The widespread and lasting effect of thinking quantitatively is truly that profound.

Quantitative thinking, at its essence, is merely but mightily our inclination to view natural and everyday phenomena through a lens of measurable events: from anticipating the day's weather, to feeling confident that medications will do their job, to sleeping soundly about investments, to being calm when boarding a plane, as well as thousands of routine and not-so-ordinary happenings in our lives. The uncertainty for all kinds of events, previously thought to be random, providential, or unknowable, is now vastly reduced because of our thinking quantitatively.

"Quantification" is the noun I use for quantitative thinking and its concomitant worldview. It implies a level of rational thought and decision-making that is forward-looking since outcomes can be anticipated with more certainty than ever before. Consequently—and significantly—with quantification, we need not impulsively react after something has happened. No longer do we perceive ourselves as beings who are at the whim of fate or providence or who do things because "That's the way it's always been," with a resigned "That's what happens." With this new viewpoint, we live with an internal sense of odds and probability. Quantification is our moment-by-moment reality.

There is a kind of freedom born from all the new choices available to us. It allows people to choose daily and long-term actions among alternatives, knowing the likelihood of a desired result. No longer is impulse or habit our only choice for action. From our quantitatively informed viewpoint, we can take meaningful, advantageous action beforehand. We react to events in new ways and interact with others differently. We are more attentive to predictive markers, as risk is gauged in our brains instantaneously. Uncertainty in outcome still exists, but now it is evaluated. We are more confident in our decisions, less doubtful.

Moreover, with quantification, we have even reformed our assessment of what we previously thought to be unbending fact. The new quantitative thinking has caused us to challenge the veracity of what was once believed to be factual, accurate, and faithful to observation. Everything we know changed because we see everything differently. That is, quantification altered our conception of reality and what is, for us, true—indeed, the Truth. The change in outlook brings an inconsistency or a kind of inaccuracy to our formerly known Truth. As a philosophical conceit, it is an error of Truth. We explore this thought in detail as we go along. In fact, this epistemological shift is interesting to trace.

The new viewpoint—summarized as "quantification"—came into being because of a momentous human achievement: namely, we learned how to measure uncertainty, a direct consequence of the invention and development of the science of probability. For the first time in history, using the methods of probability theory, we can anticipate an outcome for almost any event or phenomenon. For all kinds of things, we can describe odds and the likelihood of occurrence. By calculating correlational relationships, we understand much better how things go together.

Amazingly (yet seemingly now commonplace), with probability we can even make calculations for occurrences of events that have not yet happened: reliable predictions. For instance, we can track hurricanes and forecast their likely direction, anticipate longevity for patients with cancer, and expect particular sales outcomes based on how resources are allocated. Even planned military campaigns can be strategically evaluated in terms of their likelihood of success. Knowing the odds of an outcome is quite ordinary to us today, but it was not always so.

The science of probability has not just increased our understanding of how things work and brought to us the "odds," as if to a gambler. It has fundamentally changed our lives, in both fact and outlook. Because of it, we process information differently and hold mental pictures unlike those held by previous generations. At last, with mathematical precision, we understand much about ourselves and our surroundings that was previously unknown. For instance, because of probability, we know how our behavior and food intake relates to general health, even to specific diseases like diabetes; how school attendance and a student's motivation affect their achievement and opportunities for many years thereafter; or how neighborhood planning is related to humans' interactions.

Most modern electronic devices stem directly from developments in probability theory. The things with which we routinely interact in our daily activities—such as cars, TV sets, refrigerators, smartphones, notebooks, hearing aids, and heart defibrillators—work via the circuitry of their microchips. The binary logic of such circuitry can be traced directly to developments in probability theory. The same is true for the "Bayesian thinking" of artificial intelligence (commonly known as AI), where it formulates a response based on some input (such as with a virtual assistant like Apple's Siri or Microsoft's Cortana, both of which respond to voice commands). These Bayesian "estimations" (as they are called) are an invention of probability theory. Spam filters employ Bayesian estimation. The calculations necessary for such electromechanical reasonings have their foundation in probability theory.

Further, many things we routinely do as an unremarkable daily activity are the direct consequence of probability in our lives. For instance, the mathematical algorithms used to reliably move money between financial institutions (including credit card companies) find their basis in probability assumptions. And a Sudoku puzzle's solution is a Latin square of Fisher's devising. Machine learning and predictive analytics are also forms of knowledge that directly result from probability theory. For an especially prominent example, consider AI's momentous advance in recent years. With AI's "superintelligences," there is now the capacity for a machine to "learn" and even evolve its knowledge from its own learning. AI is Bayesian thinking in action. AI is also the very heart of control systems for the autopilot mode in airplanes. All of this is due to the invention of the methods of probability.

Throughout this book, we will see many instances like these, where the work of probability theory is now alive in our daily life. But—significantly—this book is not a mere recounting of developments in probability theory. Rather, it is a story about us: how we came to be who we are and how we adopted a quantified worldview.

Most of the events of this story happened during a relatively short time frame in world history, a period that brought forth a torrent of breathtaking advancements in mathematics and statistics (the science of making inferences), and, in particular, the ability to measure uncertainty through the developments of probability theory. It presented a cavalcade of new information. The momentous mathematical advances arguably eclipsed such achievements at any other time in history. At first, these were technical accomplishments focused on astronomy and geodesy (i.e., mapping distances across countries or the globe), but soon there arose new fields of activity altogether, particularly for applications of statistics and the quickly developing theory of probability.

Further—and especially relevant to our story—is the fact that these mathematical inventions, discoveries, and developments happened because they *could* happen—and not as the result of randomness in the universe. In all regards—social, cultural, political, educational, and even religious—the time was ripe, and right, for intellectual action, with the activity being spurred on by a number of momentous historical events. The 130 years or so in which our story occurs are generally acknowledged to be the most productive in human history, eclipsing all others up to then. During our period, Western societies made significant and innovative advances in virtually all fields of knowledge: medicine, engineering, business, economics, and education, as well as the literary and performing arts. History gave a kind of tacit assent to the quantitative developments—with profound effect.

Three specific historical events gave impetus to quantification. They are (1) the mid and late periods of the Enlightenment and its rationalism (as we see in a bit, the Enlightenment is often divided into three periods); (2) the fall of the European monarchies as a result of the French Revolution, which was caused by the Enlightenment and had an impact far beyond the borders of France, and (3) the Industrial Revolution, both the "First Industrial Revolution" in England and its re-emergence in America as the "Second Industrial Revolution." These things gave a context in which quantification *could* develop. Their meaning derives from their context. In a very real sense, understanding the times is prerequisite to understanding the phenomena of quantification.

Of course, in practice, we almost never think about actually assessing the odds by employing the statistical methods needed to make such predictions. Obviously, few of us have the training or the inclination to make those calculations. But, with quantitative thinking as our outlook, we do routinely imagine that these events are calculable in some way or another. Picturing daily events as having calculable underpinnings is so routine in our lives that we scarcely give it a second thought. Yet, what is ordinary to us now, was relatively new then—undreamed of by previous generations.

It is easy to appreciate that the new knowledge of predicting things with reasonable accuracy resulted in our altering our behavior, but the realization that this new ability has changed us at our core, far beyond actions we have taken in and of themselves, is startling. It has fundamentally transformed what we know about the world, and even what we think of ourselves, into something entirely new, giving us a quantitative perspective on all things. It's an epic shift in thinking. Yet, few recognize the complete transformation in natural outlook for what it is. Its perspicacity is stunningly subtle and eludes recognition unless intentionally examined. An entire populace changed…and did not even recognize the shift in their thinking until it was so fully incorporated into their daily lives as to seem ordinary. This was an astounding metamorphosis. It is akin to a toddler learning to speak for the first time—speech advances so simply and subtly that the child does not recognize his own learning until it is done. So, too, was it with quantitative thinking.

Further, as potent as was the influence of quantitative thinking on forming our worldview, the story of its invention and settling into our neural pathways grows more stunning when we learn that the formal methods of quantitative analysis—the measurement of uncertainty—stem from the work of a very few men and women (fewer than about fifty principals), each of almost unimaginably high intellect and most working in about the same time period, roughly from the early 1800s to about 1930. Many of the major characters knew one another personally or at least were aware of the others' corresponding work in probability theory or statistics, and more broadly in mathematics. Yet, aside from knowing one another and their reputation in intellectual circles, most of these individuals worked in relative obscurity and were not popularly known. Even today, most of these individuals are still not well known, except to biographers and scholars. But, as we shall see, their influence on us today is nothing short of astounding. In large part, this is what makes the story of quantification so interesting.

I will not list all of these individuals now because we meet them later on; but, to give a taste of what is to come, some prominent figures include Jacob Bernoulli, Thomas Bayes, Adrien-Marie Legendre, Carl Gauss, Pierre-Simon Laplace, Sir Francis Galton, Karl Pearson, Sir Ronald Fisher, and Albert Einstein. And just as important to our story are certain others who worked outside of mathematics and probability theory but still played a role in the story of quantification. These individuals are likely familiar to you but you may not have expected them to be included here—people like Benjamin Franklin, Wolfgang Amadeus Mozart, and even the poet–philosopher Dr. Samuel Johnson. As we go along, I mention others, too: politicians, composers, economists, and the like. My reason for including a given person at all is that he or she contributed to the story of quantification. Our story has a single focus, which was brought to us by many persons of varied backgrounds.

To state the obvious, this story could be expanded to include many other important people, adding to my selection the minor players and yet more who may have influenced the major contributors. But to include them all would make this book so big as to fill the Long Room at Trinity College, Dublin (a wonderful multistory, musty Harry Potter-ish room that is a huge repository of renowned classic works, including some surviving pages from the Celtic Book of Kells). I limit my selection to those whose accomplishments have had the greatest influence on quantification. And, of course, scholarship is not an insular activity; all the characters of this story had their own individual influencers. I mention some of their predecessors—from Newton to Einstein—along the way when they add to our story.

Contrast a transformed outlook for ordinary people with that held even in the century just before quantification (roughly, the eighteenth century). Although many folks lived routine and often orderly lives, the world was a tumultuous place rife with unknowable events that demanded an after-the-fact reaction. During this time, rulers (often monarchs and despots) made decisions that had direct and often daily impact. Queen Victoria, Napoleon Bonaparte, and Peter the Great were never far from people's thoughts as, seemingly, their every whim was consequential and beyond any control. The influence of the church, and particularly that of the local priest, was also profound and directed people's daily actions to a degree that, today, most would find unacceptable. Ordinary folks had no alternative but to react to what was happening around them—and, accordingly, their mindset was one of responding to events rather than anticipating them and making informed choices beforehand.

Especially when experiencing natural phenomena, the common belief was that things were directed by an unpredictable god and that anticipating them ahead of time was impossible. One natural event in particular was widely viewed as both the literal and the symbolic manifestation of this view: the Great Lisbon Earthquake of 1755. The full tragedy was actually composed of three related disasters: the earthquake itself; a devastating tsunami, which followed forty minutes later; and then a number of small fires that merged into a horrifying, citywide firestorm which lasted for weeks. All told, perhaps as many as 100,000 souls lost their lives in the catastrophe—about half the population of the city and its environs.

On Saturday morning, November 1, nearly everyone in Lisbon was in church for All Saints' Day. Lighted candles of commemoration were everywhere. People were likely listening to a sermon or sitting in quiet reverence, when suddenly there was an upset. The shaking came first to their feet, then moved up from the floor to their seats; next, it came to the building itself. People stood up and looked about with uncomprehending anxiety. It took a lifetime of seconds to realize that this was an earthquake. It lasted six minutes. Its intensity has been estimated to have been, at today's Richter magnitude scale, 8.5 to 9.0. As the buildings and houses began to crumble around them, individuals screamed and tried to find an exit. Utter panic ensued.

Amid the confusion, people had what seemed to be a rational reaction: run outside, away from the toppling buildings and toward the harbor, a large area without buildings or structures. Lisbon is located in a shallow, bowl-shaped valley that opens on one side to the sea. The shaking finally stopped, and everyone began to take stock of the devastation, doubtless looking for their loved ones. As the minutes passed, more and more people arrived in the open city center, near the harbor. Chaos still reigned, but, gradually, pockets of calm began to emerge here and there.

Then, a full forty minutes later, something happened that was yet more unthinkable: a colossal wall of water, estimated to have been from forty- to one-hundred-feet high, rolled in from the sea. A tsunami roared upon them, pushing back into Lisbon all the ships in the harbor, together with sand, mud, and all the pieces of the collapsed buildings, relentlessly sweeping the earthquake survivors back into the rubble of the fallen churches and buildings. Everything was moving. When the tsunami's surge reached the far side of Lisbon, it turned around and receded back to the sea, this time pulling debris and bodies along with it. Sea captains later reported seeing hundreds of floating bodies and wreckage dozens of miles offshore.

Next, the fires came. A million All Saints' Day candles ignited the wood sticks of the broken structures. Small fires, looking for oxygen, drew the winds from the valley's sides downward until the small fires grew into a citywide firestorm. Lisbon was now a single huge conflagration. The fire continued for weeks. It was like a scene from Dante's *Inferno*. Figure 1.1 is a copper engraving of the Great Lisbon Earthquake, showing the city in flames, and a tsunami overwhelming the ships in the harbor.

News of the Great Lisbon Earthquake quickly spread across the Western world. Everyone felt its effects. One wide-ranging consequence was that it caused people to think about their place in the world. Questions of theodicy (why a good god would allow evil) formed a major topic of discussion, not only for philosophers and clergymen but for everyone. Other questions arose, too,

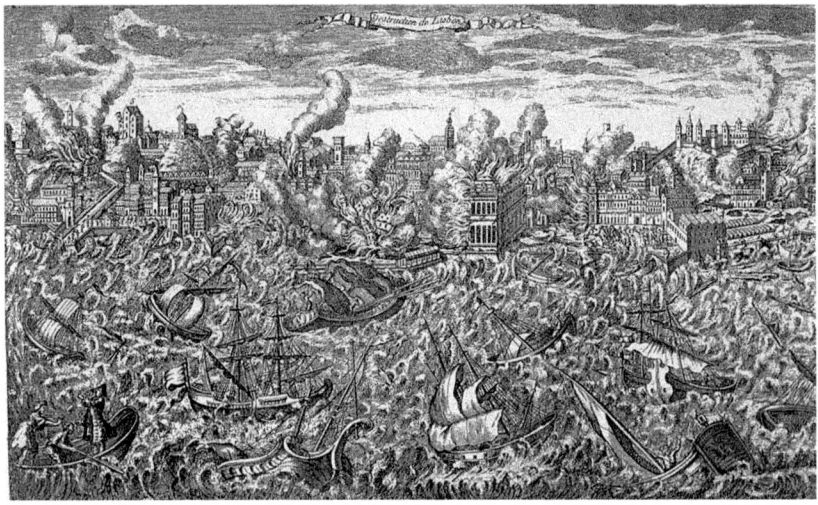

Figure 1.1 Copper engraving of *The Great Lisbon Earthquake*
(*Source*: http://commons.wikimedia.org/wiki/Category:Public_domain)

with much ensuing discussion. One such area for concern was whether there was anything that could have been done to help such a situation. Of course, people realized that no one could forestall an earthquake, but they did begin to question their perspective—not only on natural phenomena but on their internal thoughts. If one's worldview could be more based on understanding rather than reacting, one would gain a certain sense of internal control.

As we see in our story of quantification, the momentous developments in probability theory—the measurement of uncertainty—came into people's lives at this time, affecting their whole lives—soul, mind, and spirit. With this, it seemed, people gained a kind of cognitive refuge in which to place their rational thoughts. Influenced by the philosophy of David Hume and John Locke, people tried to employ their reasoning powers to understand and deal with the natural world. Certainly, with quantifying techniques at hand, something could be known about the future; it might be possible to reasonably anticipate the likely outcome of events—both ordinary ones and extraordinary ones.

The Great Lisbon Earthquake is thus often cited as giving impetus to the Enlightenment period (the Age of Reason) that followed, a time when many events in our story occurred.

Clearly, quantification contributed to a changed everything. With it, people saw that they were in control of their lives to a much greater degree than before because, in this new epoch, events could be measured, scaled, empirically analyzed, and often anticipated. No longer were people destined to react to events only afterward. All kinds of dangers, pitfalls, and unfortunate trials could be knowingly avoided. With quantification, one sees the world as controllable, to a degree previously unknown in human history. The social, cultural, political, and scientific realms aligned in importance to enable this new mindset—the mindset we hold today.

Hence, in a nutshell, this book tells an interesting story about some extraordinary people and the truly profound consequence of their work. We explore how they developed the methods of measuring uncertainty and how that (in its historical context) brought quantification to the fore for everyone, everywhere—and, particularly, how, as a result, we came to a changed perspective on what we know about the world and what we think of ourselves.

****

The German language has an especially apt word that captures what happened to our outlook by this reformation of modern thought. The word is *Weltanschauung*, meaning "a worldview," denoting the way a person looks at the phenomenon of life as a whole. The noun implies more than its literal translation. A *Weltanschauung* is also an internal sense of self. It shapes all that is seen by giving a guiding perspective to the senses. The lens and filters that we employ in taking in and interpreting sensory input constitute the crucial heart of a *Weltanschauung*.

Grammatically, the term is a contraction of *Welt* (worldwide) and *anschauen* (to look at), implying concentrated attention and discernment. It is more thought-provoking than the simple German word *blicken* (to see). There is a depth to the word *anschauen*.

While the word *Weltanschauung* may appear officious or pompous when used as common English, it is preferred in educational conversation over the simple term "worldview" because it conveys a more comprehensive idea. The *Oxford English Dictionary* cites many examples, both historical and contemporary, of using *Weltanschauung* in context.

In this book, we see, by the introduction of one noteworthy idea—namely, quantification—how humankind came to a new *Weltanschauung*. This kind of transformation of a worldview—what we know about the world and what we think of ourselves—is very rare in human history, but it has happened before. One such occurrence was in late medieval times when Copernicus published his work on a heliocentric (sun-centered) universe. He posited (with mathematical calculation) that the earth revolves around the sun, thus challenging the understanding of a Ptolemaic universe where Earth is the central and principal object in the universe, a view held as impossible to deny by the Roman Catholic Church.

The prior truth was eventually transmuted and held to be in error. Copernicus's discovery evolved humankind's thinking about what is known about the world, encompassing even what we think of ourselves. This Copernican revolution was just like the change experienced with the switch to quantitative thinking in the nineteenth and early twentieth centuries: a fresh *Weltanschauung*. It was, indeed, a rare event in history.

Copernicus's work changed people's thinking, but it is useful to realize that a restructured worldview does not appear with every discovery or invention—even the great ones with a profound effect all their own. The difference between making a new discovery or invention and changing one's worldview is the difference between advancing knowledge, on the one hand, and embracing a new perspective on what is true or real in our daily lives, on the other. It is a qualitative distinction.

Most discoveries and inventions fall into the former category; that is, they extend our knowledge and often make us change the way we *do* things, but they do not alter our way of *thinking* to a degree that modifies our understanding of everything around us. We may change our daily behavior, but we do not alter our view of the world. Typically, new inventions and discoveries offer a route to make our lives safer or more comfortable, or somehow better. Many are significant, of course, but our general viewpoint remains unchanged.

To understand this point, contrast Copernicus's work with Louis Pasteur's experiments on the relationship between germs and disease in the late nineteenth century. Pasteur momentously advanced our understanding in medicine, and his discoveries changed how diseases were treated, making cures and medical management vastly more effective. Today, as a direct result of Pasteur's studies, we understand that diseases are caused by germ-based pathogens, such as bacteria or viruses. Doubtless, his work improved lives and has saved millions of people. But Pasteur did not bring about a new way of thinking. His work advanced medicine so much that we even changed our behavior so as to avoid harmful germs and bacteria, but he did not cause the paradigm shift of a reformed *Weltanschauung*. In contrast, Copernicus's ideas *did* transform our thinking.

The new quantitative perspective applies to humankind generally—both the illiterate person and the person with many years of schooling—as well as those with talents of all forms: writers, artists, musicians, craftspeople, academics, athletes, businesspeople, and the many who simply survive daily trials by their street smarts. We act as a community in changing a worldview. Accordingly, the quantification of our lives is due to neither geography, nor language, nor ethnicity. In our story, these are not, in and of themselves, determinants.

And, of course, evolving to our quantitative viewpoint is not the result of many more people simply learning about quantifiable subjects, or more people studying statistical methods or probability theory. That did not happen. Taken as a whole, there was no increase in the number of people learning the newly developed methods of probability theory. After all, most people do not know many things explicitly mathematical, nor do they overtly care. They are simply, but meaningfully, going about their daily lives. At most, they likely use simple arithmetic only occasionally—say, at a restaurant when viewing the bill or when figuring the taxes they owe.

You may remember the classic 1986 Francis Ford Coppola movie *Peggy Sue Got Married*. In it, a thirty-something woman is magically transported back to her high school, where the exasperated math teacher implores the unmotivated students to study their algebra assignment. Kathleen Turner turns to her much younger classmates (they see her only as another student their own age) and says, "You can forget this. Trust me. You'll never use it!" In a literal sense, she is right. But the fact of not knowing mathematics is different from having a quantitative perspective on the world. Clearly, then, the tale of how an entire population came to quantitative thinking is not explained by suggesting that a large, new group of people somehow learned more math than they had previously known. Regardless, as we shall see in our story, their view of the world has been informed by a quantitative outlook on daily events—as a community, their worldview is now different from what they had before.

Realize, too, that our adopting this new quantitative outlook occurred very recently in human history—within the past two hundred years or so, only six or seven generations back: from roughly the early 1800s to shortly after the end of World War I (WWI), around 1920. For perspective, modern civilization as we know it is thought to be at least 6,000 years old, and perhaps much older. We know, too, that the species *Homo sapiens* was around much earlier. From this perspective, our transformation in thought is, indeed, recent.

This is our story, and, over the course of the next seventeen chapters, we shall see how this extraordinary, almost unknown, and certainly underreported transformation happened.

****

A few relevant notes about my approach to writing this story are as follows. Despite the fact that the tale of quantitative thinking revolves around probability and measuring uncertainty, this book is decidedly not a text on quantitative methods. I mean to tell an amazing story about a handful of extraordinary people, and I write to a general audience. In telling the story, I deliberately avoid technical description, and I do not use the jargon of mathematics. In fact, I do quite the opposite: I neither use math to describe topics nor present formulas as explanation. Instead, I stick to describing things in plain English.

However, it is sophisticated reading, and, at some points along the way, I do explain the *hows* and *whys* of particular methods when it is necessary to advance the storyline. I present a number of graphics (mostly figures and tables, but a few pictures, too) and, although some of them appear technical, they are only included to tell the story more clearly. Don't worry about following them technically. I do use the words "algebra" and "calculus" when mentioning many probability models and statistical methods, but I mean these terms only as nouns to place the description in context. No knowledge of algebra or calculus is necessary to follow the narrative—again, just the opposite is true. I write for your enjoyment, not to be didactic. I do mention some measurement terms such as inference, odds, probability, likelihood, and, particularly, least squares, with associated ideas like "regression" and "general linear model." Whenever necessary, I explain them simply. I think you will follow these terms and ideas without a problem.

To complete this thought, let me say that this book is not an account of philosophy, although I have already mentioned Truth as a large idea shaping our worldview. I do use some terms borrowed from philosophy, such as "ontology" and "epistemology," but they are easily understood, I am sure.

While I eschew mathematical description and formulas in telling the story of quantification, there *is* a set of mathematical principles behind it all. After all, things are quantifiable only by some kind of calculation. Primarily, I mean quantification to be most closely related to probability theory. This is the math of odds and likelihood, as generally expressed in ratios and distributions. It overlaps statistics quite a bit, too, because the procedures of one are often useful to the other, but they are not exactly the same thing.

Statistics as a discipline is more connected to research methodology, empirical testing of hypotheses, or modeling events mathematically. Probability is related, although less so, as it shows up in most statistical contexts, such as in descriptions of research methods and in many studies using statistical analyses. Both disciplines rely upon algebra, calculus, in particular, and, to a lesser degree, analytic geometry. I try to employ correct terminology in the telling of our tale, but the line between statistics and probability is more one of perspective and purpose than of identifying distinct disciplines.

And, finally, let me state what this book does not cover: while I recognize, appreciate, and admire the giants of early mathematics in all its forms, this book is not a history of early mathematics (for that, see *The World of Mathematics: A Four-Volume Set* by James R. Newman [Newman 1956]), nor is it an overview of modern mathematics (for that, see *The Princeton Companion to Mathematics* edited by Timothy Gowers [Gowers, Barrow-Green, and Leader 2008]), statistics (for that, see *The History of Statistics* by Stephen Stigler [Stigler 1986]), or even of our primary focus, probability theory (for that, see *Classic Topics on the History of Mathematical Statistics: From Laplace to More Recent Times* by Prakash Gorroochurn [Gorroochurn 2016a]). Quantification of knowledge and thought is a guiding theme of our story, and that happened because of the amazing accomplishments of a handful of brilliant (mostly) mathematicians, but this story is about us, and not a history of mathematics or of probability theory.

It is to the remarkable story that we now turn.

CHAPTER 2

The Context

Our story lasts about 130 years or so, beginning during the last (but most significant) years of the Enlightenment through the first third of the twentieth century, roughly the 1790s through the 1920s, covering a "long view" of the nineteenth century. The proximity and rate of recurrence of so many historical milestones in this time period has led some historians to figuratively modify the century's dates, saying it really commenced in 1815 with the Congress of Vienna (ending the drawn-out French Revolution and the era of Napoleon) and was not entirely complete until the outbreak of WWI in 1914. This is the so-called long century. The dates are approximate, since foregoing events are causal to those I describe, and, of course, societal changes leading to a transformed worldview are not cleanly bounded.

In this short chapter, we look briefly at the world during this time to understand the social, cultural, and political milieu in which our central characters were working. With a few notable exceptions, our principals are the persons who invented probability theory and developed the methods of probability estimation. They were individually brilliant and extraordinary in temperament—but, for us who are unraveling the story of quantification, we realize, too, that they lived and worked in exceptional times. We meet the first of these individuals in Chapter 3; however, in this chapter, we look at their historical context.

In every sense, the historical context gives meaning to their accomplishments. Only by knowing the contemporaneous history—and especially by realizing what these historical events connote and imply for the people living then—can we truly understand the importance of what our principals accomplished. I emphasize this point specifically because, throughout the story of

The Error of Truth. Steven J. Osterlind. Oxford University Press (2019). © Steven J. Osterlind 2019.
DOI: 10.1093/oso/9780198831600.001.0001

quantification, we will place the work of the individuals we examine in the context of a relevant history.

Geography, too, plays a role in our understanding of how quantification came about, because the influences we discuss are primarily located in Western Europe, the Nordic countries, the Americas (mostly North America), and, to a lesser degree, Eurasia. This is not to ignore or dismiss important discoveries and advancements in the Far East, the Middle East, or Africa. Those may tell a different tale, which is left now for another time.

****

The era of our story was scarcely ordinary by any measure. More change occurred during the years of the long century than in virtually any of the prior centuries, possibly since 9400–9300 BCE, when humans living in Mesopotamia developed agriculture by planting seeds, thereby enabling them to stay in one place for a long time and cease being simple peripatetic hunter–gatherers. This signature event is cited by many historians as possibly the single most momentous action in all history. After that, set aside any debate of which century saw humans change their behaviors the most, there is general consensus that the long century of our story was truly an exceptional period.

Further, it is these extraordinary historical events taking place during our period—that is, the vast societal and cultural changes—that made possible the remarkable developments in mathematics we shall discuss, the pace and scope of which are unprecedented in history. Because of them, it can be said that history gave tacit assent to quantification.

To buttress this point, a notable historian of mathematics, who spent more than twenty years researching its history and eventually compiling his effort into a four-volume masterpiece, said of statistics: "Statistics shot up like Jack's beanstalk in the present century. And as fast as the theory [of probability] has developed . . . the anthropologist must survey a literature so new and so vast that even the expert can scarcely comprehend it" (Newman 1956).

Echoing Newman's comment on the rapid growth of inventions and development in just the quantitative sciences, one may conclude that, from the beginning to today, there has never been such a cascade of numerical achievements, both in the speed with which each invention came upon the scene and in their scope, bringing numeracy to fields never before involved. Initially, the math was strictly formulaic (albeit sophisticated, with developments in calculus and trigonometry) and primarily tied to advances in astronomy and geodesy; but, soon, they broadened in scope to give a much better understanding of how numbers operate generally. It is during this period that many numerical ideas were codified into theorems (principles of mathematics that underlay a range of applications in various formulas) that are universal across all the disciplines of mathematics.

Chief among this work was the invention of probability theory, and its theorems. At first, the theory was just an application of statistical arrangements, most notably as the normal, bell-shaped curve. But soon further inventions came along, like the method of least squares, as well as density functions, regressions, and correlations. These developments grew in sophistication and were soon applied more broadly. Some examples are the formulation of conditional probabilities and widely applicable notion of "Bayesian thinking" (a systematic way of using observable evidence to modify prior beliefs). Statistical testing by inference was invented at this time, too. (Along the way, we will explore each of these interesting developments.)

Through all this, we were learning how to measure uncertainty, and, as a consequence, "quantification" was forming—we were evolving with respect to our perspective on the world. This evolution in our worldview is this history's lasting effect. Literally, it has come to help shape who we are today.

The roots of the unprecedented technical growth in mathematics during this time are usually traced back to 1680s' England, where, in the span of three years, Isaac Newton published his stunningly brilliant work *Philosophiæ Naturalis Principia Mathematica* (or the *Principia*, for short; Newton, Motte, and Chittenden 1846). This monumental three-volume work describes the theory of universal gravitation and Newton's proof for it by formulating modern calculus—two amazing accomplishments in a single work. One cannot overstate the importance of this masterpiece of scholarship. Even beyond its twin noteworthy contributions to astronomy and mathematics, it defined science generally and thereby has given direction to nearly all quantitatively oriented scholarship ever after.

While the *Principia* was never widely read (owing to its technical heft and, of course, the fact that it was originally written in Latin, which few persons could read). According to Einstein, it was the most important book ever written. In Chapter 3, we will explore this work to see more of its contents and importance, and we will see its influence pop up again and again on the work of others.

Also, throughout this book, I will emphasize the point mentioned earlier— that these achievements in mathematics leading to our quantified worldview happened because they *could* happen. They did not come about apart from their historical setting. The times were ripe for discovery and invention because of an emphasis upon reason and a questioning of the previous epistemologies. The period of our story is highlighted by the fact that it takes place during three important periods in history: (1) the last years of the Enlightenment, (2) the French Revolution and the post-Napoleonic era, and (3) the British/European/American Industrial Revolution.

Even nature itself seems to have coordinated with history to enable the developments in our story. It is only a slight oversimplification, but not at all inaccurate, to suggest that the Great Lisbon Earthquake (along with such human achievement as Newton's *Principia*) set the people of the time on a path to reason with deliberate cognitive intent and to learn as much as possible about the world around them.

Our story opens with the late Enlightenment itself, a period wherein reason and science grew in importance as sources of truth. The German philosopher Immanuel Kant summed up the era's sentiment in his powerful 1784 essay *What Is Enlightenment?* by saying, "Dare to know! Have courage to use your own reason!" (Kant and Beck 1995). The Enlightenment's important seventeenth-century precursors include Francis Bacon, Voltaire, and René Descartes, as well as the key natural philosophers of the Scientific Revolution: Galileo,

Figure 2.1 *The Weimar Court of the Muses* by Theobald von Oer
(*Source*: http://commons.wikimedia.org/wiki/Category:Public_domain)

Kepler, and Leibniz. Most intellectuals of the day were deeply influenced by John Locke's *Essay Concerning Human Understanding* (Locke and Nidditch 1987). Often, this period is called the Age of Reason.

The Age of Reason is characterized in a famous painting titled *The Weimar Court of the Muses* by Theobald von Oer (shown in Figure 2.1). In it, Oer incorporates several Enlightenment motifs, such as using bold colors in painting and depicting groups of people—men, women, and children, and nobles and ordinary folks alike—gathering in a new-found zest for learning. Purposefully, Oer includes the likeness of several important figures of the day, particularly Goethe, the German writer and statesman who promoted the belief that individuals should engage in their own learning. An oft-cited quote of Goethe is "Knowing is not enough; we must apply. Willing is not enough; we must do."

In the remainder of this chapter, I highlight a few of the most important historical, political, and social happenings of the long century in which the people who invented the mathematics that changed us so profoundly lived and worked. Obviously, the events mentioned here are noted only briefly, since extended description could (in fact, *has*) fill thousands of books.

****

During the nineteenth century, there were an astonishing number of inventions and innovations in nearly every field of human endeavor. We look at some of them here momentarily; however, for context, I will first mention some activities of governments of the day, since they had a lot to do with how individual citizens came to their inventions and innovations.

Near the close of the eighteenth century, it seems that tensions and conflicts between countries were everywhere. To start, the British state rose to prominence, with its global expansion (some call it "domination") during this time. The British Empire carried its version of modernity into far-off corners of the globe, but at a cost of subjugating native people who lost their own sovereignty. More positively, this was also the time during which the British brought more education and some improved healthcare conditions to the countries under her empire. And she brought an end to the tyranny of the Barbary pirates and thereby opened the seas worldwide to safe travel and universal trade.

In Eurasia, the Ottoman Empire and the Holy Roman Empire came to their respective ends, although fighting continued ceaselessly in the Middle East and Islamic countries—not only between ill-defined frontiers but especially among tribes and other cultural and ethnic factions. In China, the Qing dynasty obliviously exposed its corruption to the masses; as a result, unrest grew, leading eventually to the end of all Chinese emperors. (Deplorably, the Chinese communist revolutionary Mao Tse-tung [Zedong] followed, and his savagery was even worse.)

In Russia and other Slavic regions, the Romanov dynasty continued its autocratic rule and retarded its own country's development for several generations. America had its wars of independence and then its Civil War, which was followed by a westward expansion that took lands away from many Native American peoples. There was a bitter civil war in Spain, too.

Two particular inventions during this century changed the face of war across the globe. These were dynamite (by Alfred Nobel) and barbed wire (by Joseph Glidden). As soon as they were known, governments put these creations to use in their military campaigns, making their warring more intense than ever. Because these inventions brought a qualitatively different kind of fighting to wars, relationships between many countries were altered, sometimes for a common good but more often to their mutual peril and devastation. A parallel change in warfare (with a similar effect of changed relations between countries) occurred in the twentieth century, with the invention, and almost immediate use thereafter, of the atomic bomb.

During this century, the slave trade—the most horrific human injustice in all history—finally wound down to a trickle in Western societies. In 1807, Britain passed the "Abolition of the Slave Trade Act," and the other European countries soon followed. In the United States, after the US Civil War, it was absolutely abolished when the Thirteenth Amendment was formally adopted into the US Constitution, guaranteeing that "neither slavery nor involuntary servitude... shall exist within the United States, or any place subject to their jurisdiction." And, citing its impetus from the United States, Russia, too, officially abolished slavery, although slavery continued to exist underground there on a large scale well into the next century. But not all governments and countries were so quick to act, for slavery continued almost unabated in the Arab countries, across most of Africa (where the slave trade actually expanded at this time), and, to some degree, in China.

Unfortunately, there is a certain timelessness to this disgrace, because slavery and other forms of human trafficking continue today in possibly as many as 155 countries, with mostly young persons (particularly young women) being kidnapped, primarily from the Arab countries, southern parts of Africa, and southern Asia, as reported by the United Nations in a 2014 report (UNODC 2014).

But, when these wars and political rivalries subsided during the long century, much good came forth. Specifically, the living standards for most people materially improved: they lived with more conveniences, better healthcare, improved education, and more personal safety. In the sciences, it was a time of great advances, with an almost uncountable number of discoveries and inventions. As mentioned in Chapter 1, Louis Pasteur discovered that most infectious diseases are caused by germs, an idea he called the "germ theory of disease," and he invented pasteurization. John Tyndall first demonstrated the principles of fiber optics, thereby laying the ground work for all kinds of advances in electronic communication. And Alexander Parkes first made plastic, which revolutionized the scope and breadth of products that could be made economically.

In other areas, too, significant innovations materialized. Soft drinks were invented and immediately became popular worldwide. In particular, John Pemberton devised a special drink by mixing dark cola with a small amount of cocaine (derived from mashing coca leaves), marketing it as a tonic, and calling it Coca-Cola. (In 1904, the United States banned plain cocaine from all drinks. In response, the Coca-Cola company reduced the amount of cocaine it used in its drink, although Coca-Cola didn't become completely cocaine-free until 1929. Today, the recipe for Coca-Cola remains one of manufacturing's best kept trade secrets.)

As a complement, a machine was invented that would twist paper into a small tube, making the first drinking straw. While this was a minor invention, its success spurred many others to develop machines for making all kinds of ordinary things, from paintbrushes to folding cardboard boxes. Since then, simple machinery has become ubiquitous.

The century also saw for the first time a motion picture (popularly accepted almost immediately, but available then only in big cities) and a high-wheeled bicycle frame (a contraption thought by many first-time observers to be singularly weird looking. It was famously dangerous to ride: modern, low-wheeled bicycles were first called "safety cycles"). Samuel Morse devised his code for electric signals by tapping out long and short sounds, enabling long-distance communication via the telegraph. The telephone was patented soon thereafter. Electricity became widely available in cities, both across the Continent and in the United States (although its spread to rural areas was slow), spurring the growth of the steel industry, which led to the spread of the Industrial Revolution from England, where it originally started, to the United States.

In the social sciences, Sigmund Freud, working in Austria most of his life, postulated theories of personality and invented psychoanalysis. Accomplishments in literature, painting, and the performing arts at the time were just as profuse and important as those in the sciences, although these efforts were overshadowed by the powerful intellectual and artistic spurt of Europe's Renaissance.

Perhaps most important to the majority of people is the fact that with all these new developments, the physical chores of daily existence, such as when getting enough to eat, securing a safe, clean place to sleep, and caring for children, became a lot easier.

With all this, the nineteenth century was obviously one of the most productive times in human history.

The inventions, discoveries, and advancements of the twentieth century matched those of the nineteenth. For our focus on the notion of quantitative thinking, consider just the beginning third or so of the twentieth century (leaving out the computer and the Internet, because, by then, quantification as a viewpoint was already firmly set). Understandably, any list of accomplishments, discoveries, and inventions in the century would be miles long, including television, the airplane, the automobile, rocketry, nuclear power, the submarine, and antibiotics as just a few. There are hundreds, even thousands, more.

Of course, wars and conflicts raged in this century, too, most glaringly WWI and its reemergence into World War II (WWII). Perhaps more than any other twentieth-century development, this single event affected the lives of people worldwide. Incredibly, more than 3 percent of the world's population (about two billion people at the time) were killed at this time: ten to twelve million persons during WWI, and at least sixty million during WWII. Doubtless, more souls were killed during these world wars than in any other conflict in human history—a statistic so horrifying that it is difficult to grasp.

This is the milieu in which the amazing individuals who developed the procedures for measuring uncertainty lived. Their work led people everywhere to adopt a new worldview. In this context unfolds the astonishing and unexpected tale of how quantitative thinking was invented and rose to primacy in our lives.

Now, the journey of our story begins.

CHAPTER 3

Beginning in Observation

The eighteenth-century poet and philosopher Samuel Johnson began his epic poem "The Vanity of Human Wishes" with these lines:

> Let Observation with extensive view,
> Survey humankind, from China to Peru.
> (Johnson, "The Vanity of Human
> Wishes," stanza 1)

For our inquiry into how we came to a quantitative *Weltanschauung* (worldview), Dr. Johnson's words are more than poetic—they are instructive. The lines carry for us the thought that observation is key to making informed statements and conclusions. Observation forms the basis for systematic inference, the core of much scientific inquiry. In addition, observation can be applied to nearly anything. It has, says Dr. Johnson, "extensive view... from China to Peru," metaphorically proclaiming its universality of purpose. Further, one can employ such observational purpose intentionally and for a sustained period. Observation here is active, thoughtful, and deliberate, not casual. Accordingly, we learn to apply observation as logical inquiry.

Observation begins our story of the evolution of humankind's worldview from one of a qualitative perspective to one of quantification.

Common *observing*—an automatic, ordinary part of daily existence—broadened in scope to become *observation*—an empirical tool for scientific inquiry. Observation was, at this time, recognized by the greatest of scientists as a pragmatic part of their work, and it gave impetus to a sudden and immense advance in mathematics, one that sharply focused on the invention of methods to calculate uncertainty as a predictable event. In other words, gauging the odds and

The Error of Truth. Steven J. Osterlind. Oxford University Press (2019). © Steven J. Osterlind 2019.
DOI: 10.1093/oso/9780198831600.001.0001

likelihood of an event is now an evaluable problem with an empirical solution—and significantly, it all began with observation.

What sets these new methods apart from earlier forecasts and predictions was that they were based in mathematics, have estimable accuracy, and are reliable. Over time, the methods of these sciences have become ever more accurate and reliable.

Dr. Johnson seems to anticipate this operational use of observation when he instructs his readers to use observation for a particular purpose: namely, to "survey humankind." In doing so he invites scientific inquiry. Observation is now methodology. In this context, observation is both affective, as in Dr. Johnson's poem, and effective when used in scientific inquiry. G. K. Chesterton, the influential nineteenth-century journalist and social critic, called himself "an observationalist." He bridged the two disparate ways of engaging the world by saying, "The difference between the poet and the mathematician is that the poet tries to get his head into the heavens while the mathematician tries to get the heavens into his head" (Chesterton 1994).

Early astronomers, including Ptolemy, Aristotle, Copernicus, and Galileo, realized the importance of observing phenomena accurately. From the start, they took seriously the role of observation in their studies. Galileo, in particular, had profound insight regarding observation, being one of the first to appreciate variations between his measurements. He systematically studied such differences, attributing them to his varied observations. To account for this, Galileo (and these other early scientists) typically used the "best" observation: that is, the values they employed in their work were from the single observational occasion that was least influenced by immediate distractions or irrelevancies. If, say, an astronomer was observing the relationship between the moon and a given star, he might use in his calculations the numbers gathered on the clearest night.

But, after a while, Galileo and other astronomers switched from their *best* observation to using the average of their repeated observations. Now, they considered this average value to be the most representative value. Hence, the arithmetic mean of repeated observations became the number customarily employed in their calculations. The difference between the mean and any actually observed value was considered an *error* in measurement. And, thus, the term "error" came into use in their professional lexicon.

But a contemporary of Dr. Johnson, the German astronomer Tobias Mayer, realized that the arithmetic mean itself may not be the most accurate value to represent a phenomenon, because changes in the conditions under which the measurements are taken are not uniform across all observational occasions.

Even slight or subtle variations in the conditions of measurement, multiplied over repeated observations, can grow to be a substantial influence on the outcome. Because of this, the mean is often skewed and therefore can be an inaccurate representation of the phenomenon's true value. Further, when the phenomenon is itself dynamic, such as the changing orbits of stars and planets, using the mean value is yet more problematic.

From this insight, Mayer suggested an operational method for gathering data useful to scientific inquiry, in a process he called a "combination of observations." Methodologically, combining observations is different from measurement gathered with repeating observations. Essentially, he sought to reduce the error by systematically controlling the conditions under which the several independent observations were taken, and then combining the information to form a more meaningful value. At first blush, this may seem like a distinction without a difference from merely taking the average, but it turns out to be a significant methodological advance in how data is gathered for scientific analyses.

Mayer devised his imaginative approach to controlling the conditions while studying lunar variation in an attempt to accurately figure longitude. It was reasoned by astronomers that the moon's position relative to Earth would be an effective way to do this, but the problem for Mayer (and others who also studied the problem) was that the moon's position relative to Earth is not fixed. It oscillates slightly in a phenomenon known to astronomers as *libration*. Due to libration, the amount of the moon's surface visible from Earth varies, to an average of about sixty percent over time. Three types of libration—technically termed *diurnal, latitudinal,* and *longitudinal*—were known to exist. When working on the longitude problem, contemporary astronomers would follow the usual practice of taking only three measurements, one for each type of libration but irrespective of its variation.

Mayer took measurements at different hours—three, six, and nine hours apart—to account for variance in libration. For each type of libration, he took three measurements in each period, for a total of twenty-seven observations. He cross tabulated these observations for successive days to identifiable stars, like the sun and Antares (the brightest star in the constellation of Scorpius). Finally, he combined these separate values to produce the representative numbers for his design.

This approach was clever in accounting for variation in libration with multiple measurements within each one of them, and in combining values. His approach is so sensible that to us it may seem unremarkable: simply making his observations at prescribed times during each libration. As one historian of this period noted, "[Mayer's] approach was a simple and straightforward one, so simple and straightforward that a twentieth-century reader might arrive at the very mistaken opinion that the procedure was not remarkable at all" (Stigler 1986, 21). But, as we understand now, it was a breakthrough in scientific methodology.

From his data, Mayer began solving the longitudinal problem by using methods of trigonometry newly devised by Leonhard Euler (an important Swiss mathematician of the eighteenth century). But, in yet another complication, Euler's trigonometry could not accommodate Mayer's many combinations. The difficulty was that he had twenty-seven equations with three unknowns to solve: the diurnal, latitudinal, and longitudinal coordinates.

Not to be stopped at this point, Mayer then worked to advance Euler's trigonometry so as to handle the increased complexity. While hardly easy math, the longitude problem was nevertheless now one with a solution. Once he had made his calculations, Mayer published longitude tables which were later proved to be accurate within half a degree. (Later still, more precise clocks gave route to determining exact longitudes.) Figure 3.1 shows Meyer's actual tabulations.

But Mayer's approach was still a flawed methodology. Its shortcomings were not fully addressed until nearly a century later by Carl Gauss with his method of least squares, which is the approach to synthesizing observational data that we use today. Both Mayer and Gauss lived and worked at the same time as our Dr. Johnson, and while we have no record of their meeting, we can presume that they at least

[118]		OCTOBER		1772.	
Diftances of D's Center from ☉, and from Stars weft of her.					
Days	Stars Names.	Noon.	3 Hours.	6 Hours.	9 Hours.
		D. M. S.	D. M. S.	D. M. S.	D. M. S.
1	The Sun.	62. 6. 55	63. 44. 49	65. 22. 18	66. 59. 22
2		74. 58. 25	76. 32. 59	78. 7. 10	79. 40. 56
3		87. 24. 0	88. 55. 28	90. 26. 35	91. 57. 21
4		99. 26.	100. 54. 47	102. 23. 14	103. 51. 23
5		111. 7. 52	112. 34. 22	114. 0. 37	115. 26. 38
3	Antares.	33. 0. 51	34. 36. 17	36. 11. 37	37. 46. 49
4		45. 48. 30	47. 14. 40	48.	22. 24
5		58	59. 40. 36		2
6					

Figure 3.1 Tobias Mayer's lunar-distance tables

(Source: from T. Mayer, *The Nautical Almanac and Astronomical Ephemeris for the Year 1772*)

knew of one another because separately they were luminaries of the day. In Chapter 7, I explain Gauss's monumental accomplishment. As we will see there, too, the invention of the method of least squares was not solely due to Gauss.

Despite Mayer's ingenuity, his measurement strategy was not readily accepted by his contemporaries. Euler expressly rejected any advantage to Mayer's system for making observations over the traditional practice of taking the simple arithmetic mean of repeated observations.

Because of Euler's superior stature among astronomers of the day, most of them followed him and Mayer's clever approach to observation languished as an undiscovered treasure. But, by the late eighteenth century, Gauss had invented his method of least squares, complete with incontestable math proofs and careful documentation, and that technique finally prevailed. Now, we view Gauss's method of least squares as a touchstone by which other procedures for synthesizing observational data are judged.

****

Euler himself made many important contributions to early mathematics. He followed Newton with some advances in calculus and number theory. But another of his contributions is the one that we see more widely today: his codification of modern mathematical terminology and its notation, like $f(x)$. Most famously, he specified the "order of operations," which shows which step in solving a mathematics problem should be first, and then what is next. You probably recognize this mnemonic: "Please Excuse My Dear Aunt Sally," or PEMDAS, for Parentheses, Exponents, Multiplication, Division, Addition, and Subtraction. And the practice of working problems from left to right is also thanks to Euler.

Because of these numeric stipulations, he is responsible for our consistency in solving equations like this one:

$$6+\left(5\times 3^{2}+4\right)=?$$

You may have, at some point, asked your friends to solve such simple equations just to see how many answers they have. It's a fun game. Often (sometimes depending upon the number of beers consumed or the lateness of the hour), many different answers are produced. Fifty-five is the correct answer by modern order of operations. Think—and thank—Euler, because it is just practical that we all solve equations in a uniform order.

These contributions—and others—place Euler near the top of all the early mathematicians and make him a harbinger of quantification. Pierre-Simon

Laplace (whom we encounter in Chapter 4) articulated Euler's influence on mathematics by declaring, "Read Euler, read Euler, he is the master of us all."

Now, without digressing into an account of the history of mathematics (despite it being a great story, too—recall that, in the concluding paragraph of Chapter 1, I cite several works that do tell that story), I pause to acknowledge the work of another noteworthy individual who was also antecedent to our focus on quantification: Gottfried Leibniz. Initially working without an awareness of Newton's development of calculus, Leibniz would come to his own parallel invention of both differential and integral calculus, although about ten years later. And, in 1679, Leibniz introduced binary arithmetic, which—as many readers will recognize—is the elemental mathematics for directing electronic pathways in microchips.

Aside from astronomy and geodesy, there are many other fields where observation has played an inventive role in scientific inquiry. For example, Charles Darwin, working roughly a century after Dr. Johnson and well within our time frame, based his entire body of evidence for an evolutionary trail in lower animals (and then the human species) on the observations he made of animals and their habitats. His books *On the Origin of Species* and *The Descent of Man*, were accepted immediately as revolutionary. He was active, thoughtful, and deliberate in his observations, but he did not include any mathematics to support his description of species evolution, nor was he particularly careful in following research protocols, despite their being well established by his time. We explore more about Darwin and his influence on quantification in Chapter 12.

Today, we label the methodology used by Darwin as "observational research," or sometimes more narrowly as "naturalistic inquiry." Observational research is nonexperimental study wherein the researcher systematically observes ongoing behavior, while the term "naturalistic inquiry" typically suggests that the research is field based, as opposed to being conducted under controlled, experimental conditions.

****

Stemming from observation as the opening to our story of quantification, it is necessary to mention two other forerunners to the principles in our story, if only briefly. They are Sir Isaac Newton (in England) and Blaise Pascal (in France). Both men directly influenced almost all the individuals we meet in this book. Almost certainly, readers will recognize the names of these giants of early scholarship. Each made monumental achievements in mathematics, astronomy, physics, philosophy, and elsewhere. I introduce Newton in this section and Pascal in the next. Both Newton and Pascal slightly precede the timeline of our story of quantification; nonetheless, each is so consequential to the persons and events we do address that they deserve some note.

Newton, in particular, is widely credited (with agreement from Einstein) as perhaps the most influential scientist of all time. He virtually invented the physics of motion and gravitation, formulated modern calculus, and procedurally set out the scientific method. Of course, there were earlier attempts to mathematically describe motion and force (the essential focus of calculus), but Newton provided direction and solvable equations for these ideas, thereby giving a structure to modern mathematics. And, with his advanced description of the laws of gravity, he presages Einstein's theory of relativity.

The lasting impact of this brilliant individual on our lives cannot be overstated—we will see his influence throughout the story of quantification. His thinking dominated scientific inquiry for more than three centuries.

He was truly one of a kind. An early Newton scholar and translator of his work (from Latin into English) describes the man in almost reverential tones: "the author, learned with the very elements of science, is revered at every hearthstone where knowledge and virtue are of chief esteem" (Newton, Motte, and Chittenden 1846, v). Further, Newton was the first person appointed to the position of Lucasian Professor of Mathematics at Cambridge University, probably the most prestigious appointment in all academia. He held the chair from 1669 to 1702. It has been occupied by luminaries ever since; for example, it was held by Stephen Hawking from 1979 to 2009. The position is popularly referred to as "Newton's chair." Newton was widely renowned during his lifetime; his fame then was parallel to that of Einstein in the early twentieth century. Among innumerable tributes to Newton work and life is the honor of having an elite private school and a fine college named after his masterwork, *Principia*.

Upon Newton's death in 1726, the poet Alexander Pope wrote the following epitaph:

> NATURE and Nature's laws lay hid in Night:
> God said, *Let* NEWTON *be!* and all was Light!
> (Pope, "Epitaph: Intended for Sir Isaac Newton")

I wholeheartedly suggest reading something about Newton's life and accomplishments, since he continues to be important in our lives today—but chiefly because doing so places one in the company of a truly remarkable man. A fine list of books is available online, provided by the Isaac Newton Institute for Mathematical Sciences in Cambridge (see Isaac Newton Institute for Mathematical Sciences 2018).

30 | THE ERROR OF TRUTH

Another honor given to Newton is the life-size statue of him at Trinity College, Cambridge University. This statute, by Louis-François Roubiliac, was presented to the college in 1755 and has been described as " the finest work of art in the College, as well as the most moving and significant" (Oates 1986, 204). It is inscribed with *Qui genus humanum ingenio superavit* ("Who surpassed the race of men in understanding"). It is shown in Figure 3.2.

Figure 3.2 Statue of Isaac Newton at Trinity College, Cambridge
(*Source:* http://commons.wikimedia.org/wiki/Category:Public_domain)

Newton left us with many memorable quotes. A few of his best-known ones are the following:

If I have seen further it is by standing on the shoulders of Giants. (Newton, Isaac. (1675) 2017)

To explain all nature is too difficult a task for any one man or even for any one age. Tis much better to do a little with certainty & leave the rest for others that come after you [than] to explain all things by conjecture without making sure of any thing. (The Newton Project. 2011)

To every action there is always opposed an equal reaction: or the mutual actions of two bodies upon each other are always equal, and directed to contrary parts. (Newton, Motte, and Chittenden 1846, 83)

Gravity explains the motions of the planets but it cannot explain who sets the planets in motion. (The Newton Project, 2011)

About the same time as systematic observation was beginning to be recognized as an essential tool of science (that is, in Dr. Johnson's time), the *Principia* was translated from Latin into English, an act that ensured its broad influence. Almost every scholarly work on a quantitative subject has its genesis in Newton's momentous work. It forms the foundation for entire fields of higher mathematics and philosophy. From its initial publication in 1687 to today (a time span more than three centuries long), Newton's *Principia* has been considered to be an example of science at its best.

Several copies of the first edition of the *Principia* survive. Newton's own copy with his handwritten notes is located at Cambridge University Library, and another copy is at the Huntington Library in San Marino, California. One copy, which was originally a gift to King James II, sold in 2013 at auction for nearly three million dollars. Its title page is shown in Figure 3.3.

As for *Principia*'s content, Newton's work is a touchstone for modern mathematics and physics. Only a generation before, mathematics had been an immature field of inquiry mostly confined to straightforward measurements in astronomy and geodesy. Its emphasis was upon using early geometry and trigonometry to learn more about characteristics of celestial bodies, like their size and distance from where mere mortals stood. There was no developed investigation into predicting their orbits or learning why some planets move faster than others. At the time, most persons, including astronomers, did not envision such questions. The predominant thinking was that everything in existence was controlled directly by God. (Virtually everyone at the time held strong religious

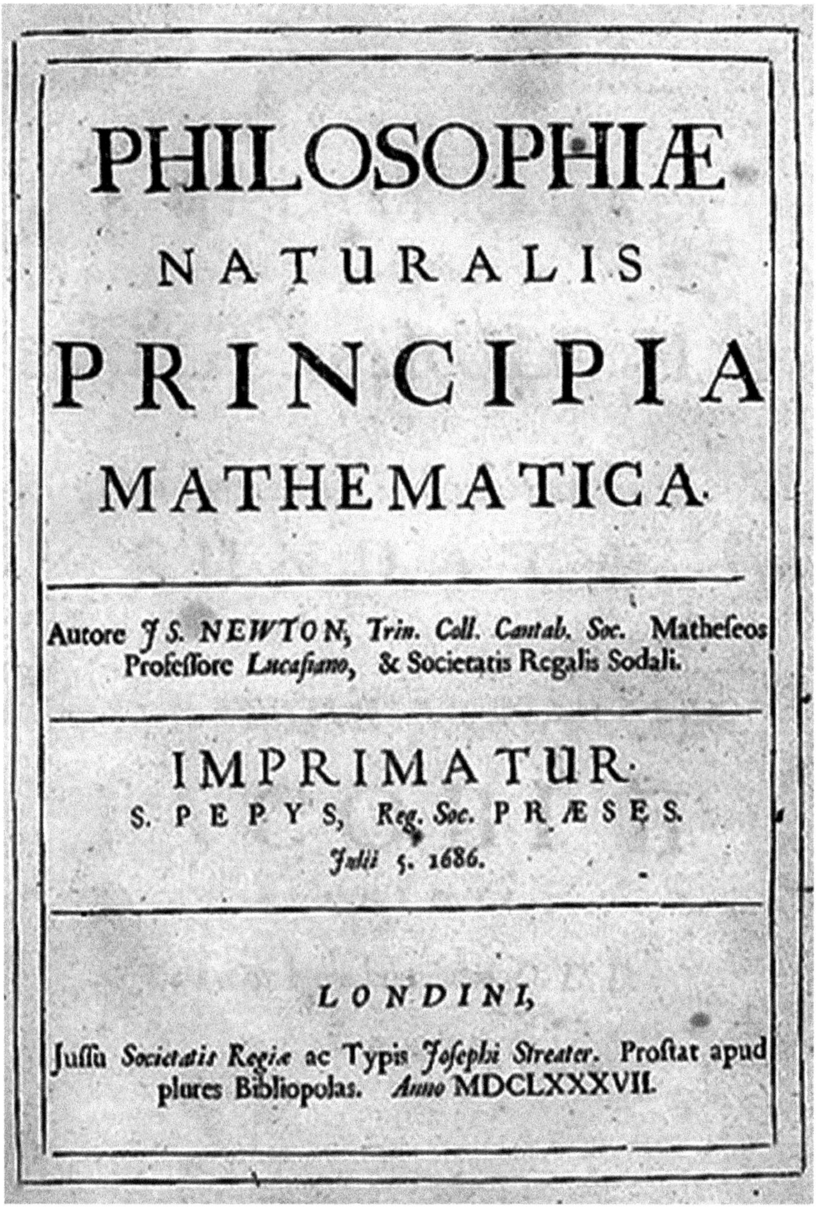

PHILOSOPHIÆ

NATURALIS

PRINCIPIA

MATHEMATICA.

Autore *J S. NEWTON*, *Trin. Coll. Cantab. Soc.* Matheſeos Profeſſore *Lucaſiano*, & Societatis Regalis Sodali.

IMPRIMATUR.

S. PEPYS, *Reg. Soc.* PRÆSES.

Julii 5. 1686.

LONDINI,

Juſſu *Societatis Regiæ* ac Typis *Joſephi Streater*. Proſtat apud plures Bibliopolas. *Anno* MDCLXXXVII.

Figure 3.3 The title page from Isaac Newton's *Philosophiæ Naturalis Principia Mathematica*

(*Source*: http://commons.wikimedia.org/wiki/Category:Public_domain)

beliefs.) The stars and planets had orbits known only to "Him" and controlled exclusively by "Him." It was imagined that nature was at the whim of God.

This was a time much prior (by about one hundred years or so) to the era of our story: quantification as a worldview was something no one dreamed of.

Then Newton came along. He was the first person to truly realize *how* things in nature behave and *why* it is so. He turned the world upside down by realizing and understanding the physics of the cosmos. Building upon Copernicus's work, Newton discovered that there are physical laws that operate across the universe, beyond just here on Earth. The most universal and consequential physical law is that of motion: specifically, gravity. Newton discovered gravity and the laws of gravitation, as specified in the three laws of motion.

He saw that gravity makes an apple fall and that it also is the reason why an object's trajectory is eventually drawn downward. If you throw a rock, for instance, it travels forward only for a bit before its trajectory begins to bend downward, owing to the silent force of gravity. With astounding insight, Newton realized, too, that gravity operates outside of the Earth and that it explains the orbits of celestial bodies. He saw, even, that the moon is held in place by gravity. As we know now, with calculable precision, that if a spaceship travels at the same speed as Earth's rotation, its trajectory will bend downward in correspondence to Earth's curvature. This is exactly how the International Space Station is kept in orbit today.

Newton not only made his observations as philosophical argument, he provided a mathematical scheme to define gravity—modern calculus. And, as readers likely know, calculus forms the underpinning for nearly all of modern mathematics, including probability theory, as we study in this book. It is simply amazing. Newton set the groundwork for all of quantification.

****

While mentioning Newton as the developer of modern calculus, I also want to reassure you about references herein to this higher form of mathematics. Throughout our story, I necessarily cite calculus as a field in mathematics; however, I will remain true to my pledge in Chapter 1 to not use any calculus to solve equations. In this book, we will not "solve" any equations, as that is not our focus. I only mention the term calculus (or algebra or geometry) to let you know where the mathematics lie. We will encounter the word often throughout this book. Just read the word *calculus* as our story unfolds—you will not be lost.

Still, a short explanation of calculus may be interesting because it is so foundational to probability theory. To grasp a very primary notion of calculus, realize that it is as much a way of thinking as a system for devising and solving equations that follow a set of mathematical principles. Calculus is a method of mathematically addressing questions about quantities that change, particularly when the change is nonlinear. Typically, in science, change is classified into two branches: rate of change (called "differential calculus"), and accumulation or change up to a certain point (called "integral calculus"). Thus, calculus itself has two complementary branches: differential calculus and integral calculus.

In both instances, the measurement is done by estimating the change many, many times, essentially breaking the measurement down into small pieces and then measuring each bit and incrementally summing them to get the useful, big picture. Knowing this, we can appreciate logically that the word *calculus* itself derives from the Latin, meaning "small stones," because it is a way to understand something by looking at its small pieces.

We can see these calculus estimates for change through an example. Consider the speed of a bullet. There is an exit velocity for a bullet as it leaves a gun barrel. The velocity soon changes due to many factors, such as air resistance, loss of inertia, and, most forcefully of all, gravity. Hence, a bullet does not move in a straight line forever; rather, its trajectory is an arc, and its rate of change continually slows, just like the rock and the spaceship. With calculus, you can compute the force or energy (e.g., rate of change, or speed) at any point along the arc.

Integral calculus can be illustrated by imagining a bell-shaped curve, the kind with which we are so familiar (actually, any shape of a curve, but this is the shape with which we are most accustomed). Now, see below the curve a straight horizontal line, the baseline. Visualize some point along the baseline—say, the midpoint. Then, take a second point some distance above it, to its right side. Project for both points a straight vertical line extending upward to the curve, thus graphically specifying a portion under the curve as between these lines. With calculus, we can determine the area under the curve that we have specified. In fact, in educational and psychological testing, this area under the bell-shaped curve represents percentiles of the tested population (in this example, since we are starting at the midpoint, it is the percent above the fiftieth percentile). In Chapters 9 and 14, I discuss this point in a more complete context (for an illustration for our current discussion, see Chapter 9, Figure 9.5).

Relatedly, by using the inverse of what I just described, calculus can be employed to determine the volume of any shape, even when it is an irregular shape. Here, think of some three-dimensional shape, like a sphere, a capsule, a cone, a pyramid, or even something less regular in its shape, such as an amoeba. Calculus, with integration, can estimate the volume of these shapes.

This is as technical as I go in this book. Obviously, I am leaving out many important and fascinating features of calculus, but, as Father William in Lewis Carroll's poem of the same name said:

> "I have answered three questions, and that is enough,"
> Said his father. "Don't give yourself airs!
> Do you think I can listen all day to such stuff?
> Be off, or I'll kick you down-stairs!"
> (Carroll, "Father William," stanza 8)

Still, beyond just explaining the essential core of calculus, it is instructive to realize its many uses. As one can imagine, calculus is so broadly applicable it is difficult to think of areas of work where it is not useful. For instance, in construction, it is vital in estimating the strength of materials for a bridge spanning a gorge; in addition, calculus is used to determine the rate at which the concrete of the bridge degrades. In medicine, calculus helps medical researchers determine appropriate drug dosages, again by estimating the rate of change. And, of course, it is employed with missiles and rockets for everything from optimizing their materials during construction to setting their trajectories. In biology, it is helpful in determining the effects of environmental factors on a species' survival. In finance, it can be used to identify which stocks would compose the most profitable portfolio. As we just saw, in educational and psychological testing, calculus is used to figure percentiles. In manufacturing, calculus aids in reckoning optimal industrial control systems . . . and on and on.

One of the most imagined examples for calculus is its application in determining the path of a celestial body such as a star, planet, or asteroid. Like the trajectory of a bullet, a celestial body does not travel in a straight line through infinite space. It, too, is pulled downward by the gravity of whatever object (e.g., the sun or Earth) is close enough to exert influence. Thus, as Earthlings, we orbit the sun; and similarly, the moon's orbit is fixed by Earth's gravity. This is so for all celestial objects in our solar system, and in others. Because not everything is exactly regular (e.g., Earth is not a perfect sphere, and it does not spin at a uniform rate), an orbital path is an ellipsoid. When a rocket's falling trajectory exactly matches Earth's rotation, the rocket remains in Earth's orbit. Astronauts could (theoretically, at least) orbit Earth forever; but fortunately, they can break the force by turning on thrusters to head back home.

At this point in our discussion of observation, some readers may think of the field of statistics, as it too relies on numerical quantities gathered from observations that change. But this term was not used until later, around the mid-eighteenth century, when Gottfried Achenwall was systematically gathering numbers for his employer (the German government) and called his system "statistics" (the original German is *Statistik* or "political state"). Some historians claim that, owing to his development of these structured methods, Achenwall is the "father of statistics," although others challenge that ascription, claiming it really belongs to Sir Ronald Fisher who, much later, virtually invented modern statistics by introducing inferential reasoning to hypothesis testing and many other methodologies. (We meet him in Chapter 14.) Achenwall, however, appears to have been the first one to use the name "statistics" to describe his organizing work.

****

The other earlier giant in mathematics who influences so many of the individuals in the story of quantification is Pascal. In truth, our story of quantification could logically begin with him, since he was instrumental in the founding of probability theory as a distinct discipline. We will see his work and influence many times in this book.

Pascal was a man of truly extraordinary intellect, and stories of his mathematical exhibitions when he was still a child abound. For one, while in elementary school, he almost fully memorized Euclid's elemental geometry treatise *Elements of Geometry*. Also, it is reported that when his teachers gave him a problem in mathematics, he often accompanied his answer with a proof. One imagines his schoolmates were not similarly occupied!

To help with his father's business as a local tax collector, Pascal invented a simple wooden calculating machine that he called the "Pascaline," or the "Arithmetic Machine." It is probably the first machine produced in any quantity to actually perform arithmetic operations. It calculated by rotating wheels of numbers which could be added and subtracted. He later made attempts to have it do multiplication but was not successful. His original Pascaline is on display at the Musée des Arts et Métiers in Paris and is shown in Figure 3.4. It has been reproduced by hand many times, in brass, in silver, and in wood. And his machine predates Leibniz's "Stepped Reckoner," which some consider to be the first calculator because it could perform all four arithmetic functions.

One of Pascal's most consequential achievements was not a physical contraption at all. Rather, it was his conception of a simple graphic triangle that displays solutions to some common expansions of the binomial. The binomial theorem is foundational to probability theory, as well as to much of algebra—indeed, to all of number theory. And "Pascal's triangle" is a clever arrangement of numbers that makes solving some of the binomial's numerical expansions quick and

Figure 3.4 Pascal's Pascaline
(*Source*: http://commons.wikimedia.org/wiki/Category:Public_domain)

easy. In Chapter 4, we meet both of these numerical inventions: the binomial theorem and Pascal's triangle.

Early on, Pascal turned his attention to studying probability, which, at the time, was not an organized field of academic scholarship but more just an interest of persons concerned about getting ahead in their gambling bets. Pascal's interest began when he learned of a special gaming scenario now called "the unfinished game," but which, for a long time, was known as the "problem of points."

The problem of points presents a simple illustration of chance, or, more to our perspective, the notion of how to measure uncertainty. In the "game," two players contribute equally to a pot of money. They then agree to an upper limit of trials for some chance event (like the rolling of a die, or the flip of a coin), at say, ten trials each, after which whoever leads takes the entire pot. This is simple enough; the "problem" comes into play when, hypothetically, the game is interrupted before it is finished.

The question for the players is how to divide the pot for an incomplete game. Obviously, it would be unfair to give the winnings to the leader midway through, since the loser would not have had an opportunity to switch places to be the winner himself. Or, one might suggest that the pot be divided proportionally according to how many points were won up to the time when the game stopped, a solution proposed by Luca Pacioli, an early mathematician who studied the problem and who worked with Leonardo da Vinci on a fascinating book on mathematics and art, *De Devina Proportione* (*On the Divine Proportion*). (We will explore this book in Chapter 17.)

But that proportional solution was thought to be unsatisfactory by two other early mathematicians, Gerolamo Cardano and Niccolò Tartaglia. They noted that if the interruption came after only one player had a single trial, it would be far too early in the game to have that division of the pot of money be fair. Over the following years, others, including even Galileo, attempted to solve the problem but also without success. The intractable problem gained attention among mathematicians (and, presumably, gamblers), in large part because it was germane to their interests but seemingly unsolvable.

Finally, in the mid-seventeenth century, Pascal stepped in to work on the problem of points. He was enlisted to tackle it by his friend Antoine Gombaud, Chevalier de Méré, who thought the mental activity would be helpful to the bright-but-frail Pascal as a way to divert his attention away from his poor health, on which he seemed fixated. Pascal agreed to take on the challenge, but he wanted some help. The Chevalier de Méré suggested an acquaintance, Pierre de Fermat. Pascal wrote Fermat a letter to inquire of his interest in joining him in seeking a satisfactory solution to the problem of points. Fermat immediately accepted the invitation, and thus began their famous correspondence. (They never met personally.)

Their five letters discussed the problem of points as a science, the first time it was so described. With immediate insight, they appreciated that the problem was one of estimating the likelihood of a given final outcome from however many points had been won up to the interruption. They each had a different approach, but together they developed a method to calculate risk, the likelihood of success or failure. As consequence, their work formed the foundations of probability theory.

The Pascal–Fermat correspondence is credited with being the first serious study of probability. Due to popular reception of their correspondence, Pascal and Fermat are jointly cited as the "fathers of probability." Their original correspondence survives and can be viewed online (see original correspondence at Fermat and Pascal 1654). It is charming reading. Moreover, the tale is fully told in a recent book whose title trumpets the significance of their correspondence: *The Unfinished Game: Pascal, Fermat, and the Seventeenth-Century Letter that Made the World Modern* (Devlin 2008). While the correspondence is significant, this retelling seems to go a bit further in interpreting its impact on probability than we see in the case of quantification at this time.

Pascal was a man of deep religious conviction, and he revolved much of his thinking about probability around religion. As one example of his linking probability to religion, he argued that a deep-seated belief in God is worthwhile for even the skeptic because, just with the slightest probability of God's existence, it is better to be with God after death than without. This proposition is famously known as "Pascal's wager." Pascal detailed his religious musings at great length in his famous Christian apologetic *Pensées*, which was unfinished at the time of his death in 1662 but published posthumously in parts over the next several years. Physically, he was unwell most of his adult life and died at age thirty-nine, probably of a stomach tumor.

In the same year as the Pascal–Fermat correspondence took place, the Dutch mathematician Christiaan Huygens published a small manual on probability: *De Ratiociniis in Ludo Aleae* (*On Reasoning in Games and Chance*), now considered to be the first published work on probability (Andriesse 2005). However, possibly even earlier, the Italian Renaissance mathematician Gerolamo Cardano (the same person who helpfully dismissed the proportional solution in the "point's problem") addressed the probability of given outcomes in rolls of dice, in a paper that has since been lost.

And work on the chances of expecting certain rates of birth and death was done by an obscure London shopkeeper named John Graunt, who taught himself calculus (quite extraordinary for then—and today). He published his investigations in a paper and was later admitted to the prestigious French Academy of Sciences. One imagines that Graunt went into the wrong line of work!

For the most part, the early scholars of probability focused their work on increasing the odds of winning in games of chance; only incidentally did they address measurement error, an essential interest in modern probability theory. Still, these efforts evidence the start of probability theory as a distinct discipline, albeit quite undeveloped at this early stage. From these labors, we can anticipate that measurement science had a bright and burgeoning future, and, alongside it, observation was finding a home. Thus, combining Mayer's idea of linearly associating observations with the early probability investigations of Pascal, Fermat, and others gave seed to a promising science.

Significantly, from its inception as a discipline, probability theory has been thought of as having practical implication. With it, problems in society, such as anticipating rates for births and deaths or calculating military probabilities in war scenarios, as well as the age-old wish to decrease uncertainty in games of chance, can be approached as having an empirical solution.

To no one's surprise, at the time, work in probability theory remained in the purview of scholars and academics, who were based primarily at universities, although a few of them were in religious orders and even fewer worked in government. Meanwhile, the vast majority of people—the populace at large—went about their daily lives, unaware of and unconcerned about these mathematical attainments. Quantification had not touched their lives in impactful ways and certainly did not constitute their worldview. The choices they made, both big and small, were done with the immediacy of their thoughts and by their personal experience. It is possible they may have discussed choices with friends or relatives and asked for advice or shared personal experiences. But empirical examination of daily phenomena was not yet anywhere in their mindset.

Thus, the seeds for quantification were being sown but had not yet sprouted.

At that time, even the educational system was limited in its impact on moving people to quantification as a worldview. For students, mathematics likely comprised just the basic four operations of arithmetic (add, subtract, multiply, and divide). Despite the fact that algebra, geometry, and part of trigonometry (along with Newton's and Pascal's newly invented calculus) were developed subjects, we know from contemporary accounts that they were not taught in most schools, owing to an insufficient number of qualified teachers and a lack of texts and other resources.

However, this form of higher mathematics was taught in at least a few secondary schools and, of course, in universities. In 1770, Euler published his *Algebra*, probably the first formal textbook on the subject. And Euclid's *Elements of Geometry* was the standard text on geometry from the time it first appeared in 1492 through the mid-eighteenth century.

One historian of the period reports that military textbooks commonly included tables of squares and square roots, something useful for military leaders when preparing orders to deploy troops, and provides this quote from a contemporary military treatise: "Officers, the good ones, now had to wade in the large sea of Algebra and numbers" (Crosby 1998, 6–7).

With "Algebra and numbers" required for military personnel (or at least "the good ones"), calls to mind a scene in Shakespeare's *Othello* (fully, *The Tragedy of Othello, the Moor of Venice*) in which Iago expresses to Roderigo his frustration with his chosen officer Cassio, who was apparently strong in mathematics but lacked practical experience for setting troops in a battlefield (Shakespeare and Rowse 1978). The Bard writes,

> I have already chose my officer.
> And what was he?
> Forsooth, a great arithmetician,

> One Michael Cassio, a Florentine,
> A fellow almost damn'd in a fair wife;
> That never set a squadron in the field.
> (*Othello*, act 1, scene 1)

Othello was written about fifty years before the first probability publication, which tell us numeracy was common even in his time, but certainly not yet part of a broad mindset.

CHAPTER 4

The Patterns of Large Numbers

Finis origine pendet, the Roman poet Manlius wrote: "The end depends on the beginning." With our story we know the end—people have a worldview characterized by quantification in thought and expression; a transformed *Weltanschauung*—before the beginning. As to its first phases, we have seen how observing—a conspicuously ordinary part of daily existence—broadened in scope along several steps, culminating in Mayer's ingenious methodology to become observation—an empirical tool for scientific inquiry.

And important antecedents have come on the scene, such as Newton's invention of modern calculus and his structural foundation to the scientific method, as well as Pascal's intentional study of estimation, leading to his invention (jointly with Fermat) of probability theory for measuring uncertainty, something they devised when studying the problem of points. And, in broad society, education was advancing through the appearance of schools in most towns and villages, effecting a slow but solid rise in the rate of literacy.

But these antecedents are just that: a preceding circumstance. With only a few exceptions (those working in the sciences and the academic peers of the individuals we have discussed), the daily lives of ordinary people had not yet been impacted. Uncertainty, indeterminacy, and unpredictability were still the mindset. More broadly across society, people had not yet considered uncertainty as a calculable part of their lives and thoughts. But how could they? After all, the mathematics of reliable estimation and prediction had not yet been invented, and probability theory was only in its infancy, as an idea. Significantly, however, the seeds for the new perspective had now been sown, if not yet sprouted.

Most ordinary people of the day (the early eighteenth century) lived ordinary lives, despite momentous historical events surrounding them. Men worked as farmers or businessmen, as soldiers or clergy, as teachers, or perhaps in the trades. Nearly all women were occupied as wives, mothers to many children, and caregivers of the elderly, who were often their own parents. It was common for them to assist their husbands in some way or another. The life of rural peasants at this time is illustrated in a famous painting by Jean-François Millet, *The Gleaners*. It depicts three peasant women bending over to gather the grain left by reapers after the wheat harvest. This moving image is shown in Figure 4.1.

All the while, the influences of quantification were in the air, suffusing people's lives in subtle and yet unrealized ways.

In perhaps the most impactful event, both on an individual level and as progress toward quantification, more Europeans were learning to read. As mentioned, almost all towns and villages had at least elementary schools, and

Figure 4.1 *The Gleaners* by Jean-François Millet
(*Source:* http://commons.wikimedia.org/wiki/Category:Public_domain)

secondary schools were commonplace. Third-level education was well established by Oxford and Cambridge in England and Trinity College, Dublin. The Paris-Sorbonne University, an edifice of the Latin Quarter, was already centuries old. In the Nordic countries, Uppsala University in Sweden was famous for its scholarship, as was Universidad de Madrid. Across the sea, in the American colonies, Harvard College was by now more than half a century old, and two new promising colleges had been founded: William and Mary College in Williamsburg, Virginia, and Yale College in New Haven, Connecticut. As one can imagine from this, the emphasis on formal schooling was broad and deep, reflecting a new-found interest in learning, not only for its practical utility but also for the sake of bringing a richness to one's existence. Learning for its own sake was gaining the appreciation of the populace at large.

Many newspapers had a widespread distribution, and local publications such as tabloids and political tracts were springing up regularly. The extensive availability of printed material contributed to spreading interest in reason and science, growing toward full expression of the Enlightenment. I mentioned earlier the enormous influence of the English translation of Newton's *Principia*, most especially its *Book I*, on this period. While unlikely to have been read (due to its technical heft), Newton's work was commonly known. During his lifetime, Newton was a renowned personality although he did not give public lectures beyond his teaching ones at Cambridge.

Virtually everyone held a deep belief in God or a higher being, and nearly all belonged to an organized religion, such as Protestantism, Catholicism, and Judaism. Only a tiny sliver of the population lived their lives apart from religion. Of a matter of course, the influence of the local pastor, priest, or rabbi on the daily life of ordinary people was weighty. Cultural and social life revolved around the local church, parish, or synagogue often literally.

During this time, too, the Italian family Stradivari (its most famous member was Antonio Stradivari) was making violins, violas, and cellos. These skilled woodworkers were known to have spent considerable time and care in selecting their woods. They used spruce for the top and maple for the back and neck, treating it with several types of minerals, including potassium borate, and applying a finish composed of still-disputed ingredients. Their craftsmanship was of such skill that it still amazes everyone who examines it. And, of course, listening to a talented musician play on a Stradivarius is an awesome experience.

Global commerce, especially by sea, was commonplace after Captain (William) Kidd, the last of the Barbary pirates, was hanged in London in 1701. (Many folks hoped that his last words would be to reveal the location of a much-rumored buried treasure, but alas, not so.) His capture and trial attracted massive public interest and was symbolic of the end of a freewheeling lifestyle, whether lived out on the sea or on land. Not only were the seas now relatively safe, but, with Kidd's unsuccessful attempt to bribe his way to freedom, public order was seen as firmly in place.

But all was not peaceful. The climate of the day—social, cultural, political, and intellectual—was fraught with disquieting influences. Forces leading to the French Revolution were building, and the colonists in America were fighting for secession from England. In the Queen Victoria period, the newly formed Great Britain (England had just incorporated Scotland and Wales into its sovereignty, following its domination over Ireland) declared war on France to stop the union of France and Spain. In Russia, the twenty-four-year-old Peter (later Peter the Great, of the House of Romanov) became the sole tsar (later emperor) of Russia and almost immediately let loose his expansionist dreams, an aggression that lasted for decades and set the course for nearly two centuries of poverty for his people and societal unrest throughout northern Eurasia. While he did bring some stability to a chaotic government, he focused his reforms on building Russia into a military power, rejecting the developments of the Renaissance and the Reformation, and he kept Russia isolated from the rest of the world. These forces inhibited intellectual development in that part of the world.

****

Although yet early in our story, we already realize that probability theory leads us through the unfolding of quantification because, primarily, this is the methodology for assessing uncertainty. As the methods of probability theory are invented and grow in sophistication and utility, we will see that adopting quantification as a worldview in ordinary folks is quick on its heels. With this as context, then, it is to Pascal and Fermat's newly invented probability theory itself that we now turn.

As emphasized in Chapter 1, I describe each mathematical achievement only to advance our story and not as didactic explanation. Readers familiar with probability theory will realize that I leave out much information about each procedure that is given in textbook explanations. Remember, our story is about people, not methods.

****

Like all things quantitative, probability theory relies upon some basic rules for numbers. Through the course of time, some of these rules have been codified into mathematical theorems, or statements accepted as fact. It's important to know that theorems are mathematically verified, typically by an algebra or calculus proof. Virtually all branches of mathematics rely upon theorems, and mathematicians, researchers, and other practitioners generally assume the theorems in their work and often cite them when interpreting their evidence for a particular finding or conclusion.

Three simple theorems—the *binomial theorem, the law of large numbers*, and *the central limit theorem*—are fundamental to how numbers operate in a probability circumstance. Although each theorem carries unique information, they are closely related, like the separate movements of a complete symphony. They coalesce to form a sophisticated mathematical underpinning for the whole theory. The binomial theorem builds to a special case in the law of large numbers, which is itself closely related to the central limit theorem.

These profound theorems are undoubtedly the most useful—and used—theorems in probability theory and perhaps in all of mathematics. They are truly that foundational. Any strong mathematician (and certainly most persons working in statistics or probability theory) will recognize them immediately. As we examine the theorems, don't be surprised if you have a sudden epiphany that they present common-sense ideas. In fact, they do! Their beauty lies in their simplicity and clarity. Here we see that the best ideas are also the simplest. Further, even though we will not look at the proofs of the theorems, know that these too are relatively straightforward. However, this is not to imply that we can gloss over them or be glib in our understanding. We should attend to their meaning with care. This is an instance where meticulousness in understanding matters.

I will explain each theorem in nontechnical language, emphasizing its relevance to quantification. As we go along, I also show the mathematical formula for each theorem—but just so you can see how they look. Certainly, we will not derive them, prove them, or even explain their math. A mere glance at them will serve our purpose.

We start with the binomial theorem, because it is the oldest, and then move on to the others.

Newton is credited with developing the modern version of the binomial theorem, although it has roots dating clear back to Euclid and probably before him. There is record of its ideas existing in early Europe, in China, and in India. Omar Khayyam, the eleventh-century Persian mathematician and poet-philosopher, is known to have worked with binomial expansions. But these early versions appear to be much less developed than Newton's formulation. Hence, we acknowledge that the idea of the binomial, and binomial expansions, has been around for quite some time, but our work begins with Newton.

Despite its long history, the binomial theorem is not stodgy old math; indeed, it is pertinent today, even dramatically so. For example (as you may be surprised to learn), the binary logic of computer code that has found application in the opening and closing of electronic pathways on circuit boards and microchips is firmly grounded in the binomial theorem. Without it, there would be no branching for computer combinatorics, the arrangement of numbers as groups of logic that move around on a circuit board.

As another example of its relevance, electronic transfers of money—the core of our monetary system, happening millions of time each day such as with credit card transactions, across the globe—use binomial expansions in their algorithms. The binomial theorem is, literally, quantification in our daily lives! We return to this amazing realization several times throughout this book.

Further, the binomial theorem appears in some other, unexpected places, like literature. One fun example is when we see it in the stories of Sir Arthur Conan Doyle, who created Detective Sherlock Holmes, of the second most famous address in London: 221B Baker Street. Fans of the amazingly observant detective know that his archenemy is Professor James Moriarty, the "Napoleon of crime." In the story *The Final Problem*, the young professor carries with him a book he wrote that is a teasing curiosity throughout the story because there are frequent hints that it may contain clues to the murder mystery. Professor Moriarty's book is a treatise on the binomial theorem! The "consulting detective" (as Sherlock is called in the stories) mentions it when describing his nemesis:

> He is a man of good birth and excellent education, endowed by nature with a phenomenal mathematical faculty. At the age of twenty-one, he wrote a treatise upon the binomial theorem, which has had a European vogue. On the strength of it he won the mathematical chair at one of our smaller universities, and had, to all appearances, a most brilliant career before him. (Doyle, Morley, and Mottram 1981)

Huh! Who would have imagined that there is a connection between the binomial theorem and Sherlock Holmes? We will see many more such gems as we go along in our story.

Now, on to describing what the binomial theorem is and how it works. In algebra, a binomial is just two terms that are either added or subtracted together. The simplest binomial expression is $a + b$. Obviously, *bi* means two, for its two terms. They can be slightly different; for example, $3x + 4$ is another binomial, as is $2a(a + b)^2$, and so forth.

The binomial theorem is a clever formula specifying how numbers can be combined in many useful ways, including the powers of sums, in permutations and in other arrangements. For explanation, in mathematics, a "combination" is when numbers are considered together and their order does not matter; a "permutation" is when the order does matter. Thus, a permutation is an ordered combination. These arrangements allow for numeric expansion, meaning an equation can be manipulated by, say, multiplying it by itself.

Now, I will show the binomial theorem as a formula, just so that you may see what it looks like. As promised, we will not work with it at all. If this is uncomfortable territory, just skip over the formula—doing so will not interrupt the story—and pick up at the paragraph below it.

The important aspect to notice in the binomial theorem is the left side of the equation: particularly, that it is just two terms added together (a binomial). The exponent n means that the two terms can be expanded (i.e., squared, cubed, etc.). This is the general form for binomial expansions (different texts vary the notation):

$$(x+a)^n = \sum_{k=0}^{n} \binom{n}{k} x^k a^{n-k}$$

Imagine solving this equation manually, a couple of hundred years before calculators were available. For the simplest binomial expansion (its square), the solution is not too difficult, but with more expansions it is exceedingly hard. The computations grow in length and complexity with each level of expansion. This is tough sledding.

Here are just three multiplying expansions of the binomial $a + b$. Note particularly that each expansion is that much longer than the previous one, and a lot more tedious to compute. The third expansion is longer yet:

$$(a+b)^2 \text{ is } a^2 + 2ab + b^2$$
$$(a+b)^3 \text{ is } a^3 + 3a^2b + 3ab^2 + b^3$$

and

$$(2a+3b)^4 \text{ is } 16a^4 + 96a^3b + 216a^2b^2 + 81b^4$$

Imagine the next few expansions of the binomial: that is, bring them to the fifth power or the ninth power. Each time it is expanded, the results grow longer and longer, ever increasing in complexity and cumbersomeness. The calculations (remember, at the time, done by hand) are certainly possible but exceedingly tedious, probably taking hours, and they would be pages long. And that is presuming no mistakes along the way.

Fortunately, Pascal invented a graphical arrangement of numbers that makes it a lot simpler to determine many expansions of the binomial. This famous graphical portrayal is Pascal's triangle. Like the binomial itself, Pascal's number arrangement has a prior (but unclear) history, but Pascal is credited with this form. Here, a picture really is worth a thousand words. As seen in Figure 4.2, Pascal's triangle is a triangular array of numbers starting with 1 at the apex and with each subsequent row starting and ending with 1. Each of the remaining numbers is the sum of the nearest two numbers in the row above.

Working with Pascal's triangle to calculate binomial expansions is rather like a logic puzzle or a game, wherein the numbers needed to solve an expansion are given as coefficients and exponents. Coefficients are numbers that go ahead of the variable and is a multiplicative factor. For instance, in 2a + b, the 2 is a coefficient stipulating that the *a* is to be multiplied by 2. Of course, an exponent is the number of expansions.

In the triangle, after the apex 1, turn your attention to its first row: 1 and 1. These values are the coefficients for the binomial's first level $a + b$; that is, this

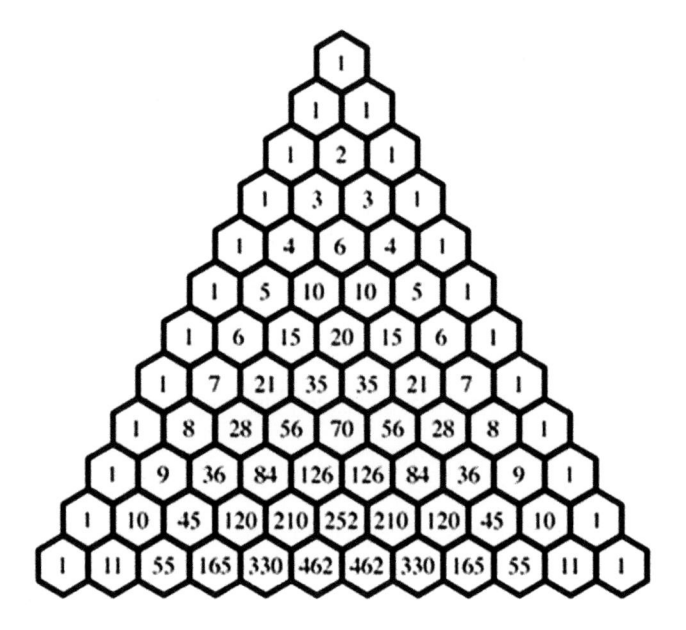

Figure 4.2 Pascal's triangle

could be written (rather unconventionally, but to make the point) either as $1a + 1b$ or as $(a + b)^1$, meaning 1 times a and 1 times b. Of course, it is more conveniently just $a + b$.

Next, consider the binomial expansion when squared: $(a + b)^2$. To solve this expansion, look to the second row in the triangle: 1, 2, 1. These are the coefficients for $a^2 + 2ab + b^2$. (For the exponents, count up the boxes above each number.) The process iterates for each level of expansion.

In advantage over Newton's formula, Pascal's triangle provides a much simpler way to solve these combinations, regardless of the number of expansions (at least up to a point). Realize, however, that Pascal's triangle is not a shortcut for all expansions and permutations of the binomial; it only works for some of them. Complications arise with negative exponents, for example. For these advanced expansions, Newton's binomial formula is still needed. Of course, today any mathematical calculation for binomial expansion (e.g., when done with calculators and computer programs) is accomplished with Newton's binomial theorem, not Pascal's triangle. But is does not matter, as the result is the same. Imaginably, most of the time, Pascal's triangle was a lifesaver to persons in our pre-calculator story.

****

The next two theorems, the law of large numbers, and the central limit theorem, concern *populations* and *samples*. Both theorems address the issue of how samples can be used to represent a population. And, commonly, but not exclusively, they are employed in research contexts. In probability, a "population" refers to the total set of observations that can be made. For example, if the interest is, say, a population of all persons with diabetes, then everyone with the disease, regardless of its type and stage, is included. Or a population can be something organic, like a plant or a microorganism, or something inorganic, like a mineral. Nearly anything can be a population here. However, it is essential in research that an exact description be determined. An imprecise description of a given population is a common reason for misinterpreting findings.

A "sample" is just those members of the population from whom information is collected, whether by observation, a questionnaire, a test, or some other record. Virtually always, the primary concern with samples is to ensure that they represent the population accurately and completely, something easier said than done. Simple random sampling is one common strategy for selecting participants, but there are many other selection methodologies, too. Later on, especially when we meet Sir Ronald Fisher, we will explore a few of them—some are rather clever.

With these terms now known, we can discuss the next two theorems.

The first is the law of large numbers. By the law, the more that is known about an event by repeating its occurrence, the more likely one can estimate the "true" outcome. The true outcome is the theoretical finding that would be obtained if everyone in the prescribed population was included in the probability consideration, and there were no errors from any source. Learning the true outcome is the goal of every such project. In reality, however, it is exceedingly rare when all persons in a given population are included. Hence, samples are taken.

Here is where the two theorems come into play. The law of large numbers tells us that the average value of many samples—in fact, ever more and more samples—will converge to be the same as the true outcome: there will be no difference between the two. That is to say, the more samples we have, the closer will be the average value of those sample statistics to the true outcome.

This idea is most easily seen through an example. Imagine a research project in which a medical researcher tracks adult women who have been diagnosed with osteoporosis. A bone-density T-score of −2.5 or below is a diagnosis of osteoporosis (−1.0 or above is normal bone density). For the theorem, this threshold value of −2.5 is the true score, in this case. The first group of subjects is comprised of twenty-five women whose average T-score is determined to be −2.8. Thus, their value is not representative of the threshold "true" osteoporosis diagnostic value. With a second group of twenty-five women, the average T-score is far above the threshold—say it is −2.0. Taken together, the two groups have a combined average T-score of −2.4—still, not exactly the true score. The third through sixth groups have T-score averages of −2.6, −2.4, −3, −2.1, respectively. Each time, a cumulative average score is computed. The point for the law of large numbers is that, over many trials, the cumulative average value of the separate sample averages will converge to the true value, despite that fact that the score of any one of the trials may not be near the true value.

This data is shown in Figure 4.3. Note that some of the values in the figure are higher than the true outcome, while others are below it, but none is exactly the true −2.5. In this example, even with only six trials, they average to −2.483. With more trials, the average would grow ever closer to the true value of −2.5.

Now, we can understand the law clearly: the law of large numbers states that the more trials we have, the closer the two values: sample observed and population expected (true). This implies—importantly and usefully—the fact that one begins an experiment with little known about what to expect, but, by taking averages of subsequent trials, the true outcome can be predicted with a high degree of certainty.

THE PATTERNS OF LARGE NUMBERS | 53

Osteoporosis T-score

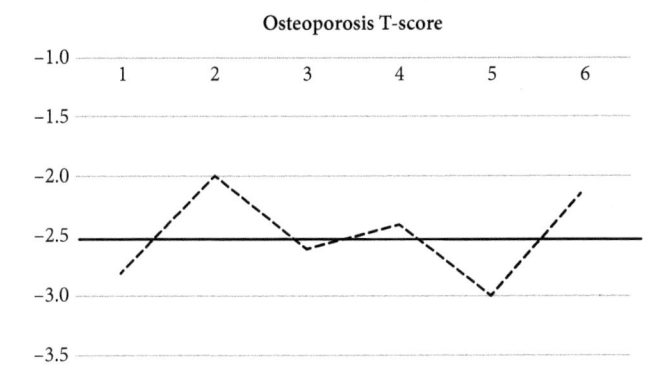

[Chart showing values from -1.0 to -3.5 across points 1-6]

Figure 4.3 Varying group averages whose mean converges to the true score by law of large numbers

The law of large numbers was invented by Jacob Bernoulli in 1713. We will learn more about him momentarily. But first we look at his own example of the law, which has become quite famous.

Bernoulli filled a jar with 5,000 pebbles: 3,000 red ones and 2,000 black ones, for a 3:2 proportion. He hypothesized that the probability of drawing samples of red- and black-colored pebbles from the jar would grow increasingly closer to the exact proportion of each color of pebble in the jar originally (3:2) as ever more samples were drawn. To prove his hypothesis, he sampled the pebbles from the jar by taking them out one at a time and placing them on the table. Each time, he noted whether the pebble was red or black. He saw that any given draw would not be predictable, but as he drew out more and more pebbles, the proportion of red to black pebbles on the table grew closer to the overall 3:2 ratio, as he predicted.

Bernoulli believed the expected (or true) value would emerge gradually in a back-and-forth convergence process wherein each successive trial would have less and less error, although, throughout, some of the errors would be above the true mean, and others below it. He labeled the entire phenomenon "reversion to the mean."

This convergence to an average (and true) value was explored more fully by Sir Francis Galton about 150 years later while he was researching intelligence in children. He established its place in social science and renamed it "regression to the mean." (We meet Galton—an interesting person himself—later on, in Chapter 12.)

As an idea, the law of large numbers is deceptively simple, leaving one to imagine that it is modest in its significance and perhaps even trivial. But there is a lot more to it than that misconception. It presents a concept that is pertinent to the very base of mathematics—namely, the *why* of an outcome. Solving a problem in probability or statistics (indeed, throughout all of mathematics) is one thing, but explaining why that outcome can be expected is quite another. Knowing why it is happening gives meaning to the problem. Bernoulli's law of large numbers provides this *why*—that is what makes it so important and consequential. In fact, it is sometimes called the "golden theorem."

The law's mathematical expression is given in a formula. As with nearly all formulas in this book, it is displayed here so that you may see what it looks like, not that I will explain it, and, certainly, we will not work with it at all. As before, do not worry about the formula itself—if you like, you can skip over it without any loss in our story. I show it in two forms: as it is most often seen in textbooks and then as a probability convergence (it has two theoretical versions, called "strong" and "weak"):

Simple strong form: $\bar{X}_n + \mu$ for $n \to \infty$

As probability: $P\left[\left|n^{-1}(X_1 + \ldots + X_n) - \mu\right| < g\right] \to 1$

It took Bernoulli nearly 200 pages to explain this idea and work through a very laborious proof. To our modern minds, this seems cumbersome, but remember Bernoulli was the first to do it, and he had many issues of mathematics and logic to work through. Over the years, our proofs have become much more efficient (with other, supporting theorems available) and modern texts explain the law in about four or five pages. Actually, the probability convergence can be proven with just a few steps, requiring only about a page to illustrate.

The law of large numbers highlights yet another important notion in numeracy: that of accuracy of measurement. This belief follows from early astronomers' debate of whether to use the single "best" observation or the mean of several observations, and then settling eventually on Mayer's method for combining observations. But because of Bernoulli's work on his law of large numbers, this debate was advanced with his developing the first principles of the calculus of variation, now referred to as "measurement error." And as we will see later on in this chapter, this is also the focus for Laplace's study.

As probability theory evolved in depth and sophistication, measurement error turned out to be an elemental concern. Measurement error accommodates variation among the samples. To Bernoulli, the whole notion behind his calculus of variation, indeed, his whole law, was just one of applying common sense, something anyone (even without the benefit of training in mathematics, he thought) could do. He churlishly said,

For even the most stupid of men, by some instinct of nature, by himself and without any instruction (which is a remarkable thing), is convinced that the more observations have been made, the less danger there is of wandering from one's goal. (Quoted in Stigler 1986, 65)

Point taken, Professor Bernoulli. Sounds like empathy was not his strong suit.

As we saw, Bernoulli initially used games of chance for his illustrations, but he soon added examples from other fields. His used his law to predict the true ratio of male to female births in a population of live births, for example. Expanding probability work beyond just gambling was novel and proved to be important in itself, helping to establish probability as a discipline with broad application.

****

Now, a bit about Jacob Bernoulli himself. He was also called James or Jacques. I note his first name particularly because he was a member of a large family that included several renowned mathematicians, perhaps as many as twelve. The most notable of the Bernoullis are Jacob, Johann, Daniel, and Nicolaus. We explore Jacob's contribution to quantification first; later, we will meet others in this famous family. Unfortunately, the family was often preoccupied with their domestic rivalry, a circumstance that historians believe lessened their potential for even greater contributions to mathematics.

Surprisingly, when Jacob invented the law of large numbers, he had not intended to work on probability per se. He was more interested in exploring Leibniz's theories on both integral and differential calculus. As mentioned earlier, Leibniz came to his own inventions of calculus independent of Newton, although his was somewhat later; and, yet further on, he advanced the original work of Newton by supplying proofs in a series of important but very difficult to understand theories. Jacob and his younger brother Johann were apparently undaunted by the mathematical complexity of these obscure proofs, and they began to study them in detail. They provided many of the missing particulars to Leibniz's work, making calculus a much more complete procedure.

Bernoulli intended to publish his law of large numbers in a treatise he titled *Ars Conjectandi* (*The Art of Conjecturing*), but, unfortunately, he died before finishing it, relatively young at fifty years of age (Bernoulli 1968). It was published posthumously in 1713 by his nephew Nicolaus I. Bernoulli.

As is obvious from the title, *Ars Conjectandi* was written in Latin, the language of most mathematical treatises of the day, but, interestingly, the early translation into English and German was not *The Art of Conjecturing*, as one would guess it to be, but *Probability Theory*. This name stuck, and thus started the name "probability theory" for this new estimation discipline. Incidentally, a first edition of this work was recently put up for a rare books sale at a beginning auction price of $40,000.

Bernoulli famously extended his work with the law of large numbers to human behavior, which he was convinced was just as important in his law as anything else. He reasoned that not all things are equal in value to everybody. Bernoulli suggested there is a moral aspect to the choices people make. For example, he conjectured that if a poor man has but one dollar, it is very important to him because it may be just enough to buy his next meal or maybe a place to sleep that night (using money's value in Bernoulli's time). But if a rich man has a dollar, it is much less important since the loss of a single dollar would not change anything for that man. By this reasoning, Bernoulli added another kind of expectation about how people make choices, which he called a "moral expectation." In Bernoulli's words:

> The determination of the value of an item must not be based on its price, but rather on the utility it yields. The price of the item is dependent only on the thing itself and is equal for everyone; the utility, however, is dependent on the particular circumstances of the person making the estimate. Thus there is no doubt that a gain of one thousand ducats is more significant to a pauper than to a rich man though both gain the same amount. (Bernoulli 1954, 26)

This moral expectation is instrumental to the story of quantitative thinking because it bridges—however early in development our overall story is at this stage—the mathematics of probability in the law of large numbers with everyday human behavior. Bernoulli used observation to formulate his law and then extended it to human behavior rather than leaving it solely in the realm of mathematicians.

The bridge between the law of large numbers and human behavior in his moral expectation was explored yet further by Bernoulli. He contrived some examples of the two effects when applied to gambling. He believed that the observed value (from the samples) will grow toward infinity (the expected value) for a rich man much sooner than for a poor man because the rich man can afford to make more bets (in effect, take more samples) and hence make more informed choices as he goes along.

As Bernoulli showed in his law, with the rich man's opportunity to make more choices, his observed values will grow closer to the expected value (a sure thing in gambling) than for the man with just one dollar and fewer samples. Thus (in his theory), the poor man will be risk adverse, since the one dollar means everything to him. Again, using Bernoulli's words:

> Now it is highly probable that any increase in wealth, no matter how insignificant, will always result in an increase in utility which is inversely proportionate to the quantity of goods already possessed. (Bernoulli 1954, 25)

This is an "inverse probability," something that we see more of later on in the story.

This dilemma between number of samples—more for the rich man but fewer for the poor man—changes the accumulated knowledge, because, as we saw earlier, having more samples means the observed values are closer to the expected value and, hence, closer to a winning bet in gambling. The question becomes: "How many samples will be taken by a random individual whose wealth is unknown?" Bernoulli reasoned that the criterion for making such a decision is naïvely done if it only considers the population expected value and not also the moral expectation. This famous problem is called the "St. Petersburg paradox."

The St. Petersburg paradox was introduced by Jacob but formally presented as a probability problem by his nephew Nicolaus Bernoulli and still later was mathematically resolved by another of Jacob's nephews, Daniel Bernoulli (brother to Nicolaus) in the *Commentaries of the Imperial Academy of Science of Saint Petersburg*, naming the host city for the conference where Nicolaus made his celebrated presentation (Bernoulli Society 1990).

Today, economists study the St. Petersburg paradox as integral to most economic theories in a classic risk-versus-reward scenario given in Daniel Bernoulli's publication. For us, it makes quantitative thinking more real, because the dilemma is not an obscure problem of integral calculus but one that includes everyday human behavior. People think quantitatively when gambling, of course, but, with the introduction of the St. Petersburg paradox, it is evermore formalized as a quantifiable entity and brought closer to ordinary folks. Risk and reward are considerations in many decisions, but when the problem can be stated mathematically to calculate likely outcomes, the dilemma is reduced considerably. Quantitative thinking for everyone was coming into being, even in those early years.

Connected to the law of large numbers are two follow-on notions: the "law of averages" and the "gambler's fallacy." Neither is a mathematical theorem with a formal logic and proof. They are lay terms used to express a belief.

The law of averages, also by Jacob Bernoulli, states that because there is a chance for all the possibilities for an event, it is inevitable that, over some very large number of trials, any one of the possibilities will happen. For instance, there are six sides to a die, and if it is rolled over and over again, in the course of time, a particular side will inevitably show as up. Suppose a gambler wants the side with two dots to show as up. If the die is rolled enough times, because two is a possibility, the die will eventually show the two dots. It is not one in every six rolls: it may be on the first roll or it may be on the millionth roll, because any particular roll is random and all roles are independent. But, be patient, dear gambler—the two dots will inevitably show! Or maybe not. The die has no memory or reasoning capacity to give impetus for the two dots to land up.

In fact, this is why the law of averages is a fallacy and has no satisfactory mathematical proof. For the person who desires a given lottery number to show, say their birthdate . . . well, you get the idea.

The other closely related notion is the gambler's fallacy (also called the Monte Carlo fallacy), although it was not developed by Bernoulli (which matters little, in this case). It is often associated, and mistakenly confused, with the law of large numbers. The gambler's fallacy is the belief in hope itself: that is, because something has not come up among many trials, it is bound to happen soon—like on the next roll of the dice or the next spin of the roulette wheel. It is the reason people stay at a gaming table. (If you are looking for a reason to leave the table, reread the preceding paragraph on the law of averages.) The gambler's fallacy keeps Las Vegas casino owners happy and, sadly, is the ruin of many relationships.

****

Now, we come to the last theorem discussed in this chapter, the central limit theorem. It is the most profound of all the theorems we have seen; indeed, in all of probability theory and maybe even in all of mathematics. It is that appreciated, well known, and used. The central limit theorem is very neat and simple; yet, it applies to nearly all probability and experimental situations.

This theorem is closely related to the law of large numbers (the two are often confused) but it is distinct. The law of large numbers tells us that the average of samples will converge to the true (or expected) value, while here—the central limit theorem—we are concerned with the shape of the distribution of the samples. It addresses the question, "Can I make a bell-shaped curve by plotting the values garnered from the sample trials?"

In 1810, Laplace was examining the law of large numbers, and he reasoned that the mean value from a series of experimental trials is not a precise statistic but actually falls within certain limits. He interpreted the true value as itself a probability, even in the infinite case (e.g., the theoretical zillion times). He saw it as a problem of studying errors, positive and negative, "about" (meaning "above" and "below") the mean. He proved that their mean result converges to

THE PATTERNS OF LARGE NUMBERS | 59

a "limit" in a precise way. This sounds technical but we can see it more simply, as follows.

The central limit theorem states that, with certain conditions (such as adequate sample size), when many independent trials are taken, their means will approach a bell-shaped curve, regardless of the fact that any single trial may itself not be this symmetrical shape. It is useful to be reminded that the central limit theorem is concerned with the *shape* of the distribution composed of many samples.

A simple example will make it easy to see the theorem. Suppose a researcher is interested in exploring the variable of annual household income. In this study, the researcher simply asks people their annual household income, as self-reported. He visits several locales in the area and approaches people at random. Each time, he gathers data from the same number of people, say, five individuals. A visit to an area is a trial, and all values for that trial are the "sample." For convenience, I will display the findings for each trial in a table (Table 4.1). Collectively, these numbers are called "sample statistics."

I use only three trials to keep things simple, but, in probability theory, about thirty trials would be a minimum to meet assumptions of the theorem (up to infinity would be better). Now, look closely in the table at the individual values for each trial, shown in the middle column, labeled "Income (in Thousands)." For the first trial, they appear to be random; in the second trial, they are relatively uniform; and, in the third trial, they are skewed by the one very high value.

The central limit theorem stipulates that—and this is the important point—regardless of how divergent any individual sample is from a normal distribution, the averages themselves (in the rightmost column) will converge to a normal distribution. Hence, plotting the values 54, 99, 90, and so on will create a bell-shaped curve, or, more technically, a normal probability distribution (called a

Table 4.1 *Annual household income for three locations*

Location of Trial	Income (in Thousands)	Average for a Trial
First trial: a shopping center (random)	20, 60, 45, 92, and 53	54
Second trial: a homogeneous affluent area	101, 92, 98, 98, and 106	99
Third trial: a modest income neighborhood including one very high-income household	41, 42, 51, 66, and 250	90

"density function" when calculated by integration—I explain this notion more fully in Chapter 9 during the discussion of Gauss's accomplishments).

As can be seen, the central limit theorem is concerned with the shape of the distribution, which is enormously important in nearly all research. Perhaps its greatest implication is that when its conditions are satisfied, research gathered from samples (individual trials) is generalizable to an entire population. After all, obtaining conclusions about populations is what makes research valuable.

In parallel with showing you the formulas for the two preceding theorems (the binomial theorem and the law of large numbers), I will show you the formula for the central limit theorem. It is given here, in two ways. First, I show it as integral calculus (this is what Laplace derived and then proved). Then, I give its essence in two parallel, much simpler forms. In textbooks, the former equation is shown in advanced texts, while the latter expressions are used in more elementary ones.

Remember, these formula displays are just so you can see what they look like and to give you a feel for what is involved. Obviously, providing an explanation would be lengthy and far off-track. You can just skip over them if you like, and pick up the text immediately below the formulas. You will not miss anything in our storyline.

As integral calculus, the central limit theorem looks like this:

$$y_n = \frac{2^{2n}}{\pi} \int_0^\pi \cos^{2n} t \, dt$$

(For the mathematically advanced reader, Laplace initially assumed that the error distribution could be ±1 with equal probability for each. And he started with the binomial $(1 + 1)^{2n}$.)

More simply—and as commonly seen in most beginning statistics texts—the idea of the central limit theorem is given in these two expressions:

$\mu_{\bar{x}} = \mu$ (the sample mean equals the population mean), and

$\sigma_{\bar{x}} = \frac{\sigma}{\sqrt{n}}$ (the sample standard deviation equals the population standard deviation).

(Note: If you skipped the formulas, here is where to pick up again.)

****

Quantification occurs in nature, too, a point not missed by Jacob Bernoulli. He made many calculations around natural objects, being among the first to recognize that natural structures can be modeled mathematically. Consider the Nautilus shell. A picture of one is shown in Figure 4.4. Doubtless, we all appreciate its

THE PATTERNS OF LARGE NUMBERS | 61

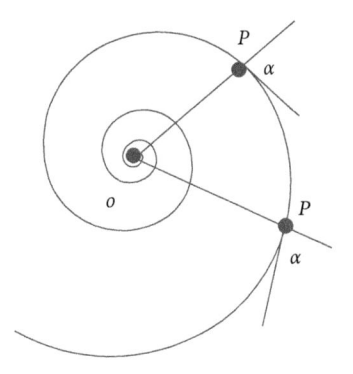

Figure 4.4 A Nautilus shell and Bernoulli's logarithmic spiral
(*Source*: http://commons.wikimedia.org/wiki/Category:Public_domain)

near-perfect shape, its symmetry, and its sheer beauty. We are drawn to it almost supernaturally. Just seeing a picture of a Nautilus shell conjures emotions for something soothing and relaxing: it is comforting, peaceful, gentle, and even magical. But what draws us in is more than just its surface beauty. When we look at the Nautilus shell, we can see that its form is regular and predictable. It is not random or uncertain. Its spirals curl systematically.

Bernoulli called this shape the *Spira mirabilis* (for centuries called the "wonder spiral" and often now called the "golden spiral") and is given credit for specifying the shell's spiral shape as a mathematical arrangement of ratios, although he was not the first to do so. A few years earlier, René Descartes had also worked out a mathematical specification.

Bernoulli was transfixed by the special spiral. He realized that the spiral shape of the shell can be construed as a series of ratios which combine to form an arc that constantly reduces along the spiral's length. Bernoulli saw it as a problem of integration (i.e., integral calculus). When so construed, one can figure dimensions for varying ratios at given points along the spiral's length. The amplitudes of the angles defined by the lines and the corresponding tangents to the shape are of a constant value specifying its size. Because of these characteristics, it is an equiangular spiral.

These diminishing ratios are best expressed in logarithmic units. (Logs are a much easier metric to use for calculating the special relationship among ratios.) Bernoulli was so taken by the logarithmic spiral that he instructed his heirs to carve it into his tombstone. Not surprisingly, the measurements for the spiral shape occur in many other places, both in nature and in man-made forms.

The starry arms of spiral galaxies often have the shape of the logarithmic spiral. And low-pressure systems of hurricanes, when viewed from space, are generally the same shape.

The ratios also relate to several other famous mathematical ratios and numbers, including Fibonacci numbers, the "golden ratio," and the "golden rectangle." In addition, it follows the form of patterns in our lives that are called "sacred geometry" patterns. We return to these special forms later on, since they do play an important role in the story of quantification as a worldview. Sufficient for now is to appreciate that, with Bernoulli's work, we know the universe can be described in numbers.

A famous example of formally employing the equiangular spiral shape in one's work is given us by Sir Christopher Wren, a seventeenth-century architect who was renowned for designing simple but elegant churches. He incorporated the spiral design into some of his structures, a few of which have survived to today. One small especially beautiful church he designed was located in a section of London that was to be rebuilt after Hitler's WWII London bombings. It was decided that, instead of the church being demolished, it would carefully be disassembled, stone by stone (with each one given an identifying number) and then transported to the small American Midwestern town of Fulton, Missouri, where it was rebuilt to its original glory.

Of course, Wren did not use the term "quantification," but, through this beautiful church's structure, we see this idea moving into the realms of experience for ordinary people, beyond math theory.

Fulton, Missouri, was selected as the site for relocating Wren's masterpiece because, a few years earlier, in 1946, Winston Churchill had visited there, brought by President Harry Truman, who wanted to show him typical Americana. The locale is certainly beautiful—rolling hills with bucolic forests and streams. While in Fulton, Churchill delivered a speech, which is officially called "The Sinews of Peace" speech, from a phrase he used, but is more popularly known as his "Iron Curtain Speech." In it he said, "From Stettin in the Baltic to Trieste in the Adriatic, an iron curtain has descended across the Continent." Churchill was warning the West about the expansionist intentions of post-WWII Soviet Union, and their closing of borders, including building the Berlin Wall to keep the people of East Germany and elsewhere under their control. It was a startling move. The world was abruptly awakened to the dangers then posed by the Communists. Many consider this speech to be the start of the Cold War, which dominated world politics for more than forty years. The Cold War began to thaw in 1989 with the fall of the Berlin Wall under the combined influence of

Ronald Reagan, Pope John Paul II, and Margret Thatcher; it finally ended in 1991 with the dissolution of the Soviet Union.

With these spiral and related curved shapes described in mathematical terms, we begin to see all shapes, regardless of how irregular, as quantifiable in and of themselves. René Descartes made a now-famous quote: "All things in nature occur mathematically." Increasingly, we see more and more fit into the evolving worldview as something both beautiful and measurable. We start to think of all things as nonrandom. We begin to take in all shapes in nature as measurable, and thus we see their forms as predictable. Through this, too, we understand how profound Jacob Bernoulli's early influence has been on our thinking.

Beyond these advances in mathematics, other contemporaneous events in broad society moved people further to an awareness of numeracy. One such happening was Benjamin Franklin beginning annual publication of his *Poor Richard's Almanack* (Franklin 1732), a book filled with quantification in its sayings, aphorisms, and predictions. Throughout it, Franklin cites mathematics as the basis for his predictions and pronouncements. The book even includes word problems and several riddles whose meaning is deduced through multiplication. Immediately upon publication, the *Almanack* became hugely popular, as people across American and Europe referred to it for guidance about predictable events. They used its mathematical information to time planting crops, when to buy and sell investments, and even when to get married. It included actuarial tables and other census data.

This kind of mathematically grounded and popular publication had not been commonly available beforehand. With it, uncertainty further receded in the lives of ordinary people as they began to look to prediction, forecasting, and quantification generally. Today, an offshoot publication, *The Farmer's Almanac*, with the telling motto "Plan Your Day, Grow Your Life," is still sold.

And just as powerful for spreading quantification to the populace was Franklin's fascination with mathematics. This side of Franklin's many accomplishments seems little noticed by many historians, but, throughout his life and in many of his writings, he stressed the importance of quantifying decision-making. In his (unfinished) *Autobiography* (Franklin and Sparks 1836), he routinely referenced mathematics as a part of his daily life.

A fine book, titled *Benjamin Franklin's Numbers: An Unsung Mathematical Odyssey* (Pasles 2007) chronicles Franklin's frequent reference to and reliance upon mathematics. While acknowledging that learning math did not come easy for him, he taught himself beginning calculus and attempted to use it in various places, such as in some predictions in his almanac.

52	61	4	13	20	29	36	45
14	3	62	51	46	35	30	19
53	60	5	12	21	28	37	44
11	6	59	54	43	38	27	22
55	58	7	10	23	26	39	42
9	8	57	56	41	40	25	24
50	63	2	15	18	31	34	47
16	1	64	49	48	33	32	17

Figure 4.5 Franklin's magic square
(*Source:* from B. Franklin, *Autobiography*)

While hardly a true mathematician, Franklin did make at least one significant contribution to the field. It was his intense interest in magic squares and circles. These are 8 × 8 arrangements of numbers with particular mathematical properties, from which patterns can be deduced. Figure 4.5 shows a magic square from Franklin's *Autobiography*.

In this magic square, he used a constant of 260. The sum of the first four numbers of any row or any column is 130 and that of a full row or a full column is 260. Also, adding together the numbers in each corner plus the four middle numbers equals 260. This is also the sum of numbers on the diagonals. Many other arrangements total to 260. Further, when the unused numbers are blocked out, symmetric and interesting patterns and geometric shapes emerge. Franklin describes dozens of them in his papers and letters. One can explore them all in Franklin's papers (a full thirty-five volumes!) at the American Philosophical Society in Philadelphia.

In his *Autobiography*, Franklin tells a funny story about himself and magic squares and circles:

> I was at length tired with sitting there to hear debates, in which, as clerk, I could take no part, and which were often so unentertaining that I was induc'd to amuse myself with making magic squares or circles. (Franklin (1706–57) 2016, 124)

Many parents know that dozens of children's books, booklets, and texts on elementary mathematics have in them magic squares and circles, sometimes asking the child to color in the missing squares. The fact that such is so ordinary gives silent testimony to Franklin's impact on bringing quantification into our daily lives.

Some readers may identify Franklin's magic squares as similar to Pascal's triangle, which we saw above as a way to explore number combinations. As with Franklin's magic squares, in Pascal's triangle, a number of clever numerical arrangements can be made, including multiplicatives, exponents, factorials, and other combinations of numbers. Plus, for the mathematically inclined, the triangles display a wide array of somewhat-complex mathematical properties. There is little doubt, due to the similarity of Pascal's triangle to Franklin's magic squares, where the good Dr. Franklin got the idea.

Similar quantification remarks as were said about *Poor Richard's Almanack* can be said for the first modern encyclopedia, published contemporaneously (with such notable contributors as Voltaire writing on history and literature, and Rousseau making contributions on music and political theory). In addition, our man of observation, Dr. Johnson, first issued his dictionary, which routinized English. All were steps toward quantification in the broader society.

Another (possibly unexpected) name in our story is Wolfgang Amadeus Mozart. Mozart was also known to be fascinated with mathematics, and his *Don Giovanni* is recognized by serious musicologists for its symmetry and mathematical precision. Listening to his musical structures naturally elicits counting, regularity, and equality among the stanzas and verses. Clearly, there is a quantifying interpretation of Mozart. Even as a novice fan of classical music, I experience this numerical feeling from Mozart.

Try it for yourself by listening to any piece of Mozart while in a quiet, contemplative mood and be involved firsthand in its quantifying effect. I suggest you will feel it, too—and will be in the company of millions of people across the globe. Mozart is perhaps foremost among the great composers in being intentionally mathematical.

Some music scholars even go so far as to suggest that listening to Mozart heightens spatial–temporal reasoning, something called the "Mozart effect" (Tomatis 1991). Discovering the Mozart effect spawned a line of musicology research on whether this phenomenon could stand up under empirical examination. Others suggest that the effect is evident but that listening to any piece of legitimate classical music would heighten the intellect. While I thoroughly enjoy classical music as a genre, as mentioned, I am not enough of a musicologist to have an opinion on this. Perhaps you do?

Playing off the Mozart effect, in 1991 the Georgia state governor allocated funds specifically for providing classical music to elementary schools, citing as justification that classical music increases intelligence. When making his case before the legislature, he played classical music in the Georgia House chambers, saying, "Now, don't you feel smarter already?" And this effect soon got swept up in the popular culture of the day. Mothers were encouraged to play classical music to their babies, even when the babies were still in the womb (!), in a fad called the "better baby." Thankfully, such silliness has now subsided.

Regardless, the connection between Mozart's music and mathematics is not in dispute, and it brings another piece of evidence to our story of the growing influence of quantification in ordinary people's lives and their worldview. Especially, it extends thinking quantitatively to a widening audience, showing its growing influence.

CHAPTER 5

The Bell Curve Takes Shape

Although off to a slow start, the roots of quantitative thinking were growing stronger across the Atlantic as well, and beyond the influence just of Benjamin Franklin. In America, on July 4, 1776, the second draft of the Declaration of Independence was presented to the Second Continental Congress at a meeting in Independence Hall in Philadelphia, where it was adopted with no opposing vote. The document incorporated many European Enlightenment ideas flowing primarily from the thoughts of Thomas Jefferson and James Madison, after substantive input from John Adams and Benjamin Franklin. Figure 5.1 shows a famous painting by John Turnbull that depicts the event. The original (a twelve-by-eighteen-foot oil on canvas) is considered a national treasure and hangs in the United States Capitol Rotunda.

In other areas, too, quantitative thinking was beginning to emerge. In design, George Washington instructed the builders of his Mount Vernon estate in colonial America to give stark corners and deliberate shape to the porticoes and columns, a style reflective of his belief in mathematical precision and consistent with themes of the Enlightenment.

Philosophers like Locke and (a bit earlier) Hume, mixed with the politics of Napoleon and those of the church (principally the Catholic Church but also the Church of England), pushed vast change onto society, moving the cultural landscape immeasurably. Their influence on the minds and outlook of ordinary people of the day cannot be overstated. The generations-old overreach of the church was passing. Its decline in influence in the daily lives of ordinary people left room for new thinking.

Hugely influential, too, was the vast rise in literacy, as described in Chapter 4. The number of people who learned to read in this period was prodigious,

Figure 5.1 *The Declaration of Independence* by John Turnbull
(*Source:* http://commons.wikimedia.org/wiki/Category:Public_domain)

making this fact alone one of the important societal markers of all time. With their new reading skills, people now had the opportunity to explore new ideas and, with that, to break from staid habits and traditions.

Still, while these forces gave momentum to the quantitative mindset transformation, they did not constitute its most significant impetus. There was an even more powerful influencer: namely, the astonishing developments in mathematics and in probability theory. This is, more than anything else, what spawned the mindset of quantification for an entire populace.

The times were so infused with new mathematical emphasis that, alone, this force could be considered almost primary to the epochal age of reason, self-determination, and mastery of one's internal environment. It brought the new ability of measuring uncertainty—*that* is what changed everything. People were now interacting with the world in new and unforeseen ways. They were beginning to employ reason and common sense informed by the ability to quantify the previously dark unknown. This new awareness was slowly, even unconsciously, making its way into people's daily decision-making. The environment was groomed for people to begin to change their mindset, their worldview—their *Weltanschauung*—to quantification.

What made the advances in mathematics, statistics, and especially probability theory so prominent was both the sheer volume of new ideas and the absolutely torrential pace at which these developments came. As asserted by one historian of the period: "Statistics shot up like Jack's beanstalk in the present century. And as fast as the theory [of probability] has developed . . . the anthropologist must survey a literature so new and so vast that even the expert can scarcely comprehend it" (Newman 1956, 1456).

We saw earlier that Newton principally set the groundwork for all of quantification. However, Newton's work did not immediately cause people to reform their thinking. As we will see, such a transformation was subtle and largely unrecognized by the populace until it had happened, which took quite a while, at least a hundred years. People did not deliberately set out to adopt quantitative thinking; rather, the environment (in all ways: social, cultural, political, economic, and educational) evolved to such a degree that they were suffused with a quantified mindset without conscious effort.

******

Because of the foundation for probability laid down by the three theorems we saw in Chapter 4 (the binomial theorem, the law of large numbers, and the central limit theorem), things really began to take off. One of the early figures in this rapid development was Abraham de Moivre. De Moivre was, first and foremost, a French Huguenot, a member of a relatively small group of Calvinist Protestants who believed that the true route to God was by believing directly in Him, who provided salvation by grace. This was in contrast to the Catholic view that praying through the church leaders and demonstrating good works was the proper avenue to God and salvation. Most especially, the Huguenots were a political threat to the French aristocracy. They became a focus for the French Wars of Religion.

At the time, the slow decline of France was well marked, and the French aristocracy blamed it primarily on the Huguenots and especially their spreading dissent to others. In fact, the French aristocracy's hatred of Huguenots was so intense that it remained strong even a hundred years after the horrific St. Bartholomew's Day massacre, when Catherine de Medici ordered French troops to kill all Huguenots and, almost indiscriminately, any sympathizers. In this horrific event and ensuing incidents, somewhere between 40,000 and 100,000 people were killed, with many slaughtered in cold blood. While the massacre actually transpired some time before our story, its effects were long lasting and profound well into the eighteenth century.

The anti-Huguenot sentiment played a large part in shaping de Moivre's intellectual efforts in mathematics and probability. He saw his energies not solely as advancing mathematics but as an extension of his belief. He said that his intention was to use mathematics to prove the existence of God. Throughout this chapter and Chapter 6, we will see that he was not alone in this perspective—it was a popular pursuit in this time of reason.

But history informs us that de Moivre was not so pure a Huguenot that he eschewed interacting with Catholics and other non-Huguenots. In reality, he was interested in advancing himself politically, which he did through seeking interaction with persons of estimable reputation. He initiated friendships with the most eminent astronomers and mathematicians of his day. His acquaintances included Sir Isaac Newton, Sir Edmond Halley (the English astronomer who precisely predicted the return of the eponymous Halley's comet), and James Stirling (the Scottish mathematician who proved that Newton's difficult computations and his invention of modern calculus were correct). August company, to be sure.

Also among de Moivre's friends were several members of the Bernoulli family. He and Jacob discussed the law of large numbers, and there are accounts that de Moivre expressed to Jacob his admiration for it. Recall, I mentioned in Chapter 4 that Jacob started to describe his theorem (the law of large numbers) in his monumental, but protracted, *Ars Conjectandi* but that he died before it was complete. Apparently, it fell to his nephew, Nicolaus, to finish the work and bring it to publication.

Nicolaus was also a mathematician, but of nowhere near the stature of his uncle or of de Moivre. Nicolaus, knowing de Moivre's admiration of Jacob's work, solicited his help on the task, but, for whatever reason, de Moivre decided against the collaboration. Perhaps this says more about the inadequacy of Nicolaus than anything else; or, possibly, de Moivre was simply too busy at the time. We do not know the reason de Moivre declined the invitation to work with Nicolaus on a project in which he (de Moivre) had expressed passionate interest.

Regardless, de Moivre was fascinated by Jacob Bernoulli's law of large numbers. He knew that, by applying it to games of chance, he could calculate the odds for a given outcome. Given that he was supporting himself by tutoring in pubs and public houses, this must have been appealing. But de Moivre's interest in the law and gaming did not stop with tutoring. He pursued it in his academic work, too. By applying Bernoulli's calculus to the underlying principle of Pascal's triangle, he demonstrated mathematically how samples of random drawings from a population would distribute themselves around the average value in a predictable manner; namely, the values from the drawn samples were always uniform and symmetric about the average, in a clear demonstration of the central limit theorem.

De Moivre quickly realized that it did not matter what object was being sampled to observe this pattern. It worked equally well with hands in a card game; political attitude; eye color; and most other objects and ideas. This led him to count all kinds of things: the height of people, the age at which they died, the distance between towns, the number of houses in them, the physical features of soldiers, and much more. Seemingly, he counted everything in sight. For each thing he counted, he aggregated his data and then mapped it graphically into what are, effectively, histograms. These enumerated elements became "variables" in his quantitative studies.

To remind you of a histogram, I present one as an example, but I am sure you have seen thousands of them and might even have made them in Excel or another software program. If you are of a certain age, you may even remember creating a histogram with a pencil and graph paper. Just to keep us on common ground, a simple histogram is presented in Figure 5.2. Histograms figure prominently in the development of probability theory.

A histogram, like the one in Figure 5.2, displays the possible values of a probability distribution as a series of vertical bars. It is useful to represent graphically the outcome of binomial events (i.e., "either-or," like coin flips of heads or tails), and histograms are a convenient way to organize data from these observations. Each column presents a frequency count, a discrete value that is less than infinity. Sometimes, these columns are called "bins," because they capture all values within that group. As a probability distribution, the separate probabilities of each column must sum to one, or 100 percent probability.

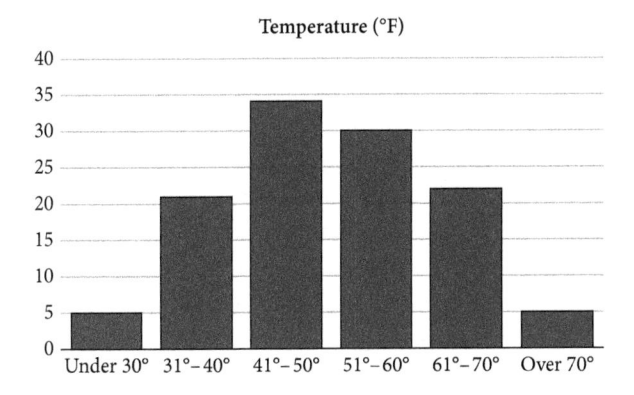

Figure 5.2 Illustrative histogram

In the figure, there are six binomial events: each is a range of temperatures, and they are shown along the horizontal scale, called the x axis. The vertical scale is called the y axis and is a frequency count for how often a particular temperature is observed. In this case, the "Under 30°" temperature was observed five times, and the "31°–40°" temperature was observed twenty-one times. Because there are several binomial events that share a common denominator (in this case, temperature), it is a "binomial distribution." We will see many examples of the binomial distribution in the remaining chapters. Later on, we will explore how this is actually a probability distribution, rather than just an assemblage of individual events.

Realize, too, that the binomial distribution is but one kind of distribution. There are many. The names for some others are "multinomial distribution," "multivariate normal distribution," "chi-square distribution," "Poisson distribution," and "Laplace distribution." Naturally, most of these are technical, and their application is nearly always associated with a given discipline or circumstance, such as medicine, engineering, or business and finance. For the curious, descriptions of lots of them can be found in a comprehensive compendium of distributions offered by Forbes et al. (2011).

De Moivre made thousands of histograms. He noticed that a great many of them were roughly symmetrical, with few instances on the left side, more in the middle, and again few on the right side, as well as the fact that this pattern emerged almost regardless of the variable under his consideration. With some insight, he connected their tops with a line that was itself almost a symmetrical curve. Apparently, he made this little wavy drawing over and over again. For de Moivre, a bright individual with a grasp of higher mathematics, the activity was not child's play; rather, it was grist for studying the shapes of his diagrams—particularly for defining it mathematically. He focused on exploring "errors" or instances that did not fit his emerging symmetrical curve.

Further, de Moivre appreciated that, by following Bernoulli's law of large numbers, each curve he had drawn was rather crudely shaped when it comprised only a few points (i.e., distinct observations), but as the number of his data points increased (i.e., more observations), the shape assumed an identifiable form. With a little hand smoothing (formal, statistical smoothing methods had not yet been invented), the rough silhouettes formed a bell shape. By this process, de Moivre had, in effect, invented the bell-shaped curve. Figure 5.3 is de Moivre's distribution for his "doctrine of chances" from thirty-six random events, with his bell-curve overlay.

THE BELL CURVE TAKES SHAPE | 73

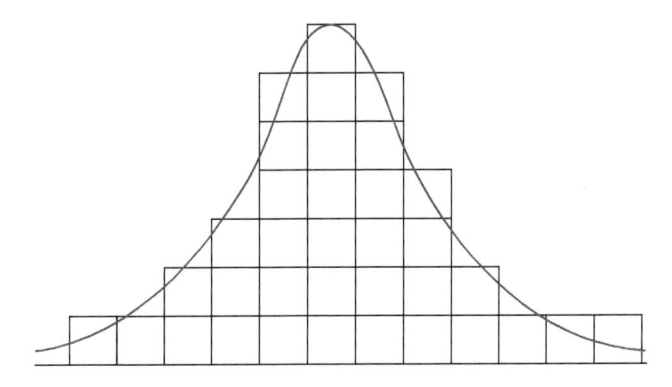

Figure 5.3 Illustration of de Moivre's distribution, from thirty-six random events
(*Source:* derived from A. de Moivre, *The Doctrine of Chances: or, A Method of Calculating the Probabilities of Events in Play*)

These observations are specified mathematically by algebraic expressions that can be derived from Pascal's triangle and are formally solved by the binomial theorem. As is readily apparent, de Moivre's illustration demonstrates the monumental importance of the theorem.

Today, we refer to de Moivre's bell shape as the "normal curve." Of course, he did not call it that, instead labeling his work a "doctrine of chances." And, significantly, his distribution was not yet specified as a density function (we will see that later on, in Chapter 9), but it did have similar properties: specifically, uniformity and symmetry.

De Moivre published his work in 1718 as *The Doctrine of Chances: or, A method of Calculating the Probabilities of Events in Play* (de Moivre 1967). The work was also one of the first accountings of a burgeoning probability theory. From its inception, *The Doctrine of Chances*, with its explanation of de Moivre's bell shape as the normal curve, has been recognized as an achievement of immense significance. For instance, it was highly praised by an eighteenth-century historian named Isaac Todhunter, who wrote perhaps the first history of probability theory in 1865, titling it *A History of the Mathematical Theory of Probability: From the Time of Pascal to That of Laplace*. In it, he said: "De Moivre's *Doctrine of Chances* formed a treatise on the subject [probability theory], full, clear, and accurate; and it maintained its place as a standard work, at least in England, almost down to our own day" (Todhunter 1865, vii).

Over the following fifty or so years, de Moivre's drawing of his "doctrine of chances" underwent several transformations, especially in the work of Carl Gauss, before finally settling into the form of the normal curve so familiar today. As another step, Jacob Bernoulli used the word "integral" from calculus terminology for the first time to specify a solution for determining the area under the curve as a density function. Regardless of these later advancements, this is where the familiar bell curve got started: with de Moivre. This was quite a step on the road to quantification.

Even beyond drawing his bell-shaped curve, de Moivre set his observations in tables of values with concomitant probabilities. For example, he invented "life tables," publishing them in an influential piece titled *Annuities upon Lives* (de Moivre 1725), which displays a bell-curve distribution of the probability of an individual's dying before their next birthday. De Moivre actually gave his work a very long full title: *Annuities upon Lives: or, The Valuation of Annuities upon any number of Lives; as also, of Reversions. To which is added, An Appendix concerning the Expectations of Life, and Probabilities of Survivorship*. Its breadth of coverage and sustained importance is described in the book's appendix, which was written by an anonymous author, who says,

> De Moivre's contribution to annuities lies not in his evaluation of the demographic facts then known but in his derivation of formulas for annuities based on a postulated law of mortality and constant rates of interest on money. Here one finds the treatment of joint annuities on several lives, the inheritance of annuities, problems about the fair division of the costs of a tontine, and other contracts in which both age and interest on capital are relevant. This mathematics became a standard part of all subsequent commercial applications in England. (de Moivre 1725, DSB IX: 454)

By the way, a "tontine" is a specialized investment plan in which individuals buy shares in a monetary investment (like an equity fund or real estate parcel) to form a pool of investors, who receive an annuity that increases whenever one of the other participants dies. De Moivre uses statistics to illustrate his examples that he had garnered from Halley's work in the 1690s. He computed tables from his bell-shaped curves to show longevity and rate of return. Today, de Moivre's tables are still used, commonly in economics and in the insurance industry but also by governmental agencies and, doubtless, others. We now call his life tables "actuarial tables." Across the globe, students today learn these features for investment based on—you guessed it—de Moivre's *Annuities upon Lives*.

****

As an ever-inventive mathematician, de Moivre did not stop at simply mapping his distributions, even though that was a significant invention in itself. He extended his work to compute yet another statistical measure, this one of dispersion about (i.e., above and below) the mean: the standard deviation. He refers to this concept as a "theory of error"; the term "standard deviation" did not come into use until much later. As we have seen already, others (notably Galileo) had previously observed variations about the mean, and even the fact of its symmetry. But they did not carry the work forward, instead merely attributing it to "error in observations." De Moivre, however, formalized the variations into a systematic measure of dispersion, a measure that could itself be calculated and studied.

Today, students worldwide in beginning statistics classes and elsewhere memorize de Moivre's formula and learn how the standard deviation operates. For curiosity only, I show the formula as used today for the "population standard deviation" (there is a slight adjustment to it when the intent is to calculate it for the "sample"). The Greek letter *sigma* (σ) is universally accepted as the symbol for standard deviation in a population. (In a sample, it is just abbreviated as SD, and μ represents the population mean value.) It is

$$\sigma = \sqrt{\frac{1}{N}\sum_{i=1}^{N}(x_i - \mu)^2}$$

De Moivre made his original calculation a function of the variance notion for any "standard normal deviate" (or "z-score"), which has the advantage of applying to any variable expressed in standardized form. This form also allows for specific expectations about the full range of a variable, which we discuss momentarily. Both forms lead to the same result; the difference is that the one given above is statistical in nature, implying that it is directed at a practical utility (as in a research scenario), while de Moivre's formula is more mathematical, and as a function implies that an algebraic or calculus proof is to follow. Actually, de Moivre developed his idea into a theorem about the variance of a distribution. Hence, this form is called "de Moivre's theorem." It is

$$f(z) = \sum_{n=0}^{\infty}\frac{f^{(n)}(a)}{n!}(z-a)^n$$

De Moivre then went on to provide the missing calculus proof. This was a huge step in probability theory because now, not only was the mean useful in understanding a phenomenon, but so too was its variation.

By formalizing the standard deviation with a calculus proof, de Moivre provided a mathematical rationale to explain the fact that, in a perfectly symmetrical distribution, approximately 68 percent of the observations fall within one standard deviation of the mean of all the observations. By extrapolation (also, more technically and exactly by integration), more than 95 percent of observations fall within two standard deviations or, similarly, any other point along the distribution.

In a syntactical sense, the standard deviation is the average error, or, in general terms, the average amount by which the observed values will deviate about a distribution's mean. Figure 5.4 shows de Moivre's distribution with standard deviations. As shown, the distribution represents a population (rather than a sample) whose mean is represented by μ and the standard deviation by σ, as we saw earlier.

As we just saw, de Moivre developed his notion into a theorem that derives from the central limit theorem. Actually, it is a special case of Laplace's work wherein the normal distribution is approximately the same shape as and has characteristics of the binomial.

De Moivre wanted to prove his theorem and used coin tosses to illustrate random events. Being a practical person, he literally tossed a coin 3,600 times (!), following Bernoulli's trials, and ultimately discovered that the probability distribution from such a tiring experiment would indeed be a bell shape. He described this work in the second edition of his by-then-famous *Doctrine of Chances* and gave credit to his forerunners: Newton, Bernoulli, and especially Laplace. As his work is an important application of the central limit theorem, today we give credit to both men, calling it the "de Moivre–Laplace theorem." It is a special case of the more general central limit theorem.

When introducing de Moivre, I emphasized his French Huguenot perspective, as it was so important to him. Recall that his primary interest in studying

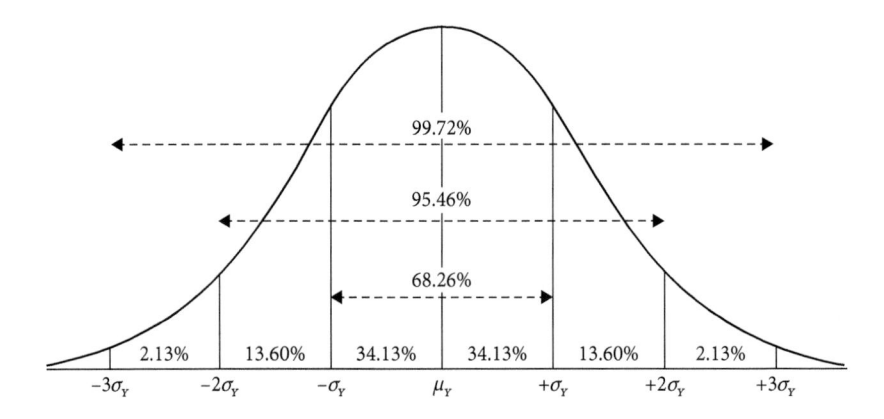

Figure 5.4 Illustration of de Moivre's distribution with centered mean and standard deviations

(*Source*: derived from A. de Moivre, *The Doctrine of Chances: or, A Method of Calculating the Probabilities of Events in Play*)

phenomenon was not so much as a theoretical mathematician but as a means to advance his Protestant beliefs and that he was actually trying to prove with mathematics the true existence of God. In philosophy, he worked from an "originalist" perspective. Reflecting the philosophical milieu of his time and his deep Protestant faith, de Moivre saw a grand design in the universality of the bell curve, which he attributed to God. Using the stilted composition of the day, he explained his idea by saying,

> Altho' Chance produces Irregularities, still the Odds will be infinitely great, that in the process of Time, those Irregularities will bear no proportion to the recurreney of that Order which naturally results from ORIGINAL DESIGN. (de Moivre (1738) 1967, 251)

In this quote, de Moivre reveals two things about himself. First, he believes that there is an order to all things, far beyond random happenings (chance "irregularities"), which is revealed in many trials over the course of time ("process of Time" and "recurrency"). And second, this order comes from God as original design. Note that, concerning the first point, symmetry exists in nature and, in man's mathematical inventions, cannot be denied.

The proposition, then, is not of the veracity of symmetry (de Moivre's first point) but of an origin for it (i.e., divine or naturally occurring), his second point. It is clear where de Moivre stands. Obviously, the bell-shaped curve is a tantalizing hint of original design for all kinds of variables.

To be sure, de Moivre's principal argument was ontological in the tradition of Western Christianity and specifically followed the ideas of René Descartes. Descartes, himself both a mathematician and a philosopher (he is often considered the "father of modern Western philosophy"), famously put forth his "proof" for God, following the logic that if the greatest possible being exists in the mind, it must also exist in reality. Descartes developed his ideas in a wealth of writings that have come to form the basis for debate about God's existence.

You likely know that the debate continues to this day, sometimes titled as "original causes" or "first causes." (A modern offshoot to this thinking is "intelligent design" or "original design.") We will see this debate resurface in the work of Bayes and Laplace, where it is integrated into probability theory as the existence of God being a likely event. With de Moivre, however, it shows the strong influence of newer thinking on Christianity—thinking that allows a quantitative approach to theological study, even to the most foundational of theological questions, is there a God?

This question has spawned an entire genre of study, in mathematics, in religion, and in philosophy, where it occupies much of the scholarship in ontology and epistemology. There are academic journals, university courses, conferences, and countless Sunday sermons devoted to the examination and debate of original design. It is an interesting topic to explore, and one that gives light to how deep the notion of quantification goes into our very souls, including our conception of the ultimate Truth: the existence of God. Of course, humankind has pursued this question since the beginning of time.

The notion of quantification in nature and its being God ordained was not only present in de Moivre's bell curve; it was moving within the broader society as well. For instance, the English poet and philosopher Alexander Pope, writing at this same time in his famous *Essay on Man*, sought to "vindicate the ways of God to Man," a variation on Milton's message in *Paradise Lost* to "justify the ways of God to Man" (Mack 1985). Written in heroic couplet (a mathematics-based poetic meter popularized by Pope in the poem), it attempts to show that "natural laws" (i.e., symmetry in nature) consider the universe as a whole and a perfect work of God. This was a popular sentiment at the time.

To our purpose, it also brings quantification to literary circles, a sphere with influence far beyond the rarefied world of mathematicians. Add to this mix the influence of religion by de Moivre, Descartes, and Pope. In their collection, we are witnessing the spread of quantification to ever more spheres of human thought and involvement. Thus, its influence is growing less obtuse, and more concrete, with respect to the daily lives of ordinary people.

With religion as his foremost thought, considering the context of the times (with still strong anti-Huguenot sentiment left over from the French Wars of Religion), life was not always easy for de Moivre. When de Moivre was just eighteen, King Louis XIV revoked the Edict of Nantes, a decree of religious tolerance granting Protestants equality with Catholics. The edict was originally intended to placate the Huguenots within France, but their obstinacy continued, as did their anti-aristocracy message. Eventually, the French government, and particularly the ruling aristocracy, decided they had had enough. The religious decree was cancelled. With no decree to protect his religious freedom, de Moivre was asked to swear his allegiance to the king, a requirement for all French citizens. But, being staunchly Protestant, he refused, and because of this, he was jailed for two years. Upon release, he moved to England (along with thousands of other early Protestants who were also persecuted by the French), where he continued his work in mathematics. It was there he met Newton and the others.

De Moivre had other setbacks, too. Despite his friendships with many important people, and notwithstanding his prodigious achievements, he seems to have been a genuinely unhappy person. Bernstein (1998) reports that de Moivre was introspective and bitter about his ever-languishing career prospect of being a university professor. More than anything, de Moivre wanted to be a professor at one of the better-known universities, but it was not to be. For most of his adult life, he unsuccessfully applied for one academic professorship after another. Needing a living, he was forced to support himself mainly by tutoring young students. Sadly, he died at the age of eighty-seven, blind and poor.

One legend about de Moivre (some biographers dispute its authenticity, but it is widely reported) is that, in later life, he began sleeping more, adding about five minutes additional sleep each night. Knowing this, he reasoned that when the additional sleep time added up to his sleeping twenty-four hours, he would be dead! From this, he computed the date on which he would die: November 27, 1754, which turned out to be his dying day. We do not know whether de Moivre took comfort in his accuracy.

****

From here on out, this kind of bell-shaped curve will reappear in our story regularly because it is basic to quantifying data and essential to its harmonizing place in our lives. To us, the famous curve looks ordinary, and we can imagine all sorts of instances where it could apply: grades or achievement test scores in school; height or weight; ages for mortality; incidences of morbidity; income across populations; and so forth. Readers may remember a popular book in the 1990s titled *The Bell Curve*, written by the respected Harvard professors Richard Herrnstein (a psychologist) and Charles Murray (a political scientist) (Herrnstein and Murray 1994). Despite their solid scholarship, the book's thesis was hugely controversial because they touched the nerve of differences in IQ. But controversy aside, for modern people so heavily steeped in a quantified world, there was no question about the title: everyone—from scholars who calculate it with integration as a density function to the illiterate individual who cannot cipher the caption—recognizes what a "bell curve" is. Anyone who sees one knows what it is. This is quantification as our worldview today.

For historical perspective, realize that the time of de Moivre was the late eighteenth century, when it was not common to count things and then arrange the values in a distribution. There was no quantitative mindset. The bell curve had not been formalized, and most folks did not even know about it. Had

Herrnstein and Murray's book been published then, its title would have been a mystery. This shows how far quantification has come.

In de Moivre's time, however, most people simply had no notion or perception of a bell-shaped distribution or, more broadly, of measuring uncertainty. They lived in a world of reacting to events only after they had happened. Decisions were made by whim, habit, or tradition. Everything happened to people unexpectedly, with little awareness that soon both everyday and not-so-ordinary future events would be predicted or forecast. But with Bernoulli, Mozart, Ben Franklin, and now Abraham de Moivre on the scene, things were changing—subtly and slowly, to be sure, but dramatically nonetheless. The influences for bringing quantitative thinking to the fore for everyone were now in place.

From Dr. Johnson's poetic insight on observation through the awesome beauty of the nautilus shell to the mathematical and statistical discoveries and inventions that form the foundations of probability, a new quantitative *Weltanschauung* (worldview) was beginning. The seeds were sown—and more importantly, they were beginning to grow in people to become a soon-to-be new perspective on everything.

****

In order to clearly understand upcoming developments, a bit of specialized vocabulary is needed. Specifically, we look at three terms: *probability*, *odds*, and *likelihood*. These terms will follow us throughout this book, so beginning with a clear understanding of their meaning is important. These interrelated terms comprise nearly the whole of probability theory because they are the object of its techniques. In other words, when a statistician employs one of the methods of probability theory (we explore some of them in the next several chapters), it is for the purpose of estimating odds, probability, or likelihood. As I explain these concepts, a few other vocabulary words will pop up, but their definitions are fairly obvious, and I doubt you will get lost.

Aristotle said, "The probable is that which happens often." Simply but significantly, *probability* is the chance of something happening, which is labeled an "event" in probability theory. An event is either a planned circumstance in research (wherein we set up an experiment of some kind, such as when examining a new medical drug to determine whether it is safe for general use) or a natural phenomenon (such as gender, height, ethnic heritage, and the like). Realize that events are constrained to just these contexts within statistics and probability theory, so not everything is a statistically relevant "event." It would be silly to think of each nanosecond of our lives as an event. However, in the probability theory context, every event leads to an outcome, although at the beginning we do not know what that outcome will be. Flipping a coin as a simple experiment to see which side lands up is an event, but initially we do not know the outcome (whether heads or tails)—only that there will be an outcome. Hence, by definition, events have outcomes. With the coin toss, there are two outcomes: heads and tails. In probability theory, this is noted as (0, 1) (read: zero, one). If we look for a given binomial outcome, the result is called "success" or "failure."

Realize that the number of possible outcomes will vary with each experiment or circumstance. It can be two, just a few (as is the case for, say, the number of children in a typical household), or several (such as the number of insect pests that feed on a given plant); or, of course, the number of outcomes can grow to millions and millions (as with stars or lottery numbers).

In exact probability applications, the theoretical domain of all possible outcomes is called the "sampling space." This is as far as we go technically with this idea, but there is much more to it.

Odds, our next term, is the expectation of a given outcome being the observed result of the experiment or natural phenomenon. Thus, odds represent the probability of success in outcome. They are calculated as the ratio between the number of successful outcomes in an event and the number of possible outcomes. To calculate odds, simply divide the former value by the latter. When predicting a given face value on a die, for example, there is one successful outcome and five unsuccessful outcomes, a ratio of one in six (the total number of sides), written as 1:6. The odds are calculated as 1 divided by 6, or 0.1666, about 17 percent. With a lottery, the odds may be 1:240,000,000 (one in 240 million) or some such. As is evident by now, in statistics, odds and probability are not synonyms.

Odds change, but not the probability of success versus failure. All instances of probability are restricted to the interval (0, 1). Think of how many times this applies to your life decisions, big and small, for success versus failure, even when the possible outcomes are many: *Did I pick the right investments for my retirement account?*, *Will holding greater inventory in my business result in faster customer fulfillment?*, and so forth. The possible outcomes for each event are many . . . and can be infinite, in fact, in the sampling space. But observing a given outcome is still a probability of (0, 1).

Our final term in this vocabulary of probability theory is *likelihood*. In statistics and probability theory, this term is used technically and means the probability of a given sample being randomly drawn from the parameters (mathematical limits) of a population. In calculations, the term "likelihood" is used rather than "odds," because of its reference to population parameters. The most probable outcome (greatest likelihood) from among all the possibilities (in the sampling space) is called the "maximum likelihood." Often, we wish to calculate the maximum likelihood; we do so by a process called "maximum likelihood estimation," abbreviated as MLE. Given that the MLE equation is a function of various kinds of information (in particular, "conditional probabilities," as described below), it is called a "maximum likelihood function."

So now we have our three terms: "probability," "odds," and "likelihood." In most scientific contexts, it is important to use them correctly. Of course, when using these terms in popular speech, things are not so fussy. Among friends, they are just words that convey approximately the same meaning.

CHAPTER 6

Evidence and Probability Data

On a clear, hot Sunday morning, July 8, 1741, in Enid, Connecticut, you sit in a wooden church pew listening to a thirty-something pastor deliver the sermon. Although normally shy and taciturn, the reverend is delivering his sermon in a loud voice, with special passion. You've long known some facts about your pastor: that he is a genius, a child prodigy who was admitted to Yale when he was only twelve years old; that he is highly regarded by the colonial leaders of early America; that he writes prodigiously about theology; and that he some years earlier married the stunningly beautiful Sarah Pierpont of New Haven, Connecticut, who bore him eleven children.

You know, too, other, important things about Sarah: that she is deeply pious, but unlike her husband she is gregarious and a remarkable conversationalist, which she uses to effect in convincing others to piety. In particular, you appreciate that she is hugely influential in shaping her husband's thoughts and writings. It is obvious to those who know the couple personally that the young pastor (who could be "difficult") is destined for greatness, in part owing to Sarah's influence on him and the actions she takes on his behalf, such as making political contacts.

Earlier that morning, before traveling to the church, you completed your routine chores of milking three cows, setting hay out for your four horses, and moving some wood inside, closer to the fireplace. Your children have done their chores, too, feeding the chickens and the like. In the pastor's home, Sarah prepared a warm breakfast and attended to the brood, getting them ready for church.

The Error of Truth. Steven J. Osterlind. Oxford University Press (2019). © Steven J. Osterlind 2019.
DOI: 10.1093/oso/9780198831600.001.0001

The reverend speaks his sermon with the intention that the parishioners not only hear his words, but feel them. He says:

> The God that holds you over the pit of hell, much as one holds a spider or some loathsome insect over the fire, abhors you, and is dreadfully provoked. His wrath towards you burns like fire; he looks upon you as worthy of nothing else but to be cast into the fire. He is of purer eyes than to bear you in his sight; you are ten thousand times as abominable in his eyes as the most hateful, venomous serpent is in ours.
>
> O sinner! consider the fearful danger you are in! It is a great furnace of wrath, a wide and bottomless pit, full of the fire of wrath that you are held over in the hand of that God whose wrath is provoked and incensed as much against you as against many of the damned in hell. You hang by a slender thread, with the flames of Divine wrath flashing about it, and ready every moment to singe it and burn it asunder. (Edwards 1741)

That morning, in that small church, the Reverend Jonathan Edwards delivered his famous sermon, "Sinners in the Hands of an Angry God." Not surprisingly, the sermon made "remarkable impressions on many of the hearers" (1741). Soon, it came to rock people far beyond just the parishioners, spreading to believers across the new American colonies. Doubtless, the sermon was influenced by Sarah, who went on to help spread its message of "Reformed Theology," holding that each human is a flawed soul with a natural predilection to wickedness and is only spared the fires of hell by a gracious God. This thinking was adopted by many early Americans as it moved quickly throughout the colonies. The Reformed Theology outlook was an emerging philosophy for the "Great Awakening," a Puritan viewpoint that rejected the Enlightenment ideas of the Continent and set God at the center of all things. Its impact on thinking by American colonists was profound and lasting.

We know already that the work of Newton and John Locke had significantly changed humankind's view of an individual's relationship to God, proposing that people could employ their own free will to reason such a relationship and choose between good and evil. This view predominated in western Europe, as we saw in the Enlightenment epoch. But, in America, many colonists moved away from the ideas of the Enlightenment, not as a return to high church orthodoxy but to the puritanical view of Jonathan Edwards. Thus, there were conflicting viewpoints: Enlightenment on the Continent, and Puritanism (that so-called Great Awakening) in the American colonies.

This conflict of philosophies had an impact beyond religious debate: it also influenced the strides of humankind toward quantification. As a consequence of the different views, the developments toward quantification took divergent paths.

I believe evidence exists to suggest that while advances in mathematics and measuring uncertainty continued briskly in Western Europe and Great Britain, Edwards's "Sinners in the Hands of an Angry God" sermon stifled parallel developments in early America and slowed the progression toward quantification there. Hence, as humankind was moving forward in its viewpoint on one continent, this transformation was muted on another. There was plenty of brain power among the colonists. Remember, the great universities of Harvard, Yale, and William and Mary were already in existence.

But, significantly, the social and cultural context for mathematical and probability advancements that were giving impetus to quantification did not come to America until much later, whereas these forces were already strongly gaining ground on the Continent. We see, then, that quantification as a worldwide viewpoint did not grow everywhere uniformly.

With quantification, people are challenged to think differently: more expansively in both time and space, and more boldly in terms of impact. We move now on to the next step for when and how this happened.

****

Meanwhile, prodigious advancements across Europe and Great Britain continued. Sometime during the 1740s—perhaps even the same year that Jonathan Edwards delivered his "Sinners" sermon—another clergyman, the Reverend Thomas Bayes, working in Great Britain, invented a simple but profound mathematical means to connect outcomes with causes. His proposition for study went something like this: is it possible to define how Event B is caused by Event A, knowing that the two things are related in some way?

This sounds like a simple and unremarkable achievement, but it turns out to be anything but simple—or unremarkable. It reaches to elemental truths, both in mathematics and in metaphysics. Bayes's work was actually twofold: a mathematical challenge requiring a numerical theorem with attendant calculus proof, and a metaphysical conceit of ontology about the very nature of existence.

From these dual points of view, Bayes's question can be stated more clearly, in a way that relates not just to a mathematical proposition but to how we process information to make decisions and reach conclusions. Now, his question is as follows:

When I learn new information about something that I already believe, how likely is it to make me change my mind?

I invite you to reread this question carefully and spend a moment pondering it, as it will occupy us for the rest of this chapter and beyond. As we will see, Bayes's question is one of the most profound of all time. Fundamentally, it presents humankind with a true challenge—one that has directed people's thoughts throughout history: how and why we change our minds. More specifically, how does one alter one's beliefs in the face of new evidence? Being aware of Bayes's challenge provides an entirely new perspective on thinking. It is that deep. It is indeed quantification in our daily thought processes.

Momentarily, we will see that his question has application across all sorts of circumstances in daily life, but first let us understand the true essence of that quintessential question. In examining it, I will introduce some terms that are useful to understanding it; these terms are commonly employed in discussion of Bayesian thinking. Note especially that Bayes's question involves two pieces of information: "new information" and "something that I already believe." Also, pay attention to the fact that there is an order to the information:

- the first consideration is "something that I already believe": this is called a *prior belief* or simply a *prior* and in logic is typically represented as Event A;
- the second consideration is "new information": this is called a *new evidence* or just *evidence* and in logic is typically represented as Event B.

An important aspect of these two elements is their relative strength. In other words, if one's prior belief is strongly held, the new evidence must itself be powerful in order to be motivating enough to change one's mind. Conversely, if one's prior belief is only weakly held, the new evidence need be only weak itself to be motivating enough for a change of mind. Imagine for each statement that there is sliding scale of confidence or adherence, with *weak* on the left side and *strong* on the right side. Reread the two bullet points but this time with consideration of their sliding confidence scales. Imagine how when one scale slides up, the other slides back, causing you to either hold fast to your prior belief despite new evidence or change your prior because of the new evidence.

With our new terminology in place and with this elaborating information, here is Bayes's question stated again:

When presented with **new evidence** *about a* **prior belief***, what is the* **probability that I will change** *my mind?*

Addressing this question methodologically is called "Bayesian thinking" or "Bayesian estimation." The latter term suggests it is a problem in probability theory, and this is exactly how Bayes addressed it. Now you can appreciate how foundational Bayes's proposition is. We do this kind of thinking all day long, instantaneously. But specifying it mathematically is quite another thing—this is what Bayes has done. It is truly a remarkable feat, and one with profound implication for our daily lives.

An additional important point is to realize how hugely innovative and imaginative Bayesian thinking is. Before Bayes brought his idea to the table, probability was a concern of counting the frequency of events. This was done by sampling a population, carefully and deliberately, as we saw in Chapter 5. By the central limit theorem, persons working the field knew the true mean of a population is ever more accurately estimated as the number of samples (or trials in an experiment) increases.

But, in a practical sense, only so many samples may be drawn; hence, the true mean is only theoretically derived and never actually known. Even when there is a high degree of accuracy in its estimation, the true mean is just a highly informed guess. There is no consideration of confidence in the value taken as the true mean; rather, by this thinking, the mean is simply accepted as true or not. Other statistics (e.g., the standard deviation) are calculated from this accepted value.

Such an approach to data is called "frequentist" because it stems from the frequency of counting samples. Frequentist thinking dominated probability theory—and was predominant among mathematicians and other probability theorists—since the beginning and continues a strong tradition today. Sometimes statisticians refer to themselves as "frequentists," meaning that they adopt this perspective for data handling.

But, as we saw in Chapter 5 (with Bernoulli's law of large numbers and Laplace's distributions of linear combinations of large numbers of independent random variables into the central limit theorem), there is another way to view the true mean of a population. This alternative—called (surprise, surprise) "Bayesian"—views the true mean not as a given value to be estimated but as itself a value that falls within certain limits. With Bayesian estimation, that probability can—indeed, should—be determined.

To see the difference between frequentist and Bayesian thinking, consider a simple example. Suppose, after many samples (everything very carefully done), a mean value of 70 is calculated. The frequentist would accept 70, be done with further refinement of the true value, and proceed to calculate other statistics and follow-on procedures. They could then happily go to lunch.

The Bayesian, however, is just getting started. This person sees the worth of modeling 70 as a probability of falling within certain limits, say, between 68.5 and 71.5, as in this expression: $P(68.5 < 70.0 < 71.5) = 95\%$. This expression means that, given that the true value actually falls with the range from 68.5 to

71.5, there is a 95 percent probability that the observed value 70.0 falls within the range of the actual true value. We return to this point in Chapter 9, where its larger context is given.

Here, Bayes presents this idea as logical thinking in probability theory. Later on, we see that Gauss defined it mathematically as a probability density function.

From this brief introduction, we now turn to learning a bit more about Bayes and then to how he solved his estimation problem—this "Bayesian estimation." We'll learn quite a bit, too, about how it fits into our mindset today. Think quantification. You may be surprised!

****

To truly understand Bayes's reasoning, it is necessary to appreciate the social and cultural context in which he was working, a point emphasized throughout this book. Bayes, remember, was a Presbyterian minister, every bit as devoted to preaching and practicing his faith as was Reverend Jonathan Edwards in America delivering his Reformed Theology in his "Sinners in the Hands of an Angry God" sermon. Thus, Bayes saw probability theorizing as religious: a study in theology put into practice via mathematics. Bayes sought to prove the existence of God. Bayes was working from a point of view called the "God argument," wherein all things in creation are original to the Creator, including mathematical theorems and all other numeracy. Humankind has only to discover them, which is itself a glorification of God.

This was the same intent proposed by de Moivre for his work that we saw in Chapter 5. Bayes had an additional advantage over his most of his minister colleagues because he was a gifted mathematician, and he knew he could employ his specialized calculus skills in his religious pursuit, too.

The thinking of Bayes and de Moivre (and many others from this era) brings to the fore a subtle but important ramification of the God argument in mathematical reporting. Some authors of the histories and chronicles of mathematics, when mentioning a new equation or math, use the term "invented," whereas others describe things with the word "discovered." At first blush, this may seem like an inconsequential difference in terminology, but it turns out there is a lot of substance to the choice of wording. It is revealing of differing perspectives. In one, everything—for example, mathematical advances, prior beliefs, and the orbits of celestial bodies—is "discovered" because it comes from God. (Recall that Newton, by introducing gravity as an explanation for physical phenomena, upset this belief.) In the other, these things are "invented" as original thoughts and ideas, stemming from an individual. For the Reverend Bayes, we clearly know his point of view.

Bayes laid out his mathematical invention in his magnum opus *An Essay Toward Solving a Problem in the Doctrine of Chances* (Bayes and Price 1963). From the title, one may imagine that Bayes intended to make only a slight comment and perhaps suggest some minor development to de Moivre's work, but his essay turns out to be much more than just a few remarks. It lays out probabilities and the whole perspective of Bayesian thinking, sprinkled liberally with religious references.

Bayes's first efforts in probabilistic thinking were to advance the three theorems of numbers we saw earlier: the binomial, the law of large numbers, and the central limit theorem. He worked specifically on solving the missing calculus of de Moivre's *Doctrine of Chances*. But as time went on, his work went much further than simply advancing the theorems. Because he was working from the God-argument perspective, he was very open in his thinking, always looking for new "discoveries" as ways to glorify God. Rather than limit his work to just numbers and equations, this broader perspective took him on a journey into the then-nascent field of probability theory.

His quest led him to explore *why* things happened, beyond the *how* of mathematics. When carried out, his efforts proved so elemental to the theory of probability that, almost inadvertently, he formulated it as a whole discipline. Although not the first to focus intellectual efforts on studying probability itself (we have already seen several folks interested in the pursuit), Bayes did bring to the field a structure and direction. Until that time, it had been considered a subordinate branch of statistics, and its developments were not organized under a cohesive rubric. Now, probability theory was itself an independent study.

Now, I draw these things together. Imaginably, Bayes's most notable single accomplishment—the one that gave form to the whole of probability theory—was to invent "Bayes's theorem" (also called "Bayes's rule" or "Bayes's law"). This is the Bayesian thinking (also Bayesian estimation) we just saw in his famous question. At essence, Bayes's theorem defines the rules for probabilistic thinking. So important is Bayes's theorem in establishing probability theory that it has been said that it "is to the theory of probability what the Pythagorean theorem is to geometry" (Jeffreys 1973, 31).

In intellectual circles from many disciplines, Bayesian thinking is widely—and deliberately—accepted. It is a near-perfect thought in science: "When presented with new evidence about a prior belief, what is the probability that I will change my mind?"

Even apart from academia, Bayesian thinking is adopted widely. In economics, as just one example, there are many Bayesians. Microsoft cofounder Bill

Gates, early in Microsoft's history, declared that he was a Bayesian in his approach to many practical computer problems. It is not unheard of for college students to wear a T-shirt emblazoned with "I am Bayesian" (definitely something for the cryptic in-crowd). It is adopted in many other contexts, too. In fact, you may use it in your thinking—with your fully quantified worldview—without even realizing that you are Bayesian, too!

Bayes did a lot of his developmental work while tutoring students in local pubs. He was a respected teacher. Taking advantage of his immediate resources (in his circumstance, a billiard table), he taught his theorem to many. Figure 6.1 is a graphic made by Bayes himself to illustrate his problem. If the figure immediately brings to mind the three theorems discussed earlier—the binomial, the law of large numbers, and the central limit theorem—you are indeed observant. Bayes's theorem is built on them.

To illustrate his argument, Bayes had two billiard balls, which he labeled W and O. Before rolling the balls, he hypothesized that they would stop somewhere between the shortest roll (the horizontal line B to A) and the longest roll (a distance of the vertical line I to i). He made this his scale of 0 to 1. He rolled the W ball down the table and measured the distance to where it came to rest. He hypothesized that 50 percent of subsequent rolls would stop to the left of his

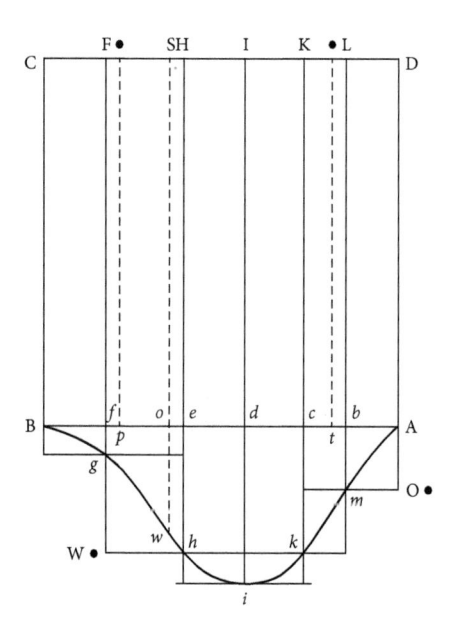

Figure 6.1 Bayes' illustration of estimating probability using a billiard table

(Source: from T. Bayes, *An Essay Toward Solving a Problem in the Doctrine of Chances*)

original toss. This hypothesis was his *prior belief*; that is, a belief before new evidence is presented. He then rolled the W ball many, many times, each time recording where it stopped. He used this scenario as a means to question his prior belief. With only a few rolls, he did not change his prior, but as more evidence was garnered, he slowly began to change his prior belief, eventually coming to a changed belief (say, that 60 percent of the time, a rolled billiard ball would come to rest to the left of his original roll).

Then he repeated the whole scenario with the O ball—but this time recorded when each ball rolled to the right of his initial O ball roll (testing again this new *prior*). His demonstration was of how a prior belief changes when presented with new information and that the change is proportional to the strength of the new evidence. Significantly, as anticipated, the distribution of the number of occurrences where the balls rolled (to the left or to the right) was a binomial, which we know from the famous three theorems eventuates into a bell-shaped distribution.

****

Now, we can really get into the Bayesian mindset, where things get interesting (don't worry—not more complicated, just more interesting!). Bayesian thinking is inductive reasoning, wherein a conclusion (which itself is usually a general statement or rule) is reached from prior observations. For example, suppose one is given a series of numbers, say 3, 6, 9, and 12, and is asked to determine the next number in the series. As seen, the prior observations are a numeric series incremented by 3. Hence, by inductive reasoning, one would conclude that the next term is 15 because it is the next increment of 3 (12 + 3 = 15). Inductive reasoning is opposed to deductive reasoning or abductive reasoning. Inductive reasoning is commonly employed in mathematics and used throughout all of scientific research.

The seventeenth-century philosopher and orator Sir Francis Bacon saw inductive reasoning as a way of understanding nature in an unbiased way, as it derives laws from unbiased observation. His thinking—done only about one hundred years ahead of Bayes's productive years—was enormously influential in setting ground rules for the scientific method and the observational techniques we saw in Chapter 3 with Mayer and the others. Now, we will use inductive reasoning in Bayesian estimation.

As we realize by now, Bayesian thinking—estimating the probability of changing beliefs based on new evidence—involves contemplating a set of probabilities: (1) the probability of a prior belief being true and (2) the probability of new evidence causing a change in the prior. Taken together, these considerations coalesce to form a central feature of Bayesian estimation: "conditional probabilities."

Conditional probability is the probability that Event B will occur given that Event A has already happened, and presuming (of course) that the events are related. Thus, conditional probabilities require three elements: Event A, Event B, and their intersection. Both events must be true and can occur independently, but they must relate in some way (i.e., have a correlational relationship, whether positive or negative).

Before going further on conditional probabilities, I pause to mention a closely related idea that is often confused with it: a "joint probability." Without realizing the distinction, either type of probability can be misinterpreted. A *joint probability* is merely the likelihood that two events are true. It does not have the added stipulations for the conditional probability described in the previous paragraph. For example, consider the two live births, where one is a girl and the other a boy. There's no dependency and no sequence—simply A and B. It brings the question, "What is the probability from two live births that one will be a boy and the other a girl?" This circumstance is a joint probability. The answer is 25 percent, because the sex probability for each birth is 50 percent (0.5), and $0.5 \times 0.5 = 0.25$. This is a legitimate joint probability question, but not a conditional probability.

Conditional probabilities commonly appear in our daily lives, even though few of us have previously classified our thoughts as such. Some examples of conditional probabilities are given in the following list. Each example of Bayesian thinking is from real life. In other words, they are nontrivial scenarios. As a feature of the theorem's richness, these probabilities can be simple or complex. Note that the scenarios meet Bayesian conditions: each event is true and can occur independently of the other event, but there is some meaningful relationship between them, and they have a sequence of one first, then the second:

- A class is given two tests. Fifty-one percent of the students passed the first test, and 25 percent of them passed both tests. What percentage of those who passed the first test also passed the second test?
- A pharmaceutical company advertises that its test for a particular disease is 99 percent accurate. A negative test result indicates the patient is free from the disease. If 1 percent of the population has the disease and you test positive, what is the probability that you actually have the disease?
- A robotic machine welds circuit boards for televisions. Ninety percent of the boards are perfect, 2 percent of them can be easily repaired, and 8 percent of them are defective. An inspector discards the defective boards. What is the probability that the inspector will incorrectly pass a board that could be repaired or is defective?

- In the general population, 0.055 percent of males die before their fiftieth birthday. Also, about 0.011 percent of males younger than fifty have cancer, with a survival rate of five years. What is the probability that a forty-five-year-old male will celebrate his fiftieth birthday? With a cancer diagnosis? With a negative cancer diagnosis?
- A survey question asked, "Is there solid evidence that the earth is warming?" The results for a "no" response were 53 percent for Republicans, 14 percent for Democrats, and 31 percent for independents. According to earlier information, 33.1 percent of voters are Republican, 34.6 percent identify as Democrat, and 32.3 percent are independent. Given that a particular voter does not believe there is solid evidence that the earth is warming, what is the probability that the voter is Republican?
- The probabilities of having a boy or a girl are each 0.5. A couple has two children, the older of whom is a boy. What is the probability that the couple has two boys? (Note: there is an ordered sequence in this scenario.)

These conditional probabilities (and many millions of others), are solved with Bayes's theorem.

The mathematics of calculating Bayesian probabilities is relatively simple, but setting up the problem requires that one understand the logic of Bayesian scenario. Here's the rub, because it is where most folks get lost. Hence, I have spent much time explaining it, from various angles. The logic is not difficult, but it is not intuitive. It requires one to diagnose a scenario and arrange its parts into the Bayesian thinking. For a quick review, recall Bayes's elemental question, which was discussed earlier in the chapter and is repeated here for convenience: *When presented with **new evidence** about a **prior belief**, what is the **probability that I will change** my mind?* Keep in mind, too, that idea of conditional probabilities. Finally, we reach a tangible point to move forward: that is, we can solve a conditional probability to give answer to Bayes's question.

Bayes's theorem is typically expressed using the set notation of logicians, as follows:

$$P(A|B) = \frac{P(B|A)P(A)}{P(B)}$$

Here, the conditional probability of Event A given B is denoted by the term $P(A|B)$, which is read as "the probability of Event A, given Event B," as seen in the left side of the equation. Like all mathematics equations, the set notation of the theorem follows Euler's order of operations, mentioned earlier. We have just seen the left side of the equation, so now move to the right side. It reads, "The probability of B given A, times the probability of A, divided by the probability of B."

The outcome term $P(A|B)$ is the final conditional probability solution. As it occurs after the problem itself, it is called the "posterior probability." The term $P(A)$ is the initial degree of belief in A (the probability of A), or the prior probability. The term $P(A|B)$, as we have already seen, is the degree of belief after having accounted for B. And the term $P(B|A)$ $P(A)$ represents the support provided by B to A, or the influence of the evidence on changing the prior belief.

Sticking to just the logic (and eschewing the math, despite its being not at all difficult), suppose some adult American smokes cigarettes regularly at the rate of one pack per day and is interested in learning the odds of being diagnosed with cancer this year. We start the short journey to solution by looking at the incidence in the population of each phenomenon independently. First we ask, "How many adult Americans smoke at least one pack of cigarettes daily?" We then ask, "How many adult Americans received a first-time cancer diagnosis this past year?" Each amount is known from health surveys. Taken separately, these questions are simple frequency counts that are usually expressed as a percentage of the population, and the probability for each is the ratio between the number counted and the population.

According to recent disease statistics, the cancer incidence in adult Americans is 455 per 100,000 men and women (National Cancer Institute 2018). Hence, one has about a 5 in 1,000 chances (actually, 4.6 in 1,000) of having cancer diagnosed, or a probability of roughly 0.005 percent, from any cause. Similarly, for cigarettes, about 15,100 per 100,000 adults smoke (an interesting side note is that slightly more women than men smoke): about 15 of every 100, or a probability of roughly 15 percent. In this scenario, thus far, there is no dependency, and the circumstance is simple to describe.

When the conditions of the two probabilities—the incidence of cancer, and the incidence of smoking—are considered together, however, both logical and computational difficulties arise. The new, combined question we can surmise from what we have learned already is "Given an adult American who smokes at least one pack of cigarettes daily, what is the probability of that individual receiving a first-time cancer diagnosis?" This dependent consideration is a conditional probability. There are some mathematical stipulations for everything to work out. And we have already seen them: namely, the events must be independent (i.e., each can happen separately), and they must relate in some way. We have long known that cancer and smoking are related. Here, we wish to consider them together, or "conditionally."

Our initial hypothesis for answering the question is just the odds of receiving a cancer diagnosis this year—which we know to be 0.005 percent, our prior information. We evaluate our prior against the evidence, that is, the known proportion of smokers in the population (approximately 15 percent). Bayes's theorem makes a ratio of these two probabilities. This combined ratio is the conditional probability. In this scenario, the conditional probability calculates to approximately 3 percent chance of receiving a new cancer diagnosis this year, given the fact that the individual smokes one pack of cigarettes daily.

The following famous problem is often used when teaching Bayesian estimation: determining the chances of a woman having breast cancer. From Centers for Disease Control statistics, about 10 out of every 1,000 women aged 40 to 50 are diagnosed as having breast cancer (about 1 percent). But the initial screening test is not infallible. A woman with breast cancer has a 90 percent chance of a positive test from a mammogram, while a woman without cancer has a 10 percent chance of a false positive result. Hence, the question refines to "What is the probability of a woman actually having cancer given that she had a positive screening test?" This is Bayesian conditional probabilities in action.

Stated in Bayesian notation given earlier, this is

$$P(cancer \mid positive\ test) = \frac{P(positive\ test \mid cancer)P(cancer)}{P(positive\ test)}$$

In other words, the left side of the equation is the question under consideration (remember, the line | means "given"). The right side of the equation is the probability of a positive test, given actual cancer, (0.90) times the probability of cancer at all (0.01), divided by the probability of a positive test, which is the four percentages of all possibilities. With numbers plugged into the equation, it is now

$$P(cancer \mid positive\ test) = \frac{(0.90)(0.01)}{(0.01)(0.91)+(0.99)(0.10)}$$

$$P(cancer \mid positive\ test) = \frac{9}{108} = 0.0833$$

Thus, the probability of a woman actually having cancer, given that she had a positive screening test, is about 8.3 percent. Incidentally, when one hundred physicians were given these statistics and asked the question, more than three-quarters of them responded incorrectly. The good news is that fewer women actually had cancer than they anticipated. With Bayesian estimation, we know the probability more exactly.

****

From the foregoing, I now present a fun problem of Bayes's theorem and Bayesian thinking. The problem I will give here is very simple, but appreciate that it is one that Bayes would find utterly incomprehensible, since it involves the twentieth-century ideas of a TV game show and cars, neither of which was invented until well more than a century after his death. Notwithstanding, this famous illustration of Bayesian thinking is called the "Monty Hall problem."

Readers may recall the very popular TV game show from years past *Let's Make a Deal*. During its most illustrious run (from about 1960 into the 1980s), a Canadian-American named Monty Hall was the host (although he has since passed away and the show now runs with new hosts and in worldwide syndication). In the show, a lucky contestant was brought onto the stage where Monty presented her with three closed doors, enthusiastically explaining that behind one of the doors was a fabulous prize (like a new car)—but behind the other two doors lay a dud prize (like a goat).

To a cheering audience, the contest would select one door, say Door no. 1, leaving Door no. 2 and Door no. 3 not selected. To prolong the suspense, Monty Hall—who secretly knew where the car was—did not immediately open the selected door (in this case, Door no. 1). Instead—and playing full on to the audience—he opened one of the unselected doors, always opening a door that revealed a goat. Say he opened Door no. 3. This left two doors still unopened: Door no. 1 (the contestant's initial choice) and Door no. 2. Then, Monty asked the contestant if she would like to change her choice from Door no. 1 to Door no. 2. As you can imagine, the audience loudly chimed in with shouts of "Change!" or "No!"

This scenario is a perfect Bayesian conditional probability. At first, there are three unknown doors. Of course, the contestant wants the car and "hypothesizes" that it is behind Door no. 1 (just a guess among the three choices at this point, but still a hypothesis with a 33 percent chance for Door no. 1). But then Monty Hall opens either Door no. 2 or Door no. 3 to reveal what is behind it, which is equivalent to presenting new evidence. We know that he always opens a goat door. Then, the Bayesian question is asked of the contestant: "Do you want to change doors?", or, in Bayesian terminology, "Modify your original hypothesis?"

Most contestants are unaware of conditional probabilities and approach the door choice as just simple odds: at first, there is a one-third (33 percent) probability of selecting the car door; and when Monty Hall takes one of the unselected doors out of consideration by showing a goat behind it, the contest thinks, "Ah-ha! With just two doors left, I now have a fifty–fifty chance, so it makes no difference whether I keep my original choice or change doors." But the contestant

Car behind		Host opens	Total probability	Stay	Switch
Door 1	1/2	Door 2	1/6	Car	Goat
	1/2	Door 3	1/6	Car	Goat
Door 2		Door 3	1/3	Goat	Car
Door 3		Door 2	1/3	Goat	Car

(Initial selection probabilities: 1/3, 1/3, 1/3)

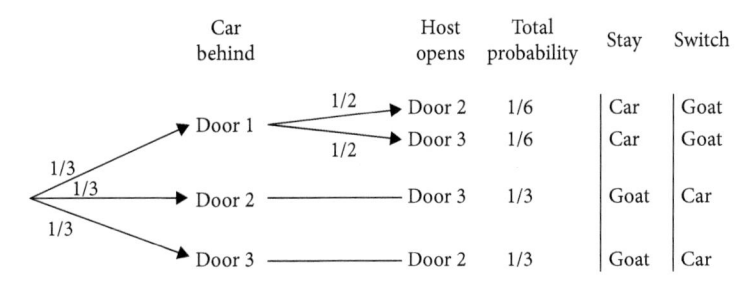

Figure 6.2 Tree diagram of the Monty Hall probability problem

is not employing Bayesian conditional probabilities. We know there is a better way to approach the door choice.

To see the problem as conditional probabilities, consider it from the beginning. At first, the hypothesis is just a one-third probability (33 percent) of choosing the car door, leaving a two-thirds probability (66 percent) of not selecting the car door. Stated as opposite, there is a two-thirds probability that one of the two unselected doors conceal the car. Now, think Bayesian: Monty Hall opens a goat door, leaving just one unselected door. But the two-thirds probability (66 percent) of those doors hiding the car has not changed. Hence, with the new evidence (one goat door known), we happily change our choice to the remaining unselected door. In doing so, we have doubled our odds. That is, by the change, the contestant has increased the probability of selecting the car door from 33 percent to 66 percent. Now you know that when confronted with this scenario, you should always change from your initial selection to the new choice.

A tree diagram of the problem, where the player initially selects Door no. 1, is shown in Figure 6.2.

A version of the Monty Hall problem uses one hundred doors, one of which hides a car while the other ninety-nine doors conceal goats. Here, the contestant initially selects any door, at random, for a 1 percent probability. From the remaining ninety-nine doors, Monty Hall then opens ninety-eight of them—all goat doors. Now, only two unopened doors remain. An unsuspecting contestant may think the odds have increased from 1 percent to 55 percent. But we know better: given the new evidence (ninety-eight goat doors removed from consideration), the odds change from 1 percent to 99 percent. Change doors—you'll almost certainly get the car!

****

There is yet another surprising fact about the Reverend Thomas Bayes's work on his theorem. After making his initial revolutionary invention, rather than continuing to advance his accomplishment, he abandoned working on it altogether. In fact, he did not labor on this problem for the remainder of his life. No one knows why he stopped working on his revolutionary approach to linking causes with evidence. Perhaps he did not appreciate his own accomplishment, although by his words and other achievements (wherein he applied the theorem and discussed his approach), this seems unlikely. Fortunately, it was rediscovered later by another mathematical genius, Pierre-Simon Laplace, who gave it its modern mathematical form and scientific application. Nonetheless, it all started here—with Bayes addressing the God argument.

When Bayes's theorem initially came on the scene, the notion of contingent probabilities was called an "inverse probability." The word "inverse" means a reverse in position: here, Event B is reversed, or reconsidered, based upon the outcome of Event A. Since conditional probabilities always calculate odds, it means that the likelihood of Event B is considered in light of the odds for Event A. Although some may consider the wording "inverse probability" awkward, it is rather descriptive. It captures the connection between causes and outcomes, expressed by a probability: the likelihood ratio.

It is almost impossible to overstate the influence of Bayes's theorem. It has surprising relevance to our lives today. We see it everywhere. Many readers will recognize that Bayes's theorem (at least the logic and solution part, often without calculations) is routinely taught in classrooms across the world, from high school to graduate school. Its ubiquity is amazing. It is used every day by people working in literally hundreds of occupations, from finance, to business, to medicine, to sports, to engineering, to law—almost everywhere. For example, many modern machine learning techniques rely on Bayes's theorem. Spam filters use Bayesian updating to determine whether an email is real or spam, given the words in the email as evidence.

In addition, many specific techniques in statistics, such as calculating P values, are best described in terms of how they contribute to updating hypotheses using Bayes's theorem. In 1992, Bill Gates announced that "Microsoft's competitive advantage lay in its expertise in Bayesian networks" and that "Bayesian theory is firmly embedded in Microsoft's Windows operating system" (quoted in McGrayne 2011, 242–3).

There is an entire specialty within criminal and civil law of probabilistic jury selection and legal defense. I recently came across an Internet site devoted to cataloging legal cases where evidence was based on Bayes's theorem. It currently lists more than three hundred cases, many of which are important and well known, including the infamous O.J. Simpson murder case. The question was

"What is the probability an individual with matched blood type, identical shoe print, and having sufficient strength, as well as with a known motive and opportunity, could have committed such horrific murders?" The judge did not allow this probability evidence because, he said, he did not understand it. Cases in which DNA evidence is used to either convict or exonerate a defendant also rely on probabilistic evidence calculated by Bayes's theorem.

A nontechnical and delightfully readable historical account of Bayesian statistics is offered by Sharon Bertsch McGrayne, a journalist who assembled numerous anecdotes of important and unexpected uses of Bayesian statistics in a book titled (ready for a mouthful?) *The Theory That Would Not Die: How Bayes's Rule Cracked the Enigma Code, Hunted Down Russian Submarines, and Emerged Triumphant from Two Centuries of Controversy* (2011). Evidently, she didn't want you miss anything. Notwithstanding the book's title, it does show the ubiquity of Bayes's theorem in our lives today.

For us, Bayes's theorem is a touchstone accomplishment to quantification.

CHAPTER 7

At Least Squares

The imminent transformation to quantification continues apace, although the journey is neither simple nor direct. The themes take a while to coalesce: the integration of nebulous societal forces with methodological advances in mathematics and statistics, highlighted by the invention of probability theory. At a very human level, it presages an intersection between the everyday life of ordinary people and the scholarly activity of some prominent mathematicians. Our storyline, moving toward a transformed worldview, follows these two interrelated and codependent tracks. And so, we follow them on the two land masses on which they had the most impact: the Americas and the larger European continent through Eurasia.

As more immigrants arrived in America, the fiercest of the Puritans were losing influence, although a more tolerant version of their belief was still staunchly held by a substantial group. Reason was settling in on both sides of the ocean. Reason gives a door to quantification in thinking and outlook.

This same sense of purpose drove the western expansion of the United States, which began with the journey of the Corps of Discovery, the party of explorers led by Meriwether Lewis and his gregarious friend William Clark, the team history just calls "Lewis and Clark." They reached the Oregon coast in 1806 with help from several Native American tribes and individuals, the most famous of whom is the young woman Sacagawea, who acted as their guide. Lewis expressed this spirit of self-determination when he wrote the following in his journal:

We are about to penetrate a country at least two thousand miles in width, on which the foot of civilized man had never trodden; the good or evil it had in store for us was for experiment yet to determine...[yet] entertaining as I do, the most confident hope of succeeding in a voyage which had formed a da[r]ling project of mine for the last ten years, I could but esteem this moment of my departure as among the most happy of my life. (Lewis and Clark (1804–6) 2005, Lewis, April 7, 1805)

In Europe, things were quite different. Notably, the French Revolution had ended only a few years earlier, bringing the ascent of Napoleon Bonaparte in the late 1790s. Although hardly benevolent, given his own brutal excesses, the young general did bring to an end the chaotic, bloody twelve-year revolution. Knowing how this period came about is useful to us because the period to follow (the beginning of the long century, from after the Congress of Vienna and onwards) is largely a reaction to it, and it is an era where many significant quantifying events happened.

During the period leading up to the French Revolution, nearly everyone in France was poor and ill-served by an absolute monarchy and a nobility-laden government that was increasingly out of touch with the people it ruled. Those in the governing class lived lives of conspicuous excess. When the people complained that they did not even have enough bread to eat, the young queen of France, Marie-Antoinette, allegedly uttered what is one of the most infamous quotes of all time: "*Qu'ils mangent de la brioche*" ("Let them eat cake"). Historians doubt the quote's authenticity, but the reality of despotic rule by a debauched nobility is indisputable.

The culture of entitled privilege by the nobles led motivated groups of determined protesters to storm a weapons fortress in Paris—the Bastille Saint-Antoine—on July 14, 1789, almost the exact day that the new US Constitution went into effect (it had been ratified one year before). The protests by ordinary Frenchmen almost immediately grew to include virtually all the peasants and anyone outside of the government. The French Revolution had begun.

The monarchy fought back with almost unimaginable savagery. They ordered French troops to carry out a bloody campaign in which many thousands of protesters were killed. Any peasant even remotely suspected of not supporting the government was brutally killed by the soldiers; many were shot at point-blank range. The crackdown's most intense period was a horrific ten-month Reign of Terror ("*la Terreur*") during which the government guillotined untold masses (some estimates are as high as 5,000) of its own citizens as a means to control them.

One of the architects of the Reign of Terror was Maximilien Robespierre, a French nobleman and lifelong politician. He explained the government's slaughter in unbelievable terms, as "justified terror … [and] an emanation of virtue" (quoted in Linton 2006). Slowly, however, over the next few years, the people gained control. In the end, many nobles, including King Louis XVI and his wife Marie-Antoinette, were themselves executed by guillotining. Some of the

Figure 7.1 Drawing of *La Terreur* in France, "Execution of Robespierre"
(*Source*: http://commons.wikimedia.org/wiki/Category:Public_domain)

atmosphere of the French Revolution is captured in a well-known drawing, which is shown in Figure 7.1; it was printed at the time with the heading "Execution of Robespierre and his conspiratorial conspirators against freedom and equality: long live the National Convention which by its energy and surveillance has delivered the Republic of its tyrants." It is not known who made the drawing originally.

The most lasting consequence of the French Revolution, however, was not ending the rule of a monarch and enshrining a popular sovereignty but fixing in people's minds the ideals of the Enlightenment, namely, reason and self-determination. This new atmosphere, cleansed of the oppressive aristocrats, gave rise to much intellectual development. A kind of cognitive release began to grow in the minds of ordinary people, leading to a mindset that viewed things quantitatively. In unseen but profound ways, people gained a sense of control over their lives. It moved from their thoughts to their daily behaviors. The tremendous mathematical advances of this period are a direct manifestation of the refreshed social environment.

****

One of the most profound developments in probability theory—indeed, in all mathematics—happened in the opening years of the nineteenth century in

France and, almost simultaneously, but independently, in Germany. This is the invention of the "method of least squares." At its essence, least squares is a data-handling technique that specifies, for plotted data, where a line may be drawn through or close to many of the individual data points. The line can then be interpreted to give information about the data variable(s). This is useful to understanding how different variables relate to one another, and it allows for mathematical prediction. As we will see, it is one of the most amazing numerical inventions of all time. It is truly *that* significant.

Least squares, as a route to regressions, is routinely employed in virtually all quantitative fields of endeavor, including statistics, engineering, aerospace, medicine, economics, and more. The invention of the method of least squares is a watershed moment in quantification, too, because of its focus on predicting and forecasting. Using mathematics to anticipate events and phenomena is given first light with the method of least squares.

It was invented twice; that is, two individuals working at roughly the same time, but independently, came to invent it, although in slightly different forms: Adrien-Marie Legendre and Carl Gauss. Its first public architect was Legendre. Immediately upon Legendre's publication of his version of least squares, its importance was recognized by scholars and academics as a significant advance in data handling. Without hesitation, it seems, they adopted the method in their own work.

Clearly, the method of least squares presaged other important developments in the field, setting probability theory on a trajectory to become a major player in mathematics. The preeminent historian of statistics, Stephen Stigler, remarked about its astounding acceptance and widespread use: "The rapid geographic diffusion of the method and its quick acceptance in these two fields [astronomy and geodesy], almost to the exclusion of other methods, is a success study that has few parallels in the history of scientific method [sic]" (Stigler 1986, 15)—impressive to think.

Inventing the method of least squares was the result of a concerted and determined problem-solving effort by several mathematicians to find a reliable way to predict the arcs (actually, moving ellipsoids) of stars and planets. They knew that such an accomplishment was possible with the new calculus inventions of Newton, Pascal, and Leibniz, as well as through Bayes's theorem, but exactly how to approach a solution was less clear. Thus, they had defined the problem and identified a working strategy—but after that, seemingly nothing—a brick wall on predicting where planets and stars would move. The more mathematicians and astronomers looked at the problem, the more intractable it seemed. No one knew exactly what to do next.

At the same time, a practical circumstance closer to the daily lives of ordinary people came to light that made the quest for solution urgent. There was a sudden and important need at hand: namely, to determine longitude. As we saw earlier, with oceanic travel and trade made relatively safe from pirates (remember, Captain Kidd was captured and hanged in a public spectacle), Great Britain and several other countries sought to expand trade via early cargo shipping.

The two quests—determine the arc of certain celestial bodies and accurately fix longitudes—were interdependent. It was widely recognized that having dependable data on the movement of celestial bodies would be of enormous use to mariners for ocean navigation, a problem of knowing one's longitude. (Precise estimates for latitude would come later.) At the time, finding one's way by the stars while in the middle of an ocean was risky business because their relative position changed each night, and estimating their trajectories was done only by eye or perhaps with an inexact sextant. This gave a rough estimate of their static location, but, before Legendre's work, mariners had no way to reliably predict the precise location of stars over time and thus learn their true longitude.

In reality, navigation was as much luck as anything else. Commonly, ships would sail off course, get lost, and travel to who knows where—often to their doom. The method of least squares provided a mathematical technique for making the necessary predictions for both the tracking of celestial bodies and then, following on that, for longitude.

Legendre studied the problem of determining arcs, which bore out his version of the method. Specifically, he was investigating the errors in measuring latitudes. He noted that, regardless of whether the errors were symmetric about the mean, he could summarize them through a process of linear combinations. Like Mayer's early method of combining observations, then, Legendre was treating the data as a unitary whole, rather than focusing on its individual elements. This insight formed the beginnings of his method of least squares.

He published his efforts in 1805 in a well-received book titled *Nouvelles méthodes pour la détermination des orbites des comètes* (*New Methods for Determining the Orbits of Comets*) (Legendre 1805). But, surprisingly, the body of his book was not about least squares; rather, it was a long description of the problem itself, measuring arcs for latitudes. The method of least squares was mentioned merely as an application to solving the arc dilemma. Only in an appendix did he describe his method.

Further, and significantly, he did not provide a mathematical proof, which would have been an expected accompaniment to such a momentous invention. This omission is all the more surprising because he was knowledgeable in calculus and certainly able to do the proof. Even further, we learn through his later writings that he realized his invention's importance. We do not know his reasons for the curious placement or incomplete description. Regardless, he is given credit for its first public account.

Meanwhile, Gauss also invented the method. In fact, he did so earlier than Legendre, but he did not publish an account of it until later. It is generally acknowledged that Gauss developed his approach first, possibly as early as 1794, when he was just eighteen years of age. But, he did not disseminate his solution until 1809, when he published *Disquisitiones arithmeticae* (Latin for *Arithmetical Investigations* but often translated as *The Shaping of Arithmetic*) (Gauss 1801, Gauss and Waterhouse 1986). I discuss this important work by Gauss more thoroughly in Chapter 9, in the context of Gauss's life and other accomplishments.

For his part, Gauss approached the problem more mathematically and provided a calculus proof to his work. His version is slightly more complex than Legendre's, and his version fits the more general case. It is the standard today.

Gauss acknowledged that Legendre published a solution sooner than he did, but he wanted Legendre to concede to him the earlier accomplishment, something Legendre refused to do. They quarreled about this attribution for the remainder of their lives, mainly in a series of letters to others. Gauss considered the method of least squares simple work, saying the reason he did not publish it sooner was that it was "so simple." He said,

> I had no idea that Mr. Legendre would have been capable of attaching so much value to an idea so simple, that rather than being astonished that it had not been thought of a hundred years ago, he should feel annoyed at my saying that I had used it before he did. (Quoted in Gorroochurn 2016a, 165)

Then, pouring salt on Legendre's wounds, Gauss added,

> Therefore, in my theory of the motions of planets, I was able to discuss the method of least squares, which I have applied thousands of times during the last seven years. (Quoted in Gorroochurn 2016a, 165)

Ouch!

In support of Gauss, there is no doubt that he employed the method several times in his work prior to Legendre's publication of it. But realize, too, that many advances in mathematics, beyond the method of least squares, were simultaneously invented or discovered by two or more people. In fact, it was not unusual for several mathematicians to independently invent or advance some procedure or another at about the same time. Moreover, at this time, advances in mathematics and allied fields such as statistics and probability theory were coming on fast and furious.

Another incident, too, led to their mutual dislike. To understand this dispute, consider a tremendously complex arrangement of numbers useful to higher mathematics, the "law of quadratic reciprocity." Although the law was used by mathematicians, its potential was not reached at first because it had no accompanying proof, making it somewhat suspect. Legendre tried again and again to provide a justifiable proof. Eventually, he gave up, saying it was too difficult. Gauss, however, at age nineteen developed a full proof. He continued to work on the problem over the years and eventually established six other proofs, and then went on to demonstrate that there are no further proofs!

There is no record of how each man reacted, but we know that Gauss was private and taciturn in his manner, whereas Legendre was quick to show anger and liked others to know about his accomplishments. I imagine that, upon learning of Gauss's proofs, he must have seethed in rage and jealousy (only conjecture, of course).

Legendre was elementary, too, in his approach to the problem of predicting the arc for celestial bodies, starting with observation. He said:

> I have thought that what there was better to do in the problem of comets was to start out from the immediate data of observation, and to use all means to simplify as much as possible the formulas and the equations which serve to determine the elements of the orbit. (Quoted in Gorroochurn 2016a, 165)

Note, particularly, that he gathers his data initially from that stalwart of our story: observation. In this account of quantification, all follows from it.

Unfortunately for Legendre, despite the widespread and lasting impact of his method of least squares, as well as the overall importance of his full mathematics canon (as we shall see, it was voluminous), he did not enjoy a high reputation while alive. But that did come later, after his death, and today Legendre is regarded as a mathematician of enormous stature. His name is one of only seventy-two names inscribed on the Eiffel Tower.

While there once, I looked for his name, but in truth it is difficult to see any of the names from the ground, as their lettering (originally in gold) is now relatively obscure and placed high up. The seventy-two inscriptions are located on the sides, under the first tower. Gustave Eiffel chose the names himself as an "invocation of science." I hope you have an opportunity to see the inscriptions, too, because viewing Legendre's name on the impressive Eiffel Tower engenders a sense of awe at his accomplishments.

Another of Legendre's achievements puts quantification in a touchable perspective. He developed a theorem relating to spherical triangles, a methodology useful for drawing triangles on the surface of a spheroid, particularly one that is elongated, like the prolate shape of a football. This can be seen by imagining using a marker to draw a triangle shape on a football—harder than it seems. Legendre's theorem defined this shape mathematically, which has application for figuring shapes and distances on many celestial bodies, including some moons and asteroids. Somewhat later (in 1790), a French astronomer with the unwieldy name of Jean Baptiste Joseph Delambre used Legendre's theorem to specify the exact length of the meter.

While many of our main characters (e.g., Newton, Bernoulli, and de Moivre) came from comfortable beginnings, Legendre had even greater childhood comfort, being born into a wealthy family. But, like de Moivre, he became infirm in later life and died poor because his pension was initially withheld (though later restored) by the French government due to his lack of support for the government-backed candidate at the Institut National, the Comte de Corbière,

Figure 7.2 Caricature of Legendre

(*Source*: http://commons.wikimedia.org/wiki/Category:Public_domain)

Ministre de L'Intérieur (Legendre 1810). Sadly, another of these most accomplished persons came to an unhappy ending.

Then, as if to add insult to injury, there is the fact that we simply do not know what he looked like. We presume that Legendre was at least ordinary in appearance; however, the only known portrait of him is a caricature in which he looks anything but normal. The drawing shows him with enormous, flailing white hair and the scowl of a junkyard bulldog. See for yourself in Figure 7.2.

It gets better: for nearly two hundred years, until 2005, books and other publications printed the profile portrait of a handsome, distinguished-looking man as Adrien-Marie Legendre. But, it turns out that it was not him at all! It was a case of mistaken identity. The two-centuries-old portrait was that of an obscure French politician named Louis Legendre—a man with the same last name but a different first name. So today, only the caricature survives as representation of one of the most famous mathematicians of all time. We know the appearance of virtually all other famous and accomplished persons within the last two hundred years or more, but not of Adrien-Marie Legendre. Alas—look again at the caricature and realize this man is an important figure in our quantification story. Dare I say, he influenced you because his achievements helped to fundamentally shape how we all view the world.

****

Now, on to describing the method of least squares. My description is rather basic, because our focus is on how the method advances quantification rather than on didactic explanation. Overall, the method's math is relatively simple, although a complete explanation includes linear algebra and the proof is by calculus.

For those interested, however, a more complete description is available via dozens of textbooks and websites, albeit with varying levels of sophistication and quality of description. One historical source on least squares is Brunt's classic *The Combination of Observations* (1931). This detailed account of the method provides a complete description, its calculus proof, and a lengthy commentary on when it may be appropriately used in experiments. While decidedly dated in approach and writing style, this work is significant and has been called "culturally important." It may be particularly useful to those interested in the method itself and its development, rather than as an introduction.

Understanding the method begins with an awareness of correlational relationships. While we all know that a mathematical correlation does not automatically prove causation, variables are commonly suggestive of one another, often to some meaningful degree. Though seemingly obvious, keeping this fact at the fore is extremely helpful in understanding exactly how the method of least squares works. Generally, the correlational relationship is only partially established. In fact, unless a correlation is perfect ($r = 1.0$), the relationship is always only partially determined. This means that when just two variables are considered, one of them does not fully correspond to changes in the other; rather, it offers only a partial explanation, and sometimes even just a hint.

When the method of least squares is employed in regression (its most frequent use), the variables involved are classified as either "dependent" or "independent." In regression terminology, the dependent is *regressed* onto the independent variable, signifying that changes in the independent variable (regardless of whether such changes are naturally occurring or the result of experimental manipulation) can explain at least some of the change in the dependent variable.

With this information set, several useful concepts come into play. One of them is to determine whether the relation between the variables is linear or nonlinear. This determination is made by considering the full range of values for the variables. A truly linear relationship signifies that two variables change uniformly throughout their full ranges. In other words, low, middle, and high values for one variable are matched by exactly corresponding changes for low, middle, and high values in the other variable: a change in one is matched by an identical (or exactly proportional) change in the other.

In real-world research, however, it is rare for two variables to behave so consistently. Following a strict criterion like this could hamper both research and other useful interpretations. Thus, in most contexts, if the relationship between two variables is fairly consistent throughout the scale (generally, with only minor deviations at the extremes), we treat them as linearly related, putting up with the slight inaccuracy. For example, most of the time, we treat people's height and weight as linearly related variables, knowing that some exceptions exist. (For the technical purist, there is an ever-so-slight inaccuracy in my description of linearity: for there to be true linearity between variables, they must also possess a characteristic called "homoscedasticity"—but that is beyond our scope to describe here.)

Linear relationships are the kind of relationships we mostly imagine in everyday thinking. Some instances are when figuring out profit over time, calculating mileage rates, or predicting which team will win a sporting event. Even without plotting the relationship on a graph, our minds instantaneously gauge a linear relationship.

Nonlinear relationships between variables, however, are typically more complex and can take several forms. In one form of a nonlinear relationship, the two variables may relate consistently for part of the scale, but then deviate at other parts. For example, consider two common educational variables: test performance and test anxiety. Research has shown that a mild amount of test anxiety can be motivating and is positively related to increased test performance. But this holds true only for the lower end of the anxiety scale. As anxiety for examinees increases to the middle and upper values on the anxiety scale, test performance varies widely: for some students, their test performance is increased, but, for others, test performance actually declines. The variables, then, are inconsistent through the scales and not linearly related.

Further, in more complex nonlinear scenarios, the relationship may be cubic or even quartic, meaning the bends in the relationship curve may go up and down more than once, to two or even three times. Of course, at some point, the relationship is so inexact that there is deemed to be no relationship at all.

Figure 7.3 displays the two kinds of ways variables relate: linear and nonlinear. Two of them are linear (positive and negative), while the other is nonlinear (in this case, random).

In research contexts, statisticians have figured out ways to deal with these situations, such as transforming the variables' scale to log or square root metric; however, such changes complicate the interpretation. These scenarios show how messy data can be, and that quantifying real-world events is never clean. As Roseanne Roseannadanna (a beloved character on the original *Saturday Night Live* TV show), when facing some perplexity, would say, scrunching her face, "It just goes to show ya. It's always something. If it's not one thing, it's another. But, it's always something!" True, true.

This is the point at which Legendre (and Gauss, but with a different approach) figured things out: namely, how to best represent these changing relations

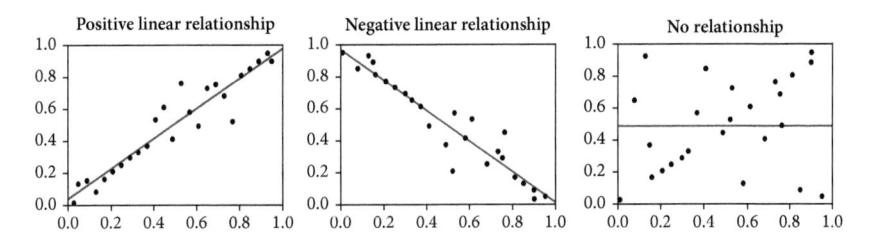

Figure 7.3 Correlational relationships: positive, negative, and none

between (or among) variables and, in particular, how to accurately extend interpretation of them even when there is no new data. Recall that they were focused on solving the two interrelated problems of (1) determining the arc of certain celestial bodies and (2) accurately fixing longitudes. Both men were innovative thinkers, and they brought to the problem an entirely new perspective. Rather than see variation in the data as errors to be minimized, they viewed the data set as a unified whole and explored ways to represent it, in toto.

Their solution was to draw a line through the plotted data to represent the many values via a single feature. This line is called the "line of best fit." The name is apt because, graphically, when all the observed data points are plotted on the x- and y-axes, a straight line can be drawn through them that is as close to all of them as possible. In other words, any other line drawn through the separate data points would not be closer to them: hence, it is literally the best-fitting line to the data. Obviously, in only one circumstance—a true, perfect correlation between variables—would the regression line go directly through all the data points. The line of best fit is the regression line. Figure 7.4 shows this regression line.

With this concept established, the practical question becomes, how to figure the line? Aha: the method of least squares! Fitting the line is accomplished by calculating two features of the line and then drawing it from these values. The two features are (1) its starting point, called the "intercept," and (2) its "slope." Both are important. For where to start (the intercept), it may seem logical to

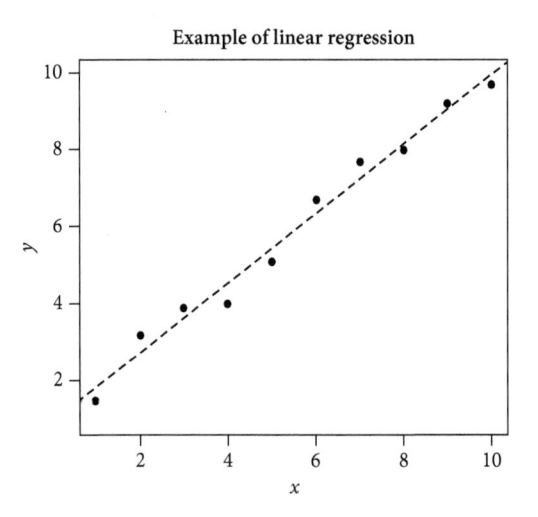

Figure 7.4 Plot of line of best fit: regression

begin at the zero point on the y-axis, but rarely does the regression line begin at zero, because, for many variables, there are no cases of zero value. For instance, with height, weight, test anxiety, and almost any belief or opinion, zero is not a realistic value. Because the start of the regression line is not zero, we must employ some procedure to figure out where it should begin. Accordingly, we define the intercept as the expected mean value on the y-axis (vertical) when the x values (on the horizontal axis) are zero. That is to say, the intercept is the average amount of y (i.e., of whatever variable that is dependent) without any consideration of x (the independent variable).

For the line's slope, we must calculate its gradient. Here is where x enters our consideration. For its meaning, the slope is interpreted to be the degree of influence of the independent variable on the dependent one. Generally, a steep slope shows significant influence, a low slope means less influence, and a flat slope indicates no correlational relationship at all.

Mathematically, positioning the regression line in the data is accomplished by minimizing the sum of squared values for each data point to the mean of all the data. It is called the "least squares solution." Because of this, regression for linear variables is technically referred to as "ordinary least squares," or often just OLS.

Again, imagine the data plot just mentioned with the data points arrayed and a straight line drawn through them. For each data point, envision a short, vertical line between it and the main regression line. For data points above the regression line, their individual short lines will extend vertically downward to the regression line, and, correspondingly, data points below the regression line will have a short vertical line extending upward. This is shown in Figure 7.5.

Next, imagine further that each short, vertical line, regardless of whether it extends upward or downward to the regression line, is one side of a perfect square. Theoretically, then, a perfect square is drawn for each data point. If there are, say, eighty-five data points, the image of the plot will show eighty-five perfect squares. Some of them are above the regression line, and others are below it, but each square will have one of its corners touching the regression line. Knowing the area for each square is useful to our progress because the smaller the area, the closer the initial data point will be to the regression line. By simple math, the area of a square is calculated by cross multiplying its two dimensions, length and height. For instance, if a square has a length of 2 and a height of 2, its area is 4 ($2 \times 2 = 4$).

In theory, in the method of least squares, the areas for all eighty-five (or however many) squares are summed to make a total squared area. This total squared

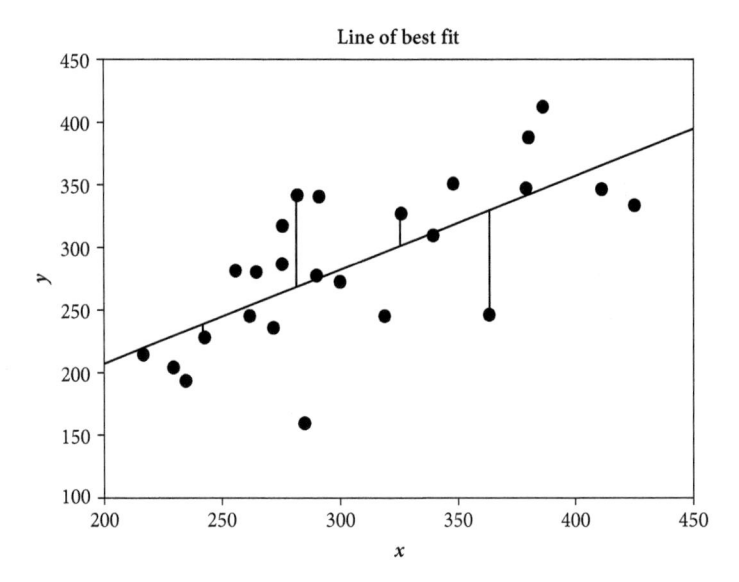

Figure 7.5 Plot of line of best fit, showing some of the drop lines

area is important to figuring the line of best fit because, in regression, the line of best fit is established when the total squared area is at its minimum. Figure 7.6 is a plot of the regression showing least squares.

In the OLS scenario, when either of the two features for a regression line (i.e., its intercept and its slope) is changed, the total squared area will be different. Obviously, there are infinite possibilities for where a line may be placed over a set of data points, but willy-nilly placement is not sensible. The method of least squares finds the best-fitting line to the observed data. This feature lets us interpret the regression line as the best representation of the data set.

That is the whole point of the regression line: to represent an array of data, gathered by observation for x and y variables. Hence, the regression is a graphic representation of the relationship between two variables. (It can be extended to include more than two variables.)

Placement of the regression line to maximize its representation of the data is easiest to imagine—and to compute—when the variables are linearly related. But even when the variables are not linear, the method can be extended to find a second-degree curve fit, or it can be generalized to accommodate even more curves. Knowing this, we turn our attention to practical features for calculating the regression line itself.

Although this book is not technical, and as mentioned several times already, I deliberately eschew formulas in it, I am making an exception for the regression

AT LEAST SQUARES | 115

Line of best fit

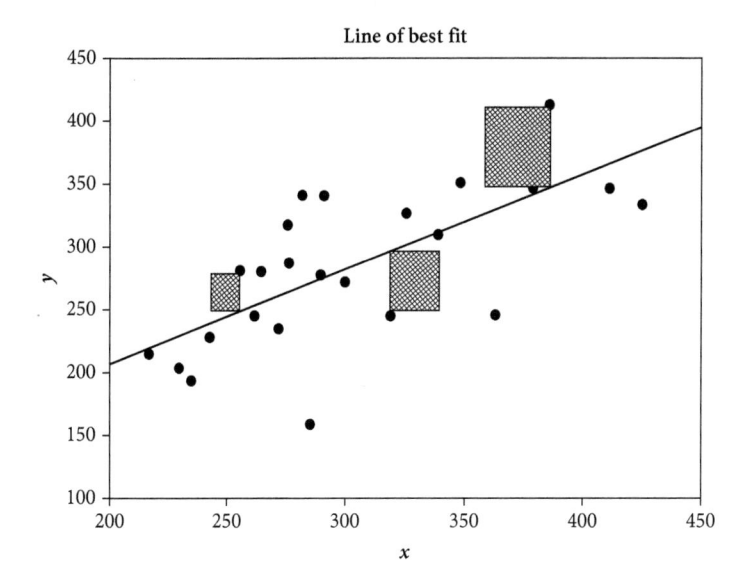

Figure 7.6 Plot of regression showing some of the squared distances

equation: Legendre's or Gauss's equation. Because this equation is so well known and the idea behind it is vital to our new quantifying perspective, it is worthwhile to examine it, however briefly. The regression equation culminates from the above explanation and is the mathematical expression of the method of least squares. It is as follows:

$$y = a + bx$$

Simply, in the regression equation, the y is our outcome, or the value we want to produce for the dependent variable. In practice, this value is a prediction for the dependent variable, given consideration of the independent variable. The right side of the equation shows how we involve the independent variable in predicting the y. Specifically, the a is our starting point (the intercept) showing where we begin placement on the dependent variable scale, as discussed above. The x is the independent variable, and the b (called a "coefficient"—remember this from expanding the binomial) is the amount of influence that x has on producing a y outcome. (Those working with regressions call this the "percent of variance explained.") This degree of influence comes from the initial correlation between the two variables. A strong correlational relationship signifies that the independent variable can be powerfully influential in predicting a given value in the dependent variable. A weak correlational relationship means it is less so.

As apparent in the equation, one must calculate the *a* and the *b*. (Realize that solving the regression equation is different from calculating its proof. As I mentioned earlier, detailing the proof requires the more sophisticated mathematics of calculus and linear algebra.)

Calculating the *a* and the *b* is not difficult. Regardless, I do not solve for them here, because that would distract us from our narrative. Our purpose is to understand the concept and see how it plays into the story of quantification. Regardless, solving the regression equation can be done by hand, or computer help is readily available on many websites. In Microsoft Excel, a macro can even be used to do the calculations. And, of course, all major statistical programs, such as SAS, SPSS, Minitab, Mathematica, and R, specialized statistical modeling programs such as Stata and Latent GOLD, and even some graphing programs like SigmaPlot can do the work in seconds. The very number and scope of these available programs illustrates how universal regression is, not only for the researchers who use these programs but also for any interested person in the broader population.

For researchers and mathematicians, there are three main uses for correlation and regression. First, in experiments, regression can be used to test hypotheses about cause-and-effect relationships. Here, the researcher explores changes in the dependent (y) variable that may be caused (at least in part) by changes in the independent variable (x). Recall that such changes may be naturally occurring (such as gender) or the result of external incidents or circumstances (such as in changing income levels), or they may be wholly contrived by the researcher (such as when an experiment requires participants to use less sugar in their diet for a week).

A second purpose for regression is to explore the correlational relationship between variables without necessarily inferring a cause-and-effect relationship. In this scenario, the variables are not identified as independent and dependent. The researcher does not manipulate one of them to examine changes in the other. Rather, naturally occurring ranges in both variables are contrasted, and the strength of association is studied. When changes are observed, the researcher infers some degree of influence of one variable on the other.

The third common use of regression is for mathematical estimation. A given value in the independent variable is set to determine the corresponding regressed value in the dependent. This is often useful for planning purposes.

From even this brief description of the method of least squares, one can appreciate its elegance as a problem-solving technique, and the profound influence it has had on the concept of quantification generally. It is both clever and sophisticated. From a math theory standpoint, it is relatively simple; with only a little study, its internal workings are easy to understand. Procedurally, the calculations are not difficult or involved. Even its calculus proof is undemanding to those with the requisite background. And, of particular importance, interpreting the results is straightforward and usually unambiguous.

Second, beyond the elegance of its use as a data-handling technique, the method of least squares exercises tangible influence over our lives, in both daily and extraordinary events. It brings prediction to a reliable, meaningful place. Just think how often it is that we predict something. Here, I do not mean *predict* as a researcher's term, but as it is used in popular speech. How common it is for us to say words such as *predict, guess, anticipate, foretell, forecast, envisage,* and *portend*, and phrases such as "I know what will happen" and "I can see that...." Certainly, you can add to this list. In popular speech, these words and phrases all indicate roughly the same thing: prediction about future events.

Regression brings to us the notion of prediction as a quantifiable thought and perspective. It is, quite simply, so broadly and profoundly influential in moving us to a quantification outlook on our world that its importance cannot be overemphasized. Thus, we have gone from observation to probability and on to prediction, through regression—quite an evolution of thought and processing of information. We will see regression throughout the rest of the story of quantification.

CHAPTER 8

Coming to Everyman

Up to now, the story of our adopting a quantified worldview has centered on advances in mathematics and probability theory as invented and developed by several extraordinary and brilliant scholars and academicians: Bayes, Bernoulli, Gauss, and others. However, they all were of a type—scholarly and intellectual—and most were associated with academia, often as lecturers or professors at universities. Now, notably, we extend the influence of quantification to a broader populace—namely, to Everyman. Most readers will recognize at least the name "Everyman" which is an allegorical figure who represents all of humanity in the fifteenth-century anonymous morality play *Everyman*. I use Everyman here purposefully to suggest that during this time, access to this new knowledge—and the concomitant worldview—was transiting from the intellectuals to ordinary people, who were occupied with their daily doings and not with mathematics.

Thus, in our storyline, quantification is finally coming to ordinary folks. Across the social order, quantification is slowly but inexorably seeping into the daily life of Everyman.

Further, my argument for quantification is that the incredible mathematical achievements of these times happened in large part because the unprecedented historical events of the era set the stage for rapid intellectual growth. But, as a theme for the times, it was a reciprocal arrangement. These mathematically inspired ideas themselves contributed to spheres of thought of the Enlightenment, such as the overthrow of absolute monarchs in France, the decline of dominant states such as imperial Great Britain, the emergence of individualism in the new United States, and the soon-to-come industrial revolutions in England, the Continent, and America.

The Error of Truth. Steven J. Osterlind. Oxford University Press (2019). © Steven J. Osterlind 2019.
DOI: 10.1093/oso/9780198831600.001.0001

Reciprocally, history supported mathematics, and its development advanced history. The effect on Everyman was to move toward a quantified worldview.

Specific inventions, too, contributed to quantification. One such invention was itself nearly a touchstone in changing people's worldview. In fact, it alone contributed to a change in how people thought of the world and about their place in it. This important invention was the chronometer, a device for keeping accurate time in all locals and circumstances. The chronometer's real value was that it worked regardless of whether it was held still or in motion, and whether it was at sea level or high on a mountain.

Of course, timepieces and clocks had been invented long before the chronometer came along, and by the early eighteenth century, they were quite accurate . . . within limits. But their exactness held only so long as they were on steady ground, and they often did not work in thin mountain air with lower air pressure. On a rolling ship, early clocks simply did not work at all, because their pendulums swung around erratically. With the invention of the chronometer, however, mariners and mountaineers alike could accurately know the time.

Obviously, knowing the time was important to mariners. But the urgency of this need is highlighted by the fact that, with their new knowledge of accurate time, they could now determine their exact location as longitude and latitude. In modern times, since at least the early 1960s, of course, latitude and longitude are both determined by one's position relative to satellites, and not directly by using time.

Another problem for time was also widely recognized then: the need for a common reference locale. Up to then, time often varied from place to place, and several rulers, like the tsar in Russia, demanded that wherever they were standing be the place from which time should be measured. High noon in Paris was different from high noon in St. Petersburg, for example. This was more than a problem of ego, too, because railways were coming onto the scene and they desperately needed a standardized time system.

While people recognized the problem and knew that the solution was to fix a certain place as the referent for timekeeping, there was debate about where that place should be. Finally, in 1884, an International Meridian Conference was held in Washington, D. C., and the choice was made: the village of Greenwich in England. Politics aside, this was a natural choice to be the center for time keeping because the Royal Observatory—the place from which official time observations were taken—was located there. At exactly noon on November 18, 1884, telegraph operators sent a coordinated signal to major cities across the world as the start of modern time. From that second to today, everywhere across the globe, time is relative to its start in the London suburb of Greenwich—Greenwich Mean Time (GMT). The international dateline, too, is determined from GMT.

For mariners, having GMT meant they could determine their own longitude as the difference between a time interval and the ship's position relative to the Greenwich meridian. This was a huge advance for many obvious reasons. For us, it signifies that quantification is finding ways into the ordinary necessity of commerce, and thus into the life of Everyman.

Another surprising event in the mid-eighteenth century also advanced the notion of quantification: Europeans suddenly had sugar! Sugar—made cheap and readily available around this time—is more than just symbolic of the rise of a middle class; it is a great leveler in society, a shared experience for the populace, accessible to both Everyman and the nobility. Prior to this time, sugar had been affordable only to persons of means and had been a signpost luxury that signaled belonging to the upper ranks in a well-established societal class system. The same leveling effect was caused by tea, coffee, and chocolate (which were also entering daily commerce about the same time).

If tangible things could be held in common, then certainly intangible thoughts—worldviews—could be shared, too.

In other areas of life, too, there was a plethora of quantifying ideas coming into frequent use by all people, ordinary folks and intellectuals alike. For example, actuarial tables for estimating longevity were commonly used in life insurance plans now; in business, market-planning estimates were growing in sophistication and gaining widespread acceptance as tools for making financial decisions. State-sponsored lotteries ("lotto" in most countries, even then) were up and running (apart from gambling itself, which, of course, has been around forever). These ordinary phenomena rely on mathematics, statistics, and probability theory operating in the background.

Quantification as a worldview had been sown and was growing; its branches were beginning to spread.

****

As some daily experiences, like pleasurable cooking with coffee, sugar, and chocolate, were becoming more commonly shared across all social classes, one intellectual intentionally sought to bring the measurement of uncertainty—probability—to the everyday affairs of ordinary folks. This individual was Pierre-Simon Laplace. Unlike most of his predecessors, and even his contemporaries, Laplace did not stay behind the stony walls of academia. He wanted to make the notion of probability public and commonplace.

To Laplace, quantification was something beyond formulas and theory—it could be a part of everyone's daily experience. He took several specific actions to bring his intention to realization. As one effort, Laplace moved the study of probability theory beyond its then almost-exclusive application to games of chance. He applied it to problems in architecture, engineering, geodesy, and many other areas. This action was imaginative, and although he was not the first scholar to apply probability outside of gaming, he was the most deliberate in pursuing its broader application.

To our modern minds, Laplace's aim seems very unimaginative, but at the time—in the context of intellectual activities staying generally with peers and inside ivory towers—what he did was almost groundbreaking. Remember, ordinary people of the day had no contact with such intellectual activities. However, to truly appreciate Laplace's efforts, it is useful to first examine the work of Joseph-Louis Lagrange, a man who influenced him greatly. Examining his work will aid in our understanding of Laplace's contribution to quantification.

Lagrange, like Laplace, was a solid mathematician, and both men were fairly renowned. Lagrange was only a few years older than Laplace, having been born in Turin, Italy (then, Piedmont-Sardinia) in 1736. His exact birth name is unverified, but it is mostly reported as either Giuseppe Lodovico Lagrangia or Giuseppe Ludovico de la Grange Tournier. We will just call him Lagrange, since he is the only one of that name in our story. A sad fact of his family history is that, of eleven children born to his parents, he was the only one who survived beyond infancy.

As one might imagine, as an influencer of the gifted Laplace, he made substantive contributions to quantification himself. While a member of the French Academy of Sciences, he helped to establish the metric system there during the period of the French Revolution.

Lagrange worked primarily in analytical and celestial mechanics and made important contributions in these areas, principally adding to work originally done by Newton, as was similarly done by Laplace. He is best known, however, for extending the work of Euler (whom we met briefly in Chapter 3) to found a very technical study called the "calculus of variations."

The calculus of variations seeks to find (read: mathematically define) the shortest distance between two points. On a flat plane, obviously, this is just a straight line, something taught in beginning analytic geometry or algebra courses as a set of points whose coordinates satisfy a given linear equation. We saw this earlier in the description of least squares in Chapter 7. But in geodesic space (as with a sphere, say, Earth or other celestial body), the figuring is not so simple.

Look at the sphere displayed in Figure 8.1, and imagine how to find the best-fitting line between the two end points of the bolded line, A to B. It is a

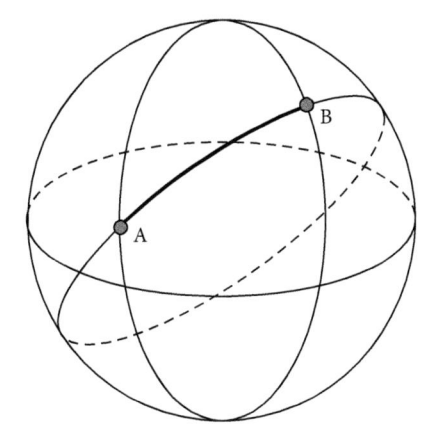

Figure 8.1 Projection for the orthodrome problem
(*Source*: http://commons.wikimedia.org/wiki/Category:Public_domain)

problem of special curvature. In analytic geometry, this is a famous problem known as the "orthodrome problem." As can be seen, in three-dimensional space, a straight line from the starting point would just extend into infinite space and never reach the second point, so something different is needed.

The solution is to draw a line tangential to the curved surface to create a defined space that can then be measured. Because the arc along this curve is constantly changing, a large number of tangents are needed, each considered a variation. Mathematically defining these spaces leads to a set of what are called "Euler–Lagrange equations." Lagrange based them upon initial work by Euler. The shortest distance can now be evaluated through repeated estimation by these equations. The entire process is called "Lagrange's calculus of variations." Its description can be lengthy; in fact, there are entire books devoted just to explaining Lagrange's calculus of variations and its application in various fields. We will not go there—thank goodness.

The orthodrome problem, with its solution by Lagrange's calculus of variations, is given life today in long-distance travel. It is easiest to visualize this in a scenario on an azimuthal projection (viewing the earth from distant space) that considers a "great circle." A great circle divides a sphere into two equal hemispheres, making it easier to see actual distances. In a given scenario (e.g., see Furuti 2012), suppose an airline flight is scheduled to travel from São Paulo, Brazil, to Tokyo, Japan, with a required refueling stopover. If the route planner used an equidistant cylindrical map and drew a straight line between the cities, it would seem that Hawaii would be the logical stopover point. But this would be a mistake. As seen in the great circle, a stopover somewhere along the northeastern coast of the United States (e.g., Boston, New York, or Philadelphia) would be the preferred choice. Figure 8.2 makes this point clear.

124 | THE ERROR OF TRUTH

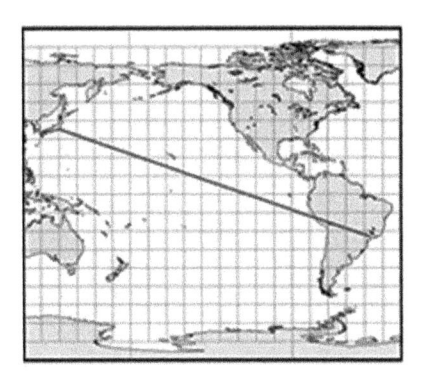

Figure 8.2 Azimuthal orthographic projections illustrating the orthodrome travel problem
(*Source:* http://commons.wikimedia.org/wiki/Category:Public_domain)

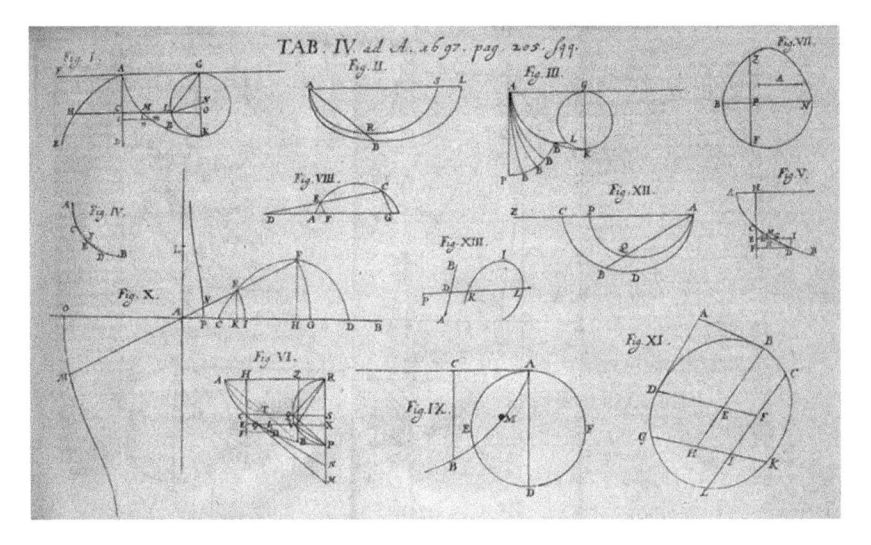

Figure 8.3 Galileo's notes on attempting a solution to the orthodrome problem
(*Source:* http://commons.wikimedia.org/wiki/Category:Public_domain)

Although Lagrange provided a solution to the orthodrome problem, he was not the first to tackle it. Just a few decades earlier, Galileo, without the benefit of calculus, had also tried to solve it. Actually, at the time, Galileo had not been focusing on this particular problem; rather, he had been trying to figure out what caused the tides to ebb and flow. In trying to solve the surf problem, he realized that the curvature of the earth was somehow involved and that tidal change was related to the distance of any particular spot on the earth's surface to a given point on the moon. Without being aware of it, he was tackling a version of the orthodrome problem. Although he was on the right track, he was not successful in finding a satisfying solution.

Remember, he did not have the benefit of Newton's calculus. One famous historian of mathematics noted that since even a genius like Galileo could not solve the orthodrome problem, then you should not feel too bad for not solving it either (Gorroochurn 2016a).

To our benefit, notes from Galileo's musings have survived, and it is interesting to view them and imagine his thinking, as shown in Figure 8.3. Now, knowing what we do about Lagrange's mathematical accomplishment with his calculus of variations, it is little wonder, then, that even an accomplished intellectual like Laplace was so heavily influenced by him.

****

By all accounts, Laplace was astonishingly bright. In fact, he is sometimes called the "French Newton"—a high compliment indeed. Laplace lived about three-quarters of a century after Newton and admired him greatly, both in his personal life and for his work. He studied Newton's calculus creations and realized that they were potentially applicable to a much broader array of disciplines than just astronomy and geodesy, the arenas in which Newton primarily worked. In particular, he understood that although Newton had provided its broad outlines and set an overall structure, many of its parts were still undeveloped. Laplace aimed to finish it, and in doing so, make both differential and integral calculus complete disciplines that could be integrated into other fields of study.

Remember, Laplace had the intention of employing quantifiable activities in areas where it would be influential to benefit ordinary people. The fact that he opened up calculus for application to a much broader range of problems than was heretofore known is some evidence of that. This influence lives on. Today, Laplace's work is especially well known in mechanical and civil engineering applications, such as for building bridges, dams, or skyscrapers—really, for making almost any large structure. As a result of bringing calculus to such a wide range of new applications, his work is considered a watershed accomplishment.

He described his deeds in a five-volume treatise, titled the *Traité de mécanique céleste* (*Celestial Mechanics*; published from 1799 to 1825), a pivotal and inspiring work. In it, he also set out his nebular hypothesis regarding the origin of the solar system. Catching the scientific world by surprise, he postulated the existence of gravitational collapse and the notion of black holes—quite astonishing!—explaining how some stars could have a gravitational pull so strong that even light could not escape, and it would be sucked back into the star. He provided

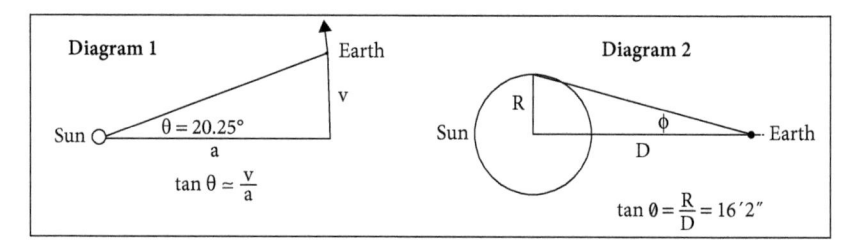

Figure 8.4 Laplace's diagram used to explain black holes
(*Source*: from S. Laplace, *Traité de mécanique celeste*)

a series of calculus and geometry theorems to prove his claim, which I discuss momentarily.

Figure 8.4 depicts a diagram Laplace drew to explain his theory of black holes. Although the idea of black holes, and even a time–space relationship, had antecedents in incomplete indications by others, many modern astrophysicists credit Laplace as their discoverer. Einstein credited Laplace with having tremendous influence on his own work in developing his theory of relativity. We will revisit this important work in Chapter 15, during our discussion of Einstein.

Beyond the legitimate black hole explanations to astrophysicists and others, in our time, of course, black holes (and sister notions of wormholes, a tenth dimension, warp drive, etc.) have been great silage for Hollywood moviemakers. Dozens of movies and TV shows have been fashioned around them, although most make no attempt at scientific accuracy. My quick Google search brought up these titles: *Interstellar*, *The Void*, *The Theory of Everything*, even *Godzilla vs. Megaguirus*, and many more. All stem from Laplace's original quantification work, attesting to his influence on us today. To think, Laplace could scarcely have imagined his theories showing up (however inaccurately imagined) in movies designed for popular audiences, some even made for preteens! In serious astrophysics work, black holes are very complex, and describing them is difficult. In Chapter 15, we meet a contemporary scholar, Professor Kip Thorne, who proffers a description of them in the context of modern advances in astrophysics.

While explaining his ideas for black holes and the universe, Laplace invented a mathematical process by which a very difficult calculus formula may be solved through the application of a much simpler one. The easier expression, it turns out, applies not only to a single more-complicated formula but a whole family of such expressions. Thus, Laplace has developed a "transform." In mathematics, a transform is a function that converts numerical information to an equivalent but different form. This is called the "Laplace transform."

$f(t)=\mathcal{L}^{-1}\{F(s)\}$	$F(s)=\mathcal{L}\{f(t)\}$		$f(t)=\mathcal{L}^{-1}\{F(s)\}$	$F(s)=\mathcal{L}\{f(t)\}$
1. 1	$\dfrac{1}{s}$	2.	e^{at}	$\dfrac{1}{s-a}$
3. $t^n,\ n=1,2,3,\ldots$	$\dfrac{n!}{s^{n+1}}$	4.	$t^p,\ p>-1$	$\dfrac{\Gamma(p+1)}{s^{p+1}}$
5. $\sqrt{t}$	$\dfrac{\sqrt{\pi}}{2s^{\frac{3}{2}}}$	6.	$t^{n-\frac{1}{2}},\ n=1,2,3,\ldots$	$\dfrac{1\cdot 3\cdot 5\cdots(2n-1)\sqrt{\pi}}{2^n s^{n+\frac{1}{2}}}$
7. $\sin(at)$	$\dfrac{a}{s^2+a^2}$	8.	$\cos(at)$	$\dfrac{s}{s^2+a^2}$
9. $t\sin(at)$	$\dfrac{2as}{(s^2+a^2)^2}$	10.	$t\cos(at)$	$\dfrac{s^2-a^2}{(s^2+a^2)^2}$
11. $\sin(at)-at\cos(at)$	$\dfrac{2a^3}{(s^2+a^2)^2}$	12.	$\sin(at)+at\cos(at)$	$\dfrac{2as^2}{(s^2+a^2)^2}$
13. $\cos(at)-at\sin(at)$	$\dfrac{s(s^2-a^2)}{(s^2+a^2)^2}$	14.	$\cos(at)+at\sin(at)$	$\dfrac{s(s^2+3a^2)}{(s^2+a^2)^2}$
15. $\sin(at+b)$	$\dfrac{s\sin(b)+a\cos(b)}{s^2+a^2}$	16.	$\cos(at+b)$	$\dfrac{s\cos(b)+a\sin(b)}{s^2+a^2}$
17. $\sinh(at)$	$\dfrac{a}{s^2-a^2}$	18.	$\cosh(bt)$	$\dfrac{s}{s^2-a^2}$
19. $e^{at}\sin(bt)$	$\dfrac{b}{(s-a)^2+b^2}$	20.	$e^{at}\cos(bt)$	$\dfrac{s-a}{(s-a)^2+b^2}$

Figure 8.5 Some Laplace transforms
(*Source*: http://commons.wikimedia.org/wiki/Category:Public_domain)

The Laplace transform is very important and useful—perhaps second only to the Fourier transform in its utility for solving differential equations. Individuals today working on engineering problems commonly use Laplace transforms. In fact, there are many Laplace transforms (at least forty common ones), but a precise number is difficult to state because there are various solutions depending upon which mathematical arrangement is employed, and each can be considered a transform application.

Even computing Laplace transforms directly can be fairly complicated; hence, mathematicians usually just use a table of already prepared transforms in their differential calculus work. These tables are commonly reprinted in textbooks on engineering, computing science, and other areas of mathematics. Figure 8.5 shows an example of a table of Laplace transforms.

****

Near the end of the eighteenth century, Laplace refocused his energy away from orbital calculus toward general statistical analysis and then on to probability theory. Probability theory absorbed him for most of his productive years, primarily the twenty-year span of the last decade of the 1700s and the first decade of the 1800s. His work in this field was especially significant. In Chapter 6, we saw Bayes's invention of the logic that became generally known as "Bayesian," and particularly its application through conditional probabilities in Bayes's theorem. There, too, we learned that, soon after inventing his now-famous theorem, Bayes mysteriously abandoned further work on it.

Laplace, to our great advantage, recognized its importance and began work to advance it. He reimagined Bayesian statistics to virtually invent the modern Bayesian interpretation of probability, a huge triumph and one that we will look at in some depth, in both this and later chapters. One important writer said: "Laplace's contributions to probability are perhaps unequaled by any other single investigator" (Newman 1956, 1322).

Laplace shares credit with Bayes for establishing probability theory (and its calculus foundation) as a distinct field of mathematical study. Their work in establishing the discipline is set apart from the correspondence between Fermat and Pascal first suggesting the field, including outlining some preliminary probability concepts.

Especially germane to our interest in quantification is that fact that Laplace resolutely sought to make the main ideas of probability theory accessible to a general audience. For Laplace, probability theory was a whole deterministic philosophy that should be brought to ordinary folks because it went to the heart of "the whole system of human knowledge" (Laplace and Dale 1995, 107). We will return to this quote momentarily because it explains much of Laplace's thinking. From this initial beginning, we can see plainly that it was also his philosophy of life.

It is the intersection of these two accomplishments—(1) modernizing Bayesian approaches to probability theory and (2) popularizing his philosophy about the theory—where Laplace contributes most to advancing the quantification perspective.

Before moving forward in our story of Laplace, however, we pause to look broadly at his work and life. Even while acknowledging his contributions to the study of probability, readers should not take away the impression that this was Laplace's single focus for all of his productive life. He worked in several disciplines and was prodigious in his output, especially in astronomy and mathematical physics. To cite his full body of work would require a very long recitation of accomplishments, many of which are not easily understood, much less concisely explained. We will stick to just the portion of his work that advances our interest in bringing quantification to a worldwide outlook: probability theory.

However, know that his reputation as the French Newton is indeed well deserved. In testament to his scholarship, Laplace is among the honored few mathematicians whose name is inscribed on the Eiffel Tower, along with our earlier friend, Adrien-Marie Legendre, the one with the scary caricature as his only surviving picture.

Apparently, Laplace had a natural facility with higher mathematics, grasping complex topics with an ease far beyond that of even most other mathematicians. This talent was noted early on by his teachers and then by others throughout his life. Biographers of Laplace routinely state that his acquaintances and colleagues mentioned him as a quick study. I can report firsthand that Laplace's natural ease for immediately understanding difficult concepts of higher mathematics is exceedingly rare, having spent much of my professional career in the company of mathematicians, some of whom are (or, sadly, were) exceptionally bright. Laplace, it seems, could grasp a tough concept right away, regardless of how complex, technical, or involved. Ah, to have such a gift of quick understanding!

There is another aspect to Laplace that is worth mentioning at this point. Of all the stunningly brilliant individuals we see throughout our story, Laplace is perhaps the most difficult to read and understand. There are two reasons for this. First, his work was exceedingly complex and technical, accessible to only the few with a deep knowledge of calculus and its theoretical base. Fortunately, over the years, scholars and theoreticians have been able to bring out his findings in more comprehensible forms, making them useful to us today.

We saw earlier that the Laplace transform is one example of this; another instance is his work in astronomy, where he was possibly the first person to have hypothesized the existence of black holes and the notion of gravitational collapse.

The second reason Laplace is tough to follow is his writing style. When he wrote about his numerical inventions, he was positively byzantine in his composition. His official translator to English, Nathaniel Bowditch, described it as follows: "Laplace made no concession to the reader. The style is extremely obscure and great gaps in the argument are bridged only by the infuriating phrase 'it is easy to see'" (quoted in Newman 1956, 1321). Bowditch added that while translating Laplace, whenever he came across the phrase *aisé de voir* ("it is easy to see"), he knew he had hours of hard work before him. Thank goodness, we will not read Laplace verbatim but only read about him!

Sadly, too, for Laplace scholars who wish to tackle his writings in their first form, many of his original writings were lost in a house fire at the home of his great-great-grandson, the Comte de Colbert-Laplace, in 1925. Nearly all the surviving works of Laplace have been collected in the fourteen-volume edition *Oeuvres complètes de Laplace* (*Complete Works of Laplace*) published posthumously between 1878 and 1912 by the French Academy of Sciences.

Laplace and Carl Gauss (already Gauss's name keeps popping in and out of our story; we finally meet him fully in Chapter 9) are considered contemporaries in their work, despite a twenty-eight-year age difference. Laplace was older. They corresponded regularly, and each made helpful suggestions to the other, given that their worked covered the same domain of scholarship, but they reported their work very differently. Not only did they have dissimilar writing styles, but their approach to explaining things was vastly different.

The reclusive Gauss was careful, thoughtful, and detailed in his explanations. He considered it mandatory to include a proof or two, or three, or even seven, in at least one paper. Sometimes, he even added commentary on why no further proofs were possible! Moreover, Gauss did not typically disseminate his findings until years after he had done the work (as we saw in Chapter 7 on his dispute with Legendre over the method of least squares). Routinely, he put things aside for a while, sometimes years, before returning for further consideration. Only then, after he was fully satisfied, did he publish his findings. We can say he was slow, deliberate, and extremely thorough, but, today, Gauss is widely praised for the thoughtful scholarship in his writings, despite their technical difficulty. His proofs are the standard for exacting scholarly study.

Laplace, on the other hand, was quick to publish his findings, cranking out scholarly papers at an astonishing rate. During his first three years as professor, he published thirteen substantive papers, a rate that is considered to be highly productive (and is the envy of every assistant or associate professor today coming up for promotion or tenure—I can attest to this, having reviewed the publishing record of many dozens of individuals over the years).

But, as we see from the comments of his longanimous (ever-patient) translator, he was haphazard in his writing, often jumping from thought to thought without any transition and leaving great gaps for the reader to fill in. Also, he borrowed liberally from the writings of others, sometimes even employing their phraseology, although he usually acknowledged his source. Today, this practice would be considered sloppy authoring, at best. Be that as it may, we know Laplace's work was original, even if getting through it is tough.

In terms of personality, too, the two men were quite different. Gauss was reclusive and academic and worked almost unceasingly in his study. We know he did not like traveling away from his hometown. Laplace, by contrast, was gregarious, always seeking contacts with a wide group of people, from academics to politicians. It is ironic that Gauss—the more careful writer and deliberate person—intended his publications for scholars and other mathematicians, whereas Laplace—the more difficult to read and understand—wanted to make his work accessible to a wide audience both inside and outside academia.

Each man contributes to our story of inventing and then spreading the notion of quantification, but in very different ways, obviously. Perhaps, then, Laplace and Gauss were not too different from any two people with distinct personalities.

Laplace was born to good parents of comfortable, but not rich, means. His father worked in agriculture before turning to selling wine in the Normandy region. As a boy, Laplace attended a Benedictine school in a small village in Normandy, where he showed exceptional math talents but did not appear to be a prodigy. His father, also named Pierre, wanted his son to become a priest. At sixteen, Laplace entered the University of Caen to study theology, intending to follow his father's wishes, but two professors, Cristophe Gadbled and Pierre Le Canu, noticed his talent in mathematics and encouraged him to study that rather than continuing toward the priesthood. They mentored Laplace throughout his higher education. He grew increasingly interested in mathematics, especially astronomy.

At some point, Laplace changed his study from theology to mathematics and announced his intention to become a professional mathematician. Laplace never adopted a formal religion fully, as his father had wished. Instead, he sought a kind of refuge in scientific determinism, something we will explore shortly.

While a university student, he wrote *Sur le Calcul integral aux differences infiniment petites et aux differences finies* (*On the Integral Calculus with Infinitely Small Differences and Finite Differences*), a paper confirming Lagrange's interpretation of Euler's assumption of a constant difference, a longtime problem of mathematical astronomy (Laplace 1878–1912). Quite astonishing for such a young person who had just been introduced to mathematics! He was given the opportunity to read this paper before the august French Academy of Sciences, which proved fortuitous to his future. Soon after the reading, he left the university (without graduating) to pursue his new interest in mathematics in Paris. But, before his move, he procured a letter of introduction from his university mentor, Le Canu, to the preeminent Paris-based mathematician Jean le Rond d'Alembert.

This d'Alembert was the editor of the first modern encyclopedia, which was published at the same time as Benjamin Franklin was publishing his *Poor*

Richard's Almanack (as mentioned in Chapter 4). By all accounts, this accomplishment was historic. Earlier encyclopedias are considered nascent in comparison to d'Alembert's. Its many editions (the 1911 edition is generally thought to be the best) have included original articles by a long list of important persons, including Voltaire, John Muir, Bertrand Russell, Sigmund Freud, James Dewer, Julia Child, and Einstein. When I was very young, a boyhood friend showed me his parent's copy of their 1911 edition of the *Encyclopedia*, but it was lost on me at the time. Regardless, it must have impressed me on some level because I still remember him showing me the set of books.

From its inception to today (the last print edition was in 2010, but it remains active online), the *Encyclopedia* has a reputation of strong scholarship. Much scholarly information given back by Internet search engines references the *Encyclopedia* as its basis. For quantification, the *Encyclopedia* is a veritable gold mine of valuable material for literally hundreds of fields. Perhaps as much as any other single publication, the *Encyclopedia* has promoted a quantified outlook for people on the whole of information retrieval.

Wikipedia, in contrast, is an unsourced offshoot of it to which anyone can contribute, and accuracy is mostly unchecked until challenged by others: a democratic, if not slightly anarchistic, update on the careful scholarship of the original.

Otherwise, d'Alembert made only minor contributions to his primary fields of mathematics and classical mechanics, and none on the scale of our central characters like Bernoulli or de Moivre, and certainly none nearly so important as the achievements of Newton, Gauss, or even Laplace. Thus, I mention him primarily as a person who was important to the life and work of Laplace, advancing his efforts.

In terms of character, d'Alembert was reportedly an opinionated and difficult person, with a reputation for being stubborn and pugnacious in his frequent arguments with other mathematicians. He made outrageous pronouncements and then attacked anyone who disagreed. For example, he declared, "The true system of the World has been recognized . . . everything has been discussed, analyzed, or at least mentioned" (quoted in Cassirer 1951, 46–7). Such a declaration— whether by d'Alembert or anyone else—certainly would invite disagreement and debate. For d'Alembert, however, it was routine grist for his argument mill. Incidentally, his preposterous assertion closely channels another infamous claim, this one attributed to Charles Duell, the US Patent Office commissioner in 1899, that "Everything that can be invented has been invented." Talk about inviting debate!

Despite d'Alembert's unpleasant personality, he was a prominent mathematician of the time, and he both influenced and helped Laplace, especially during the early years after Laplace had left the university.

Laplace, once in Paris and using his letter of introduction, promptly sought out the intractable d'Alembert. True to his reputation, d'Alembert (according to a widely reported but undocumented legend) derisively gave the young man a tough problem to solve. Laplace solved it almost immediately, surprising d'Alembert with how quickly he had grasped the problem's complexity. Laplace, we know, was a quick study, and here it paid off for him.

Another version of the legend is that d'Alembert gave Laplace a difficult mathematics text to read. Laplace returned in just a few days, ready, willing, and able to carry out a detailed discussion of the book. Regardless of which version of the legend is true (or even if neither), we know enough about both men that the tale is credible.

Regardless, evidently d'Alembert was impressed with Laplace, for soon he helped the young man secure an appointment as a professor of mathematics at École Royale Militaire in 1771, while just twenty-two years of age and sans university degree. Imagine doing that today—unlikely, to say the least.

Laplace, we know, was ambitious both in his career as a professor and in wider spheres, especially in politics. Almost immediately after his appointment to the faculty, he sought membership in the respected French Academy of Sciences, the same society where he had been so impressive as a promising student. Despite being young and inexperienced, he thought he deserved entry into the Academy since his paper had been well received a few years before. Alas, he was unsuccessful—at least on his first two attempts. Finally, in 1773, he was admitted on his third try.

But the ever-ambitious Laplace wanted more. He cultivated numerous acquaintances outside of academia, particularly with those in politics or who had some riches. His ambition was not without risk, for this was just prior to the French Revolution, and the times were highly charged with social unrest. He doggedly infused himself into government circles, seemingly to gain advantage in the contemporary disorder by knowing the right people.

As the French monarchy decayed, things grew even more chaotic. Everyone, it seems, was seeking some kind of refuge from this societal storm. Many fled to England or emigrated to America. Other prominent individuals retreated from public life altogether, seeking safety in anonymity and seclusion. It seemed that everything was in turmoil—it was, in fact, a time of mortal danger for those on the wrong side.

Robespierre came to power in a coup but was soon replaced and eventually executed. (Recall that we saw a famous depiction of this in Chapter 7.) Not long thereafter, Napoleon Bonaparte took over the reins, first as emperor of France (1804–14), then as king of Italy (1805–14) and simultaneously as protector of the Confederation of the Rhine (1806–13). By now, the ambitious Laplace was well known among politicians. He even knew Napoleon personally. It turns out that Laplace had had earlier contact with the emperor. In 1784, when Napoleon had attended the École Royale Militaire, Laplace, as professor, had been Napoleon's examiner. Of course, Napoleon was then a student, and his future as the emperor (with an aggressive expansionary goal via military means) had not yet been envisioned. Imagine being Laplace examining Napoleon on that day! (I'm guessing Laplace passed the young Napoleon.)

While emperor, Napoleon apparently saw Laplace as someone who had both good connections in governmental circles and a strong academic reputation. Napoleon, it is widely reported, was favorable to scholars and academics, even fancying himself one of them. We will see in the Chapter 9 that Napoleon revered Carl Gauss, even sparing the town of Göttingen because Gauss resided there. Perhaps because of this idolization of scholarly achievement, Napoleon appointed Laplace to be the minister of the interior. However, this was a big mistake.

Almost immediately, other government officials realized Laplace's insufficiency as minister and complained loudly about it. Laplace, although brilliant in his scholarship, was simply not suited to be an administrator. It is reported that he nitpicked everything in the government, both within his area of responsibility and beyond, apparently driving everyone around him nuts. After only six weeks, Napoleon dismissed Laplace, saying that Laplace had attempted to "carry the spirit of the infinitesimal into administration" (quoted in Gorroochurn 2016a, 3). Napoleon then appointed his brother to succeed Laplace as minister.

But despite his dismissal from his government post (and true to his appetite for politics), Laplace became a count of the empire in 1806. A few years after that, in 1817, after the Bourbon Restoration, he was named a marquis, a title of nobility, permitting him to be addressed formally as Marquis de Laplace. Reportedly, he liked the title very much and asked people to use this address for him whenever it was even remotely within protocol.

Laplace's role in government service did not end with his earlier dismissal, though. He was appointed to head a commission to standardize weights and measures. This position alone (apart from his prodigious achievements in calculus, mechanics, and especially probability theory) places him as a central character in the saga of bringing quantification to Everyman.

There is a widely reported story of an interaction between Laplace and Napoleon that needs telling. Napoleon, we have seen, had an inclination to be around academicians, especially the great mathematicians of the day. At one meeting of state, Laplace presented Napoleon with a copy of his famous work on the solar system (recall, it is *Traité de mécanique céleste*). Napoleon must have been briefed on the book beforehand, and particularly that it did not mention God, for at the presentation he said to Laplace, "They tell me you have written this large book on the system of the universe, and have never even mentioned its Creator." Laplace gave a quick reply that has become famous: "I had no need of that hypothesis." Later, Napoleon amusedly told Lagrange about the encounter, including Laplace's reply. Lagrange exclaimed to Napoleon: "Ah, it is a fine hypothesis; it explains many things!" Napoleon retold the story many times, citing it as one of his favorite encounters with the great mathematicians (Bell 1986).

****

Laplace lived and worked in tumultuous times: through the French Revolution (1789–99) and the Napoleonic Wars (from 1803–15). We have already seen that he didn't simply seek to survive (the norm for many others during this time) but tried to prosper, mostly though making political contacts. Yet, another world event was happening during this time, and it, too, came into his life. This event was the drafting of the United States Constitution in 1787 and its subsequent submission to the states for their ratification. These happenings were followed widely throughout the Continent and, it is highly probable, by Laplace, given his penchant for contemporary politics.

Despite the uncertain outlook across Europe, the success of the American Revolution gave people on both sides of the Atlantic a sense of self-determination; this was, arguably, something that had been unknown in all of the preceding history. The new Americans, as colonists in pre-revolutionary times, were viewed by Continentals as a group of rebels who were either zealously religious or sundry misfits.

Now, however, with America gaining stability amid the European tumult, the colonists did not seem so radical after all. In fact, they were suddenly folk heroes of a sort. In a shift of attitude for many, going to America meant a kind of release from centuries-old social and cultural constraints—a paramount desire for people caught up in the French Revolution, which in no slight manner seem designed to keep a noble class dominant, albeit with different rulers, again and again. America was a new and welcome respite from all that. It was, after all, the land of opportunity. It soon came to mind that going there was a sensible choice.

It was also a chance to escape from the oppressive guild system, wherein merchants and crafts persons could not advance by their own talents and energy but only by union-controlled channels. America represented a chance to start anew, with personal responsibility and freedom of self-reliance as strong inducement. Almost everyone knew someone, and often a whole family, who joined in this new attitude by either wanting to go to America themselves or supporting relatives and friends who actually made the journey.

This spirit of self-determination, embodied in the American experience, came to dominate Laplace's thinking and soon showed up in his writings. For Laplace, self-determination was the outgrowth of his adoption of a philosophy of determinism: a belief wherein all events are the result of human choices and decisions with sufficient causes. He was passionate about his determinism, and he sought to share his viewpoint. Even more, he wanted others to adopt determinism as a philosophy of life as he had. Accordingly, he used his writings as his most effective means for bringing this about. Thus, his publications had a twofold intent: first to communicate his very important and genuine mathematical inventions; and, second, to influence others toward his philosophy. With this in mind, we examine his probability writings toward these ends.

In 1814, Laplace published his seminal work on probability, *Théorie analytique des probabilités* (*Analytic Theory of Probability*) (Laplace 1878–1912). In it, he presents an immense number of new ideas. He describes his creation of inverse probabilities, defines his idea of using both prior and posterior information in predicting future events (thereby evolving ideas by Bayes into a form now called Bayesian approaches to probability), and develops a special case for using such posterior information in forecasting. Although each invention is substantial in and of itself, they are conceptually interrelated.

Further, in a break from his forerunners' studies of probability, Laplace describes his new concepts using applications outside the usual venue of games of chance, in areas such as vital statistics. No other collection of ideas has advanced probability theory so far, so fast. As the leading mathematician of his day, he thrust probability theory to the center of intellectual activity. Through Laplace's work, previous developments in probability by Bayes, Bernoulli, Legendre, and Gauss were highlighted anew. With probability finally well established as a distinct study, and with Laplace's substantive contributions, there was a remarkable spurt of research activity that focused on probability theory.

There is yet another far-reaching aspect to this work, apart from its mathematical developments. In the treatise, Laplace articulates his ideas on scientific determinism. He does this by adding to the book a defining essay as a sort of stand-alone introduction. He even gives the introduction its own title, *Essai philosophique sur les probabilités* (*A Philosophical Essay on Probabilities*), in the second edition of his *Théorie*. The essay was itself was accessible in language and tone to lay persons and hence immediately popular with general audiences. It helped to make the whole book more widely read and known.

Laplace's essay on probability theory is noteworthy because it is not the typical introduction to a scholarly book. It is not technical and does not specifically presage the book's contents, as is done in most introductions. Rather, in it, Laplace sets a philosophical context for the mathematical tome to follow. He explains his views on probability to the general reader and grounds them as a part of his philosophy of determinism. In particular, Laplace aims to tell the general reader why probability is important in everyday life and does so with a perspective integral to his ideas on determinism. In other words, for Laplace, probability is reified determinism. It has a life all its own. The historian of statistics Stephen Stigler translated the introductory essay into English and described its influence as "immense."

Laplace kept revising this essay again and again until the sixth edition of his main *Théorie*, adding to it new concepts and revising earlier ones. Later, he published it separately.

One phrase in the introduction has become especially famous. He said, "Probability theory is nothing but common sense reduced to calculation" (Laplace and Dale 1995, 79). This is pure Laplace, popularizing probability. Acting on his belief that probability is fundamental to the lives of ordinary people, he continues, "It is remarkable that a science which began with the consideration of games of chance should have become the most important object of human knowledge" (Laplace and Dale 1995, 123). Presented this way, it represents a wholly new notion of the study of probability theory, suggesting we view it as accessible to everyone and as "the most important object of human knowledge."

Such thinking was revolutionary at the time: the most famous mathematician of his day was espousing something completely different and new to popular readers. The academic subject of probability theory is being espoused by an important scholar as central to the lives of ordinary people! To think that he said that while knowing, feeling, *believing* that this most important object of human knowledge came from scholarly work ("a science," as he calls probability) yet stemmed from a nonscientific source, namely, games of chance, is truly astounding.

Further, expressing his belief that probability encompasses the most important questions of life, he elaborates his opinion by proclaiming that people have a "sort of instinct" to know an "exactness" for their lives, by describing his book on probability theory as follows:

> Here I shall present, without using Analysis [mathematics], the principles and general results of the *Théorie*, applying them to the most important questions of life, which are indeed, for the most part, only problems in probability. One may even say, strictly speaking, that almost all our knowledge is only probable; and in the small number of things that we are able to know with certainty, in the mathematical sciences themselves, the principal means of arriving at the truth—induction and analogy—are based on probabilities, so that the whole system of human knowledge is tied up with the theory set out in this essay. (Laplace and Dale 1995, 1)

In recognizing this innate aspect of people's experience, Laplace put his finger precisely on their pulse by identifying the concept as central to their existence: "the whole system of human knowledge." For the first time, Laplace brings to the study of uncertainty an epistemological perspective (i.e., the way we know things) that may be taken up by ordinary folks.

His argument now moves to describing his scientific determinism. He says that all persons have within them an "infinite intelligence" (his words) that enabled them to have full knowledge of past and future events with complete certainty. To achieve such full knowledge, however, one needed to know everything that there was (and is) possible to know about a given event, something Laplace thought was perfectly reasonable to expect from human beings. He reasoned that, with complete knowledge of an event, people could deduce patterns of thought, making sense of the event's manifold causes.

Laplace extended his logic to say that if someone had the advantage of complete knowledge (referred to in some writings as "infinite intelligence") for making such deductions, then the science of probability theory—indeed, all of statistics—would be unnecessary.

This approach to empirical questions is now referred to as "scientific determinism." Laplace was the first—and probably the most famous—proponent of scientific determinism. But he did not invent determinism; indeed, the idea can be traced back to Socrates.

Still, the notion of infinite intelligence was a deep-seated truth for Laplace. It was his philosophical determinism, his epistemology. To Laplace, infinite intelligence was innate to all humans, and so complete that a given individual would know every atom and force in the universe for each event in his life. With this complete knowledge, one could subsequently use the laws of Newtonian mechanics to calculate the past and future location and momentum of everything else in the entire universe. It is the ultimate free will. Every speck of knowledge is knowable and can be acted upon by laws of mathematical mechanics to make perfect and complete predictions of what is to come.

He referred to this innate capacity as a "demon." And, throughout the ensuing years, this deterministic perspective has come to be called "Laplace's demon." In what is now his most famous quote on his demon, he remarked:

> We ought then to consider the present state of the universe as the effect of its previous state and as the cause of that which is to follow. An intelligence that, at a given instant, could comprehend all the forces by which nature is animated and the respective situation of the beings that make it up, if moreover it were vast enough to submit these data to analysis, would encompass in the same formula the movements of the greatest bodies of the universe and those of the lightest atoms. For such an intelligence nothing would be uncertain, and the future, like the past, would be open to its eyes. (Laplace and Dale 1995, 2)

For Laplace, with such infinite knowledge—that is, knowing everything there is to know of one's past—one could accurately and fully fathom the future. Consequently, and by definition, then, Laplace obviates all of statistics or probability. There is no need for regression equations or Newtonian mechanics to make predictions, since it all lives within the individual's infinite knowledge. For Laplace, there is no uncertainty, no indeterminacy—even no choice. Everything is predetermined.

One writer, commenting on the ascent of science during this same period, contextualized Laplace's demon this way: "Laplace had taken Newton's science and turned it into philosophy. The universe was a piece of machinery, its history was predetermined, there was no room for chance or for free will. The cosmos was indeed an ice-cold clock" (Silver 1998, 234). Laplace was certain of it.

This is reinforced in his famous six principles of inductive reasoning for probability, which he labeled *General Principles for the Calculus of Probabilities*. True to Laplace's writing style, in his original wording, they are difficult to make sense of. Here, these principles are rephrased so as to capture their essence:

1. Probability is the ratio of the "favored events" to the total possible events.
2. The first principle assumes equal probabilities for all events. When this is not true, we must first determine the probabilities of each event.

Then, the probability is the sum of the probabilities of all possible favored events.
3. For independent events, the probability of the occurrence of all is the probability of each multiplied together.
4. For events not independent, the probability of Event B following Event A (or Event A causing Event B) is the probability of Event A multiplied by the probability that both Event A and Event B will occur.
5. The probability that Event A will occur, given that Event B has occurred, is the probability of both Event A and Event B occurring, divided by the probability of Event B.
6. Three corollaries are given for the sixth principle, which amount to Bayesian probability.

As one can imagine, by sheer force of his reputation and obstinacy, Laplace influenced others to adopt scientific determinism. For the next generation of scientists, scientific determinism was firm and entrenched.

But, by the mid-1800s, thinking began to change. A more reasoned approach considered the impossibility of an infinite knowledge. And Laplace's demon began to crumble. Obviously, to some degree, past events influence future ones, but the true extent of human comprehension to know all and, from that, infer the future does not match Laplace's imaginings for it. One simply cannot know the full extent of every breath of a being and realize the movement for all of its atoms. Nor are such things static in time and place. There is a dynamism in the world that is inherently uncertain and thus unknown, or unknowable.

Scientific determinism has a theoretical appeal in that every aspect of life and experience is controlled. But, as understood today, it does not follow absolutes in nature. Simply, scientific determinism does not comport with reality.

Rejecting Laplace's determinism, by most accounts, is best done by citing the physics of thermodynamics. Quite simply, the laws of thermodynamics refute scientific determinism. There are three basic laws of thermodynamics. Later on, a fourth law was added. Interestingly, since the initial three laws were so well known before the fourth was added and it comes logically before the others, it is teasingly referred to as the "Zeroth Law." These complex laws describe how atoms, and even subatomic particles, act in nature.

The first law states that energy (the basic stuff of the universe) can neither be created nor destroyed, although it can change form. This leads to the second law, which is relevant to rejecting scientific determinism. It states that when energy changes form (something that requires heat), some of it is thrown off as waste, and this waste cannot be recreated. This fact of thermodynamics is considered by physicists to show the probabilistic nature of quantum mechanics, because Laplace stated he needed to know the position and momentum of every atom; but this cannot be possible because waste energy is lost to us forever. By this reasoning, then, Laplace's determinism does not stand up to empirical inquiry, at least as evaluated by the laws of thermodynamics. It shows that Laplace's demon is not a demon after all. To further this end, most people incorporate some degree of religiosity in their worldview. God plays a role, too, in determining one's life. My reason for rejecting determinism is more modest in thought. I assert that a worldview is human and not calculable in other terms, even when those terms are Newtonian mechanics. Hence, one's worldview cannot be deterministic.

Regardless, Laplace remains one of the most influential thinkers, philosophers, and mathematicians of all time. He died in 1827. He was seventy-seven years old.

CHAPTER 9

Probably a Distribution

Here, at last, is our Carl Gauss: the *Princeps mathematicorum*, Latin for "the foremost of mathematicians." His full name was Johann Carl Friedrich Gauss, but he is commonly known as either Carl Gauss or, using his original first name, Friedrich Gauss. Today, he ranks among the greatest mathematicians that the world has ever known, along with Archimedes, Isaac Newton, and Leonhard Euler. Quite remarkable.

Even his signature displays a certain eminence, as seen in Figure 9.1. A noted biographer of Gauss describes his handwriting as "beautifully clear" (quoted in Bell 1956, 339). His signature and handwriting are a kind of metaphor for his clarity in composition and lucidity in thought: carefully considered and with attention to detail, not hurried.

Carl Gauss (his image is seen in Figure 9.2) was, quite simply, a man of amazing accomplishment, and he was pivotal in advancing quantification, both through his numerous mathematical discoveries and for his spreading quantification to an ever-widening audience. He was a man of unquenchable curiosity. Although he worked almost exclusively in mathematics, the breadth and scope of his quantitative achievements advanced astronomy; calculus; statistics and probability theory; number theory; algebra including matrix theory; geometry; geodesy; geophysics; magnetism; mechanics; and even optics—quite a list. It all represents a range of accomplishment that is simply astounding for a single person. One noted historian of mathematics summed up Gauss's career by stating, "He lives everywhere in mathematics" (quoted in Newman 1956, 339). But he had a special interest in number theory. He said, "Mathematics is the Queen of the Sciences, and Number Theory the Queen of Mathematics" (quoted in Bell 1986, xv).

The Error of Truth. Steven J. Osterlind. Oxford University Press (2019). © Steven J. Osterlind 2019.
DOI: 10.1093/oso/9780198831600.001.0001

Figure 9.1 Carl Fredrick Gauss's signature
(*Source*: http://commons.wikimedia.org/wiki/Category:Public_domain)

Figure 9.2 Image of Carl Friedrich Gauss
(*Source*: http://commons.wikimedia.org/wiki/Category:Public_domain)

Gauss loved mathematics to the point of obsession. Beginning at a young age and continuing until his dying day, mathematics was his life, occupying his every thought. When informed that his years-long-ailing wife was dying, he is reported to have said, "Ask her to wait a moment, I am almost done" (quoted in Bell 1986, 201). Don't be too quick to judge him, however. In truth, he was a compassionate man who spent many years as the care giver to his mother and then to both of his wives during their respective long-term illnesses. Regardless, for Gauss, mathematics was *everything*.

Even at a very young age, Gauss was more than just good at mathematics—he was a child prodigy. The tales of his amazing feats in mathematics while he was still a juvenile are the stuff of legend. His childhood years are recounted in many stories; the full truth is often unverified, but the amazing tales are commonly repeated. His documented accomplishments, however, are so numerous and impressive that the anecdotes are certainly believable. By any account, his genius was on par with some of the greatest, such as Pascal (who wrote a treatise on vibration at age nine, putting its proof on a wall with a lump of coal), Mozart (who composed a full symphony at age five), Chopin (who was composing at age six), or William Sidis (the youngest person to enter Harvard, at age eleven).

There are many stories of his photographic memory, such as being able to recite long passages from books he had read only a single time, years previously. He remembered the name of everyone he met, even if only in passing, and could recall verbatim their exact conversation. We saw earlier that Gauss developed one of the most important mathematical inventions of all time, his method of least squares, when only eighteen years old. Further, he later provided the calculus proof of this method when others could not. He developed a special instance of reaching solution in some kinds of calculus integration called a "Gaussian quadrature," and a particularly useful form called the "Gauss–Hermite quadrature" wherein the values for integrals can be approximated. For persons working with calculus to solve all kinds of real-world problems (e.g., in engineering), these Gaussian methods are of enormous practical value.

A popular story is that, at age three, he noted an error in his father's reckoning and, meaning to be helpful, pointed it out to him. Instead of showing amazement or appreciation, however, the elder Gauss who was an uneducated, poor laborer, berated young Carl harshly. Apparently, he didn't like being upstaged by his own son. Throughout Carl's childhood, his father resented the boy's aptitude for mathematics and his general intelligence and often showed the sensitive child a cruel and inhibiting hand. His mother, however, presented a kinder approach in raising Carl and saw to it that he got a good education, despite their poverty.

By fortune, the Duke of Brunswick, Carl Wilhelm Ferdinand, learned of the young prodigy; and, upon seeing young Carl's genius, the duke was so impressed that he assumed patronage, financially supporting Gauss's university education and even offering assistance for some years after graduation. The duke continued to support Gauss until his own death by gunshot. Their meeting was a happy accident—one that changed the course of history and whose influence lives on in us through our quantified outlook.

A popular anecdote of young Gauss's feats includes his mother. She was illiterate and did not record his birthday. When he asked her about it, she could only remember that it was on a Wednesday, eight days before the Feast of the Ascension (also known as Holy Thursday, it commemorates Christ's ascension into heaven three days after the crucifixion). Gauss knew that the Christian holiday occurs 39 days after Easter; but, as readers likely know, Easter is not on a fixed date but calculated anew each year: it is the first Sunday after the fourteenth day of the lunar month that falls on or after the vernal equinox, which is usually around March 21. Gauss calculated the date for the Feast of the Ascension for his birth year—and, from there, he figured out his own birthday: April 30, 1777. One hopes a cake with the correct number of candles on it was then presented to him each year thereafter.

Not stopping with calculating his own birthday, Gauss went on to derive formulas for finding any date, past or future. Gauss's annular formulas are still used today—if you have a smartphone, it probably has the formulas imbedded in a chip. Albert Einstein is said to have remarked that not only was Gauss the only one who could have come up with these formulas, he was the only one who would have thought that it would be possible to come up with them in the first place (Schocken 2015).

His scholarship at school was legendary. Once, when he was eight, his primary school teacher asked the class to add the numbers 1 to 100 ("What is the sum of the first one hundred whole numbers?"). The students dutifully set out to do the addition, but Gauss reasoned that using a formula would be faster. He split the numbers into two groups: 1 to 50 and 51 to 100. When he added the first digit from Group 1 to the last of Group 2, it summed to 101. Next, when he added the second digit from Group 1 ("2") to the penultimate number in Group 2 ("99"), it also summed to 101, and so on for each pair. He reasoned that adding 1 to each successive number in the first group and subtracting 1 from each number in the second group would give him 50 pairs of 101, which multiplies to the correct answer: 5,050; in other words, the equation $50 \times 101 = 5{,}050$. What's particularly interesting is the elegance of Gauss's formula: $1 + 2 + \cdots + n = n(n+1)/2$. Remember, he was only eight years old!

Gauss completed grammar school early and, at just fifteen years old, entered Collegium Carolinum (now renamed University of Braunschweig–Institute of Technology), intending to study languages, but he soon came to pursue mathematics. He graduated three years later with a degree in that field. Thereafter, he transferred to the University of Göttingen to further his studies. He returned to his hometown of Brunswick and completed his doctoral studies at Helmstedt University in 1799. In his dissertation, Gauss provided a proof for the so-called fundamental theorem of algebra, a difficult complex of numerical relationships, whose proof had eluded many professional mathematicians. (For mathematically curious readers, the full proof can be found at various online sites (e.g., see Cain 2017)). Gauss did this difficult work when he was barely twenty-two years of age.

Many of Gauss's discoveries came early, some while he was still a teenager. When just fifteen years old, he grew interested in finding the answer to a problem that had eluded even the best mathematicians of the day: finding a pattern in an array of apparently random prime numbers. After some study, he realized that, given these numbers increase by ten, the probability of a next prime reduces by a factor of approximately 2. He had found a hidden pattern. However, his pattern was not exact and thus he could not produce a proof, something Gauss believed necessary for completion. Because of this, he kept his solution secret for much of his life. Since then, however, others have given Gauss credit for coming closest to a solution.

In another problem, when he was just nineteen years old, Gauss demonstrated that a regular polygon of seventeen sides can be constructed by a ruler and compass alone. This was a major discovery in the field of mathematics, as construction problems of this type had baffled mathematicians for centuries. Astonishing work for a student—but then, remember this student was Carl Gauss! We are, indeed, grateful to the duke for ensuring that so deserving a young genius as Gauss got the education necessary to take advantage of his talents.

One of Gauss's most important quantitative advances was in number theory, an achievement he codified in 1798 but did not publish until two years later when he was twenty-one, as part of his magnum opus *Disquisitiones arithmeticae*. (Readers will recall that we saw this book in Chapter 7 when we were discussing Gauss's version of the method of least squares, and its comparison to Legendre's version.) It has been said that this book is as important to number theory as Euclid's *Elements of Geometry* (1491, Euclid et al. 2010) was to geometry. Both works are foundational, not only to their respective fields but to mathematics and philosophy generally. Some of his other achievements are the Gauss unit (in magnetic field theory), the Gauss–Markov theorem, Gauss quadrature, and Gaussian correlational inequality.

Later in the same year that *Disquisitiones arithmeticae* appeared, Gauss published *Theoria motus corporum coelestium* (*Theory of Motion of the Celestial*

Bodies), which is his most important work on applied mathematics (Gauss 1871). His theory introduced the Gaussian gravitational constant. It remains today a cornerstone of astronomical computation. Every NASA mathematician, engineer, and physicist knows about this constant and uses it in their calculations.

Throughout his life, Gauss was awarded numerous prizes, including the prestigious Lalande Prize for his contributions to astronomy, the Danish Academy of Sciences for his study of angle-preserving maps, and the admired Copley Medal for his inventions in magnetism. There are more than one hundred theorems, processes, laws, and proofs named after Carl Gauss, to say nothing of all the other things named after him, showing the truly dizzying scope of his achievements in astronomy, physics, probability theory, statistics, calculus, and other applied and theoretical mathematics. The professional society International Mathematical Union convenes a mathematical congress once every four years to discuss achievement in the mathematical sciences. At the opening ceremony, it awards the Carl Friedrich Gauss Prize for Applications of Mathematics in his honor. It is one of the most respected awards that the field has to offer, on par with the ultra-elite Fields Medal in Mathematics (also awarded only once every four years). (Notably, there is no mathematics category in the annual Nobel prizes.)

In addition, Gauss was offered several other awards and even professorships at important universities (including Berlin University, around 1817), but he routinely declined them because he was loath to leave his hometown of Göttingen. The affinity he had for his native town, in the pressures of so strenuous and exacting a career, is a bit of a surprise, but he had his reasons. His mother fell ill, and he took her into his house to care for her throughout her remaining years, more than twenty. Later, in succession, both his first and his second wives also became infirm for many years, and he cared for them, too. So, despite the earlier quote, it seems Gauss was a sensitive caregiver.

As one can imagine, Gauss's fame was worldwide, particularly in his homeland of Germany, but he was also known throughout the Continent and even into Eurasia. He eschewed the notoriety, however, preferring to work in the relative inconspicuousness of his study. One story about his renown is that, during the Napoleonic Wars of 1803–15, as Napoleon's troops were approaching Gauss's hometown, Napoleon ordered them to go around Göttingen and spare it because "the greatest mathematician of all time is living there" (quoted in Bernstein 1998, 168). Certainly, Napoleon made the right call that time.

Perhaps a more significant recognition of Gauss's genius and his accomplishments, however, was paid by Laplace, the acclaimed French Newton. After the

Napoleonic Wars, the victorious French demanded money from their vanquished foes, asking the enormous sum of 2,000 francs from Gauss. Living on the modest means of a university professor, Gauss simply could not afford that amount; plus, he simply objected to the French demand and refused to pay it. Laplace, a man of relative wealth, paid Gauss's fine, saying that he did so because Gauss was "the greatest mathematician in the world" (quoted in Bell 1986, 242). Coming from the eminent Laplace, this was a supreme compliment.

Gauss, typical of his careful nature and thoroughness, often did not publish his findings as he worked on a problem, sometimes waiting years before making his work public. He frequently kept his ideas to himself and returned to a given notion many times over the years, almost always changing and adding to it before finally releasing it. His motto was *Ut nihil amplius desiderandum relictum sit* ("that nothing further remains to be done"). When he did publish one of his ideas, he was elegant and detailed in his explanations, always including a thorough proof. Many times, he gave several proofs, and more than once he even offered explanation of why no additional proofs were possible.

All told, Gauss published more than 155 meticulously researched and written papers. There are several modern collections of them. One of the largest collections of his publications and personal papers is held at the Cammie G. Henry Research Center at Northwestern State University of Louisiana, thanks to the thirty-year collection effort of a determined professor (see Louisiana 2018). Also, the Gauss Society at the University of Göttingen is active in commemorating his work, and they maintain a collection of Internet links to Gauss's works and life (see Gauss Society 2018).

Gauss was married twice. With his first wife, he had three children. She passed away after a long illness. He then married her best friend and had three more children before her demise, also after a lengthy illness. He spent most of his adult years caring for his mother and then his wives. In older age, he grew increasingly infirm himself, and, in the last years of his life, his daughter Theresa cared for him. He died at age seventy-seven, in 1855.

Now, we turn to describing a few relevant terms, and then specifically to how this remarkable man's accomplishments advanced the story of quantification.

****

We reach a point in our story where it is necessary to employ some terms—not too technical, but it is important we use them correctly, to avoid confusion. Earlier (at the end of Chapter 5), we saw the definitions for "probability," "odds," and "likelihood." Here, we expand this list of terms to include *sample space* and *random variable* and, specifically, two types of random variables: *discrete* and *continuous*.

"Sample space" is the number of possibilities in an experiment or research context. For example, if the experiment involves rolling a single die, the sampling space is just 1, 2, 3, 4, 5, and 6: all the possible outcomes. Of course, typically the sampling space is much larger. Say, an experiment focuses on an adult's weight—then the sampling space is all the possible weights (usually within reason): say, from 80 to 500 pounds. While possible weights are above and below these limits, such extremes are so rare as to be ignored. In some experiments, there may be more than one sampling space, such as when drawing cards from a standard fifty-two-card deck. One sampling space could be any card at all, while another sample space may be restricted to face cards only. It is defined by the researcher.

A "random variable" is not quite just any number at all. Rather, a random variable quantifies an outcome. For example, if the variable is rainfall within the last twenty-four hours, then a random variable is the amount of measured rainfall. It could be any number that falls within the sampling space (that is reasonable), say, rainfall from 0 to 50 inches. This means, too that a random variable has a probability distribution. We saw this earlier in Chapter 5, when looking at a histogram. As we know from the central limit theorem, with more trials (i.e., samples), this distribution will tend toward the normal (bell) shape.

Variables themselves can be either "discrete" or "continuous." A discrete random variable is one which may be any distinct value within the sampling space, such as 0, 1, or 102. A continuous random variable is one which can be any of an infinite number of possible values. These are usually measurements, such as an achievement test score, the time taken to run a marathon, or height or weight. While (in theory) these values can be $\pm\infty$, it is usual to truncate the range to just reasonable values. Also, the numbers are limited by the scale on the dependent-variable measuring instrument. For example, if a test score is measured in whole numbers, then a fractional value would not be considered in the sample space.

****

Now, with this background, we return to describing the accomplishments of the remarkable Gauss, as they contribute to quantification. The noted dispute with Legendre over authorship of the method of least squares has a more confused—and confusing—context than simply a debate over "who got there first." Recall, I described a simple version of the procedure in Chapter 7 and identified it as OLS (ordinary least squares) regression. Legendre claimed the invention as his own, and it is indisputable that he published a version of the method first. But Legendre's original description was relatively elementary, proffering only an algebraic explanation of adding up residual values of the difference between the observed data and the closest point on the regression line. He did not provide a rigorous rationale, nor did he give a calculus proof, usual requirements for a newly invented procedure. What he presented was important, but not sophisticated in the form given.

Further, at the time of its publication, Legendre did not seem to give it much weight, as he included it only as part of an appendix to another publication, and his text was limited to just the simple explanation and one example. It was not until after Gauss had put forth his much more sophisticated version that Legendre very publicly claimed authorship.

Gauss did not dispute that Legendre had published a summing procedure first, but he claimed that he had, in fact, invented the method of least squares years ahead of Legendre. To buttress his assertion, he produced earlier papers where he had employed it several times without mentioning it as a new procedure. He said that the Legendre version was so simple that he had used it for years, as almost routine.

In the end, Gauss got the better of the dispute, both rhetorically and in technical achievement. As for the argument itself, it was mostly carried out in a series of letters to others. Recall that he expressed surprise that Legendre was claiming credit for something Gauss saw as so straightforward that it was barely even worth mentioning. This is scarcely the attitude of a gentleman with the personal empathy he showed in caring for his ailing mother and his infirm wives. But then, Legendre was not such a good guy either, wanting all the credit for something that was an accumulated achievement and making a fuss only after Gauss presented his more sophisticated version. Gauss, as we now realize, invented the generalized method with a more rigorous rationale and a calculus proof. Legendre's accomplishment is now viewed as very useful, but as just one application of the method of least squares.

Recall, too, that this dispute with not the first between the men. Earlier, Legendre tried unsuccessfully to address a complex problem of astronomy but eventually abandoned his effort. But Gauss was only nineteen years old when he successfully solved it. His work is now adopted as the law of quadratic reciprocity.

In terms of today's version of the method of least squares, Gauss defined the sample space of a combination of observations as a density function, requiring estimation by calculus. He thereby extended the one-case example of Legendre, allowing it to be appropriately used to estimate population parameters. This step is what distinguishes Gauss's method from the simple summing procedure of Legendre. I explain the notion of a "density function" momentarily, after completing this bit of history.

As a consequence, thanks to Gauss, the method can be made broadly generalizable and may be used in many scientific contexts. Today, his rationale and proof are the accepted versions. This was a major advance for statistics and immeasurably advances quantification in our storyline.

Gauss demonstrated his least squares method by using it to solve a now-famous problem in astronomy called the "Ceres estimation." On January 1, 1801, the Italian astronomer Giuseppe Piazzi discovered a new and important asteroid he named Ceres. He tried to figure out its orbit, tracking it for forty days before he could no longer see it because of the glare of the sun. He wanted to know where Ceres would emerge from behind the sun to be seen again. But he did not know how to calculate this—nor did other astronomers. At the time, it was a well-known mathematical Gordian knot. Actually, this particular problem was part of a larger unknown for determining the orbits of celestial bodies.

The problem was vexing to all astronomers and physicists until Gauss, at just twenty-four years of age, applied his version of the method of least squares. Using only three of Piazzi's observations, he correctly calculated the when and where of Ceres's reemergence. Gauss published his solution in his *Theory of the Combination of Observations Least Subject to Error* (Gauss and Stewart 1995). Fortunately for us, Gauss's original sketches of his work on the problem survive today at the University of Göttingen, shown in Figure 9.3.

Young Gauss's calculation was an amazing mathematical feat. His reputation was set by it, and, just as important, he established the method of least squares as verifiable. There are several good reads (although some are mathematical in nature) of how Gauss determined the orbit of Ceres (see Teets and Whitehead 1999, Le Corvec et al. 2007). One especially interesting historical record is by Johann Pfaff in his *Programma inaugurale in quo peculiarem differentialia investigandi rationem ex theoria functionum deducit*, a work in which Gauss scribbled notes of his proof—maybe for the first time—on a back page and signed the frontispiece. Pfaff was also Gauss's dissertation advisor. A recent first printing of this work was made available for sale at an initial auction price of $75,000, the size of which was attributed mostly to the book's inclusion of Gauss's handwritten notes and signature.

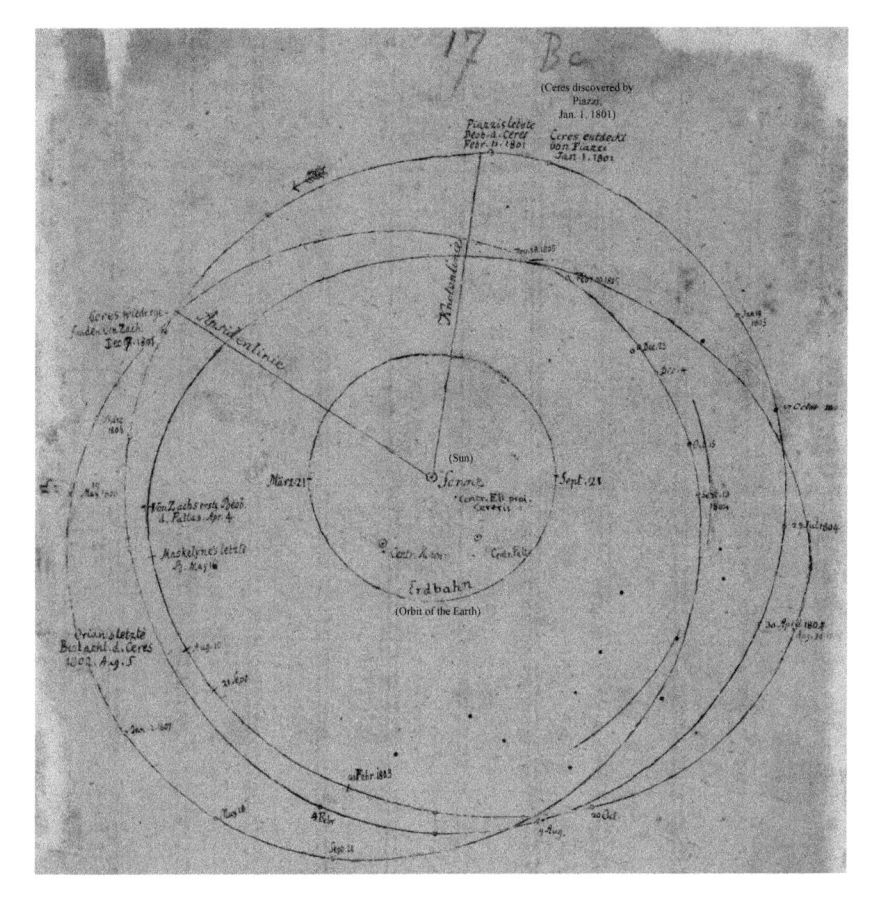

Figure 9.3 Gauss's sketch of Ceres's path
(Courtesy of SUB Göttingen. Cod. Ms. Gauß Handbuch 4, p. 1)

In determining the orbit of Ceres, Gauss realized that an underlying harmonic motion that helped to explain the changes observed in the positions of the planets was also useful in estimating large distances on the earth. From this observation, he began his work on geodesy, the science of determining the distance between two points that are far apart, such as the distance between cities or between continents, where the curvature of the earth must also be considered. During his life, he was quite well known for this pursuit. He was commissioned by George IV of England to survey the Kingdom of Hannover, a hilly area whose dimensions had previously been unknown. He set about this work by first making a remarkable invention: the heliotrope. This device is a sort of specialized small telescope that measures angles on the land topography from rays of sunlight. From this, distances can be estimated with reasonable accuracy.

Figure 9.4 Gauss's heliotrope, about 1822
(*Source*: http://commons.wikimedia.org/wiki/Category:Public_domain)

While small in comparison to an observatory's telescope, the heliotrope stood nearly four feet tall. There are stories of various measurements across the countryside taken for the first time using Gauss's equipment. A picture of his invention is shown in Figure 9.4.

In particular, the specific achievement of accurate measurement of long distance brings quantification closer than ever to the daily lives of ordinary people. For example, people could now anticipate how long it would take them to travel between cities. This seemingly ordinary reckoning brought common distances into the daily lives of folks, contributing to quantification as a mindset.

****

Up to this point, we have traced the development of the normal distribution (and especially its bell shape), as seen through the work of Laplace and others. They were captivated by the fact that this distribution occurs for everything measurable everywhere in nature (e.g., height, weight, test scores, opinions, rainfall, the size of pumpkins). But, their efforts were not merely to satisfy a simple curiosity. They realized an empirical interpretation, too. Such a scientific realization was advanced by de Moivre, who explained through the central limit theorem *why* this distribution occurs so regularly in nature. But, these efforts, while significant, were constrained to mere frequency counts, and development of the normal distribution had seemingly come to a standstill.

Then, Gauss, with his characteristic ingenuity and persistence on a problem, realized that the bell shape can be a function rooted in probability. And that this probability function need not be limited to just the bell-shaped distribution. Any distribution, regardless of its resultant shape, can be a function of probability.

By this reasoning, he envisioned the normal distribution as a density function, which is more formally called a "probability density function." This was a momentous realization and led him to devise one of the most important formulas of all time: the Gaussian probability density function. A probability density function is a mathematical specification for a continuous random variable (as opposed to the discrete random variable that we saw earlier) of the probability that a given value within the sample space falls within certain limits. In other words, when a phenomenon is seen as a likelihood ranging from zero to one, the problem arises to determine where a given value lies.

Gauss developed a formula to determine the probability that any point within the sampling space can fall within a range on the dependent variable's scale. At foundation, using calculus integration, the area under a given curve can be determined for specific upper and lower limits.

This can be seen by examining the normal curve shown in Figure 9.5. For any two points along the dependent-variable axis (the *x* axis, at the bottom of the graph), the area can be estimated. Note that the area between the two dotted vertical lines can be calculated by taking the integral from the lower bound to

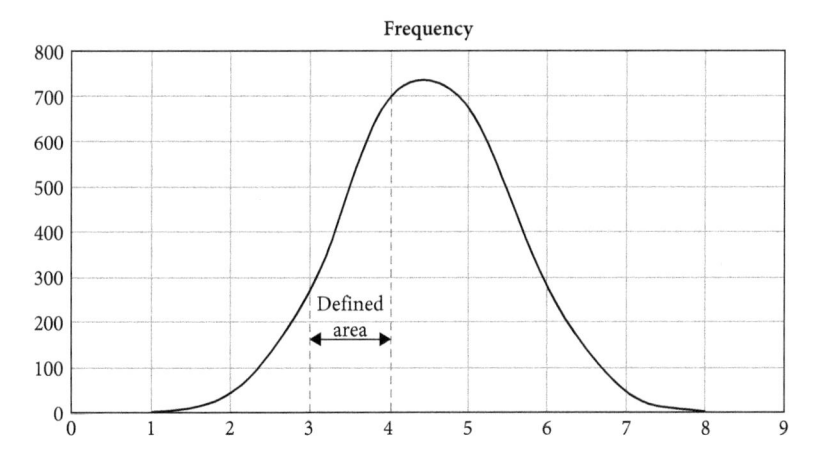

Figure 9.5 Probability density function area

the upper bound, in this case from 3 to 4. Gauss's formula can determine the likelihood that a given value falls within these limits.

This description also shows why it is called a "density function" a given region (such as that shown as the "defined area" in Figure 9.5) is "filled." An amount is specified from the lower limit to the upper limit and then filled to constitute an area.

An example of a density function with which you are likely familiar is percentiles, from the first to the ninety-ninth, sometimes written as "1st %ile–99th %ile." The lower limit is the point at which the dependent variable begins on the left. The upper limit is any percentile chosen. Say that a researcher wishes to determine the midpoint in a distribution. The area is bounded on the left by the lowest score (or value) within the sampling space, and the midpoint is the upper limit. Here, the left half of the area under the curve would be filled. By integral calculus, this area can be determined. In this case, it would be 50 percent of the total area.

When a researcher concludes "statistical significance" or NSD ("no significant difference"), it is also interpreted off the Gaussian normal density function. In many research scenarios with hypothesis testing, the significance criterion is at the fifth percentile, to give the 0.05 level of significance. The Gaussian probability density function yields the likelihood that a given value (say, a *t*-score or an ANOVA *F* ratio) falls within this region.

Thus far, the only probability density function we have examined is the normal distribution, which, when considered as a density function by Gauss's formula, is called the "standard normal distribution." The standard normal distribution is defined as the circumstance wherein the population mean is 0, and the standard deviation is 1. Often, statisticians express this distribution as $X \sim N(\mu, \sigma^2)$. This specification (data is normally distributed with a population mean and a variance of 0 and 1, respectively) yields the bell-shaped curve—far and away the most common curve and nearly the only one we will consider in this book.

But, in what is a masterful stroke of understanding the characteristics of a distribution, Gauss's probability density function is not limited to just the standard normal distribution with its familiar bell shape. It applies to any continuous probability function, regardless of the resultant shape. In fact, the feature of its generalizability is one of Gauss's density function's greatest strength. Hence, it can apply in many other circumstances. Figure 9.6 illustrates some other curves are that are not the standard normal distribution but still fit as a Gaussian density function.

To illustrate the difference between the probability density function and the standard normal distribution, I present two formulas, one for each construction.

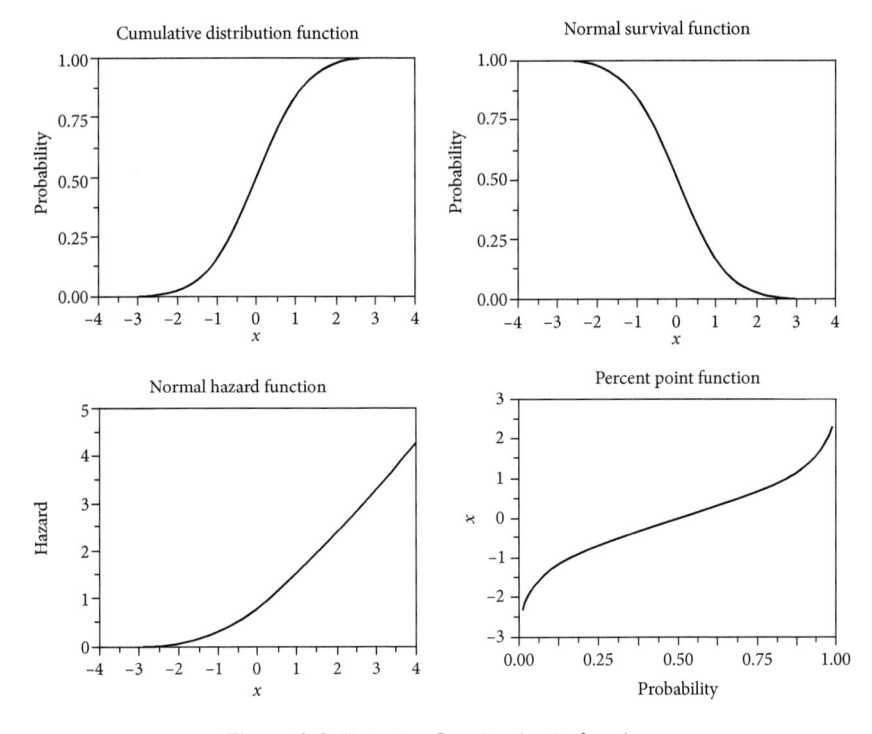

Figure 9.6 Illustrative Gaussian density functions

The first formula can specify any probability function, such as the curves shown in both Figures 9.5 and 9.6. The second formula applies only to the normal curve of Figure 9.5. As always in this book, if formulas are not for you, just skip over them. You will not lose any of the storyline.

First, we see the more general form for any instance of a probability density function:

$$f(x) = \frac{e^{-(z-\mu)^2/(2\sigma^2)}}{\sigma\sqrt{2\pi}}$$

Next, we see the form that occurs when the population mean and the standard deviation (technically, the "location parameter" (μ) and the "scale parameter" (σ)) are 0 and 1, respectively:

$$f(x) = \frac{e^{-x^2/2}}{\sqrt{2\pi}}$$

This is our more familiar standard normal distribution: the bell curve. As is apparent by comparing the two formulas, the second version, while closely related to the first, is a simpler expression. By this, one can see why Gauss's probability density function is a monumental piece of scholarship.

****

It is a pity that the name Carl Gauss is known today primarily only within academic circles, for, through his work, Gauss is actually connected to the daily lives of nearly all of us. At first blush, such a connection may seem a stretch. After all, his work was almost entirely in mathematics and characteristically relied on calculus, a difficult form of mathematics that is completely unknown to the vast majority of people. But the connection is not because of the underpinnings he provided for higher mathematics; rather, it is due to his work's application to common, everyday situations—the kinds of things we run into while going about our daily lives, like making predictions and forecasts, or determining a relationship between variables. Here, the connection is real and strong.

The commonplace-to-us aspect of Gauss's work is seen particularly in two of his developments: (1) his method of least squares and (2) his invention of the normal curve. We see outcomes of each procedure almost daily (or at least regularly), although, of course, we do not label them as such. Two examples of seeing Gauss's equation in our daily lives are the bell curves for test scores at school (we'll see that one in Chapter 13 when discussing Karl Pearson), and in weather projections, for example, when tracking the paths of hurricanes. An interesting weather-related incident, tracked by Gaussian projections, is seen in Figure 9.7, which shows NASA's forecast for the solar eclipse of 2017.

On August 21, 2017, North America experienced a total solar eclipse—with possibly the longest path of a total solar eclipse across Earth ever, and certainly the longest across the continent in ninety-nine years. The last eclipse over the United States, in 1979, passed over only a portion of the continent. The NASA trajectory shown in Figure 9.7 for the sun–moon–earth alignment was made using Gauss's methods, the same ones he invented for tracking the asteroid Ceres. On that Monday in August 2017, the path of the eclipse swept across the continent with startling effect.

You may have seen the eclipse, in either total or partial coverage, depending upon where you were during the event. By happenstance, I was in Columbia, Missouri, that day and experienced a complete covering of the sun. For that brief time, day was night . . . literally. It was both exciting and terrifying. Realizing what was happening and knowing beforehand that the event would last a mere two minutes, forty seconds at my location—or about forty-five total minutes, including the coming and going phases—made it exhilarating to witness.

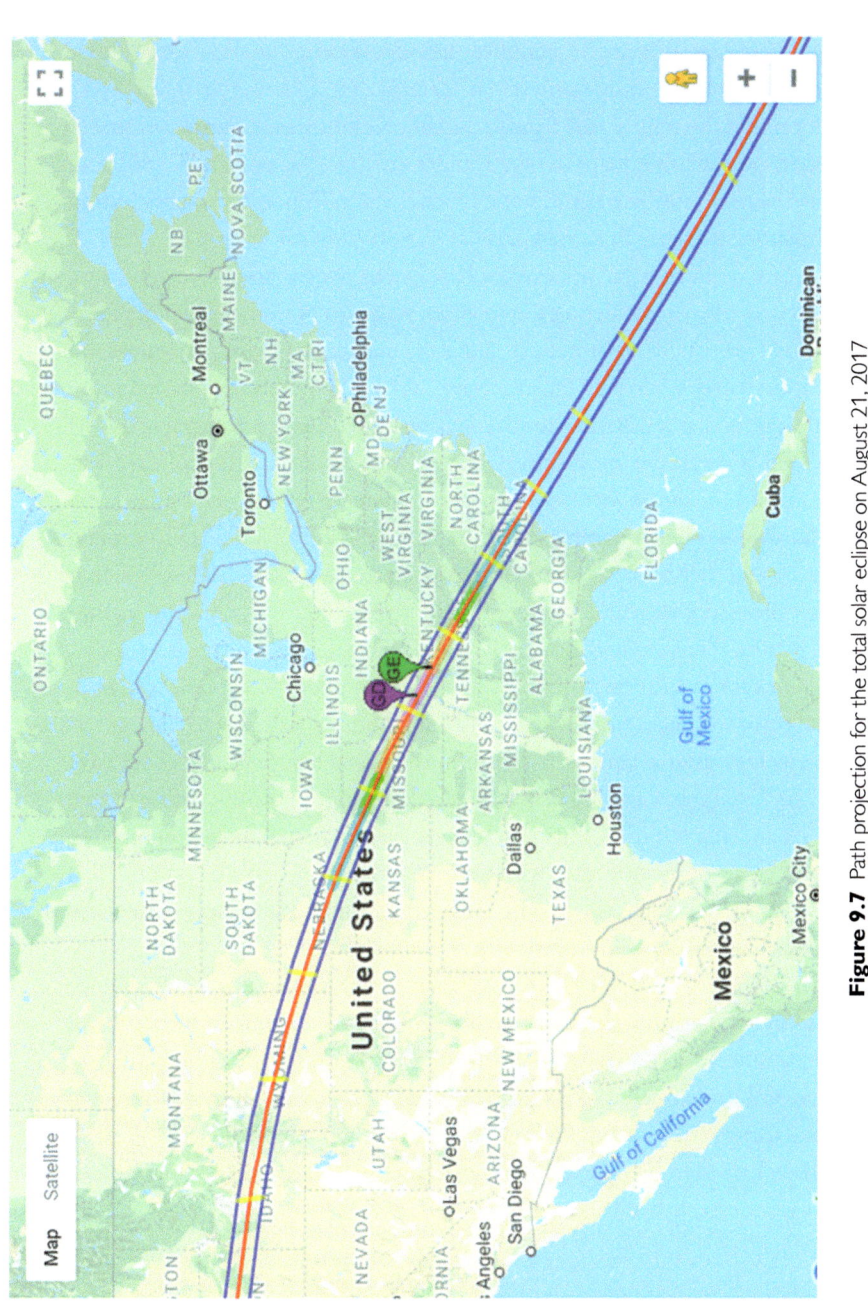

Figure 9.7 Path projection for the total solar eclipse on August 21, 2017

(Source: http://www.weather.gov/lmk/eclipse_2017, accessed August 21, 2018)

To be in the presence of such a phenomenon brought to mind the infinitesimally small stature of Earth's inhabitants compared with astronomical events. For me, there was a feeling of complete lack of control. I do not know about others, as I kept this feeling to myself at the time. Yet, I knew, too, that NASA's prediction that the eclipse was a passing phenomenon was accurate, because it was based on Gauss's equations—you know, the ones he proved in 1801 and that have been reliably used innumerable times since. Hence, my uneasy feeling was transitory. I quietly thanked Gauss.

The normal curve is far more prevalent in our lives than just in reporting school-based tests or predicting the weather. As a way of showing a data distribution, it is seemingly ubiquitous in our daily encounters: with print news, on TV, in advertising, when labeling groceries, and much more. Obviously, it is not always presented as a bell-shaped distribution. A pie chart, a bar graph (histogram), and pictographs of all kinds are actually representations of distributional statistics. Almost without effort, we can create myriad such graphs, by using Excel and other programs, or by just drawing one on a napkin to explain something to a friend when having coffee at Starbucks or Costa Coffee. Whether we draw these graphs with a pencil or using a computer program, the mathematics behind them comes from Laplace and Gauss.

The point for us in seeing these as quantification of our worldview is to realize that, prior to Gauss, things were not so methodologically based. Of course, pie charts, histograms, and other pictographs existed before Gauss, as we have described throughout this book (recall that Galileo, in the sixteenth century, recognized distributions of stars and planets and presented them graphically). But these were simple drawings without mathematical support or proofs—just conjectures at the time. They could be used to immediately represent some observed quantity, but, without an underlying architecture, such drawings were not reliable. They could not be used to track stars for mariners' navigation or to accurately determine the distance between towns, countries, or oceans. It would be silly (nay, dangerous) if NASA's projections for the flight path of manned and unmanned rockets were done using a seat-of-the-pants drawing. To make things truly useful, the underlying, provable mathematics is needed. Only then can the calculations become reliable estimates with known limits. Gauss's work established this foundation.

****

Yet another real-world example of Gauss's influence on our daily lives is seen in the devising of driving routes used by carrier companies such as FedEx and

UPS. Of course, these companies must deliver consumer goods to the correct address hundreds of thousands of times every day. They employ thousands of drivers, and each one has a unique routing sheet, giving the best street route for that day's deliveries. Imagine if the drivers' routes were established by drawing lines on a street map. Chaos! Clearly, a methodology for calculating all the routes is a business necessity. Fortunately, a formula is available from which they can program computers to do the job—a formula invented by Gauss. Even traffic flows can be factored into the algorithm in real time to update each driver on the most efficient course throughout the day.

The story of the American express delivery industry's founding is legendary. In 1965, Fred Smith, a Yale University undergraduate, got the idea while studying Gauss. He wrote a term paper describing his business model, calling his express delivery company "Federal" because he believed the word evoked patriotism. He received a high mark on the paper—mostly for ingenuity, not usefulness, because his professor thought the idea impractical. Upon graduation, Smith entered the military, but the idea stuck with him. He continued to refine it while in the service; when he completed his tour of duty in 1971, he immediately started up Federal Express, now FedEx.

In the early years of these delivery companies, they used "operations research" (or "OR") to formulate routes, with variables, such as type of package and time of delivery, being entered into multiple regression equations. Since then, they have evolved their methods to more sophisticated modeling, including a special type called "hierarchical linear modeling" (or "HLM"). All these statistical modeling approaches stem from the foundational work of that sage, Carl Gauss.

Even on a technical level, the relevance of Gaussian methods is still strong. Recently, I was invited to a routine colloquium devoted to "consideration of the computational and statistical issues for high dimensional Bayesian model selection with Gaussian priors." I am sure Bayes and Gauss would be pleased that their procedures are still the subject of active discussion.

As we see in this chapter, quantification is not just a matter of establishing the formulas—it's entering our daily lives more and more thoroughly. It has taken a quantum leap forward with the imaginative and groundbreaking work of the semi-reclusive genius of Göttingen, Carl Gauss.

CHAPTER 10

Average Man

The advance to quantitative thinking moved in sync with the happenings of an era in which the technical developments in probability theory and statistics were abundant, amazing, and profound. Where we are now in the story, the early nineteenth century, was a time of transition, both for its historical events and in terms of people's thinking. In this chapter, I first describe the contextual history (picking up from the description in Chapter 7 of events leading to the French Revolution) and then move to the mathematical innovations that prompted people toward quantification at this point.

In 1815, Napoleon suffered his infamous defeat at the Battle of Waterloo. Although he resigned as emperor shortly thereafter (for a second time; he had previously resigned, had been sentenced to exile on Elba, and later escaped and returned to France), the battle itself did not end the Napoleonic Wars or immediately close the Napoleonic era. The defeat is significant in world history primarily because it placed France in a weak position during the ensuing peace negotiations with the coalition countries (principally Great Britain and Germany) at the Congress of Vienna in 1815.

This was the most comprehensive treaty in European history, and it set the terms for much of Europe's political interaction, from its signing to the outbreak of WWI in 1914. But the Congress of Vienna was itself as tumultuous as the times. Some ministers began their negotiations before others had arrived, and it was unclear as to who could vote on what. After its acts were signed, nearly everyone criticized them, with some saying the terms were too reactionary against the European (mostly French) aristocrats, while others saw it as too liberal. Regardless, this important event did finally end the frightful and exhausting period of the French Revolution and the Napoleonic era.

The Error of Truth. Steven J. Osterlind. Oxford University Press (2019). © Steven J. Osterlind 2019.
DOI: 10.1093/oso/9780198831600.001.0001

It also initiated the slow ascendency to power by Germany, which, at the time, was helpful to many mathematicians and other academics. The German people already had a culture of hard work and intellectual pursuits, which was further encouraged by their new freedom. With the constant wars and military campaigns now behind them, ordinary people could at last go about their daily lives with a sense of personal security, and even tranquility. Consequently, the period was ripe for adopting a new view of things. This is yet another place in our story where a reformed worldview advanced because the social, cultural, and political milieu fostered such change. We know by now that this expression is a metonymy for "the times supports its developments."

One outcome of the peace was a redirection of public attention and resources toward improving the lives of the many forgotten people who lived in poverty and subjugation to an inflexible class system; indeed, it was almost an unspoken caste system. Writers, in particular, were influential in developing this new social consciousness. Charles Dickens first gained fame by writing comic characters in *The Pickwick Papers*, but as he saw the inequalities among classes persist, he turned to writing more serious works, including *David Copperfield*, *Oliver Twist*, *A Tale of Two Cities*, and *Great Expectations*. These works brought a shared sense of social responsibility to people and helped to unify them in a purpose larger than self-preservation. When thinking of Dickens, who can forget the wonderful array of names he created: the Artful Dodger, Uriah Heep, Mr. Micawber, and Sergeant Buzfuz. I have a friend who called his dog the Artful Dodger, and many of Dickens's colorful names have been usurped by unimaginative celebrities and pop musicians.

Herman Melville wrote the most famous opening line in literature by using just three words: "Call me Ishmael." Although *Moby Dick* is outwardly a whaling story, through its heavy symbolism it tacitly brings to ordinary people an exploration into the "self" and finding purpose in life. This, too, is part of a new way of thinking. Remember, Freudian psychoanalysis had not yet been invented.

In America, Ralph Waldo Emerson penned his essay *Self-Reliance*, which put forth a distinctly American approach to living and opened the philosophy of the New England transcendentalists. His ideas were tremendously influential not only on the general public but on other authors, including Henry David Thoreau, Margaret Fuller (an early feminist), Walt Whitman, and John Muir (founder of the Sierra Club and credited by some as the first environmentalist). Frederick Douglass, a respected abolitionist speaker, wrote his autobiography, *The Narrative of the Life of Frederick Douglass*, which was popularly received and contributed to growth in the number of people opposing slavery. Once again, we see a new way of thinking and imagining things.

This time in history is sometimes called the "Romantic period," but the designation is most often linked with accomplishments in music. The Romantic composers include Beethoven, Schubert, Schumann, Chopin, Mendelssohn, and a few others. However, unlike Mozart, with his affinity for mathematical expression, these artists were almost the opposite. For the most part, their masterworks had flowing rhythms with little mathematical symmetry; there is almost no syncopation or staccato. The point–counterpoint of the several Bach composers (the most famous of whom is J. S. Bach) is noticeably absent. Nonetheless, their accomplishments did show an individualism of expression that is linked to the self-determination of the age.

Also in this era, two curious inventions had an unexpected effect on advancing quantification: the stereoscope and the slide rule. Both were great levelers of society, bringing together experienced mathematicians and ordinary people. Just by hearing the names of these clever inventions, readers of a certain age will immediately recognize them and may even have owned one or both—or perhaps used them in school a long time ago. Although I did not own a stereoscope, I loved to use the one in our elementary school (whenever it was not already being hogged by other students!). I did own a slide rule and used it regularly throughout high school and into college, when the first four-function calculators appeared. Today, it sits proudly on a shelf in my study, like a piece of petrified wood, interesting only because it is so old and obsolete.

As for the stereoscope, it was "virtual reality," only happening about 180 years ago! In 1838, a British scientist named Charles Wheaton discovered a way to mimic binocular vision, the normal way humans and almost all other animals see. In our binocular vision, each eye takes in a slightly different perspective on whatever is in our field of view. The brain combines them into a single image, giving depth— hardly technical, but this is the core of how we see. Wheaton discovered that the illusion also works when two nearly identical pictures, each taken from a slightly different perspective, are presented to a subject. The viewer perceives depth in the picture, and voilà—a three-dimensional (3D) image.

Working from Wheaton's discovery, British scientist David Brewster made an invention to put the idea into practice. In Brewster's "stereoscope," two photographs of the same scene, each taken from a slightly different perspective, are held in a wooden frame a few inches away from someone's eyes. To the viewer, the picture appears to be 3D. The invention was an immediate hit, and the London Stereoscopic Company produced stereoscopes by the thousands.

The company still exists. People were quickly making their own "two-pictures" and having a lot of fun in viewing them. Folks had stereoscopic pictures of their families, famous paintings, illustrations from stories, landscapes, landmarks such as cathedrals—all sorts of things. I recall from my elementary school days having scenes from stories, as well as—no kidding—famous quotes! All were presented in 3D. Figure 10.1 shows a typical stereoscope.

The stereoscope contributed to quantification in an unexpected way. According to one historian of these novelties:

> The device crossed all cultural and class boundaries: intellectuals used it to ponder the mysteries of vision and mind, while kids merely goggled at the cool views. . . . You could stay at home and go to France, to Italy, to Switzerland and China, and you could visit all these places by your fireside. (Thompson 2017, 21)

Further, the historian asserts:

> Stereoscopy began to transform science. Astronomers realized that if they took two pictures of the moon—shot months apart from each other—then it would be like viewing the moon using a face that was the size of a city . . . The technique indeed revealed new lunar features. (Thompson 2017, 21–2)

This leveling effect helped to make experiences common to people of all classes, occupations, and educational levels. Today, we have a similar device but with a different name: "virtual reality," or "VR." Present-day VR headsets, of course, are much more sophisticated in their technology, allowing you to have an entire 360-degree experience, but the idea is the same as in Wheaton's discovery and Brewster's

Figure 10.1 Early stereoscope
(*Source*: http://commons.wikimedia.org/wiki/Category:Public_domain)

stereoscope. And the leveling aspect inherent in the viewing is still extant. Today, astronauts practice space visits with VR headsets, and many youngsters with VR headsets shoot to kill the bad guy around the next corner (it is shameful, though, that the scenarios too often have graphic detail that is disturbing in its realism).

The second invention at this time that brought still more common ground between academics with their mathematical skills and ordinary folks (in this instance, mostly students) was the slide rule. For those who don't already know (likely anyone who grew up after the advent of Apple and IBM personal computers), a slide rule is a small slip stick with several logarithmic scales printed on each of three slides. As one rule slides past the other, the scales align, and the results to a problem can be read on one or another of the scales. These little devices made many arithmetical operations easy, especially multiplication and division. That was not all they could do: they also allowed calculations for square roots, exponentials, logarithms, and trigonometric functions. Remember, before this invention, such calculation could only be done manually—in *longhand*, as it was termed. This is how Newton, Euler, Bernoulli, de Moivre, Gauss, and all the others did their daily work. Figure 10.2 shows a popular version of a slide rule, used in the nineteenth and well into the twentieth century.

To get the idea of what can be done with a slide rule, try solving this simple problem manually: (192 − 88) ÷ 8 = ? This is certainly an easy problem but it will take a few minutes. (The correct answer, by the way, is 13.) Next, try this slightly more difficult problem, again by longhand: $\sqrt{49^3}$. (The answer is 343.) Again, solvable manually (presuming one remembers—or was ever taught—how to manually compute cube values and square roots) but, obviously, it would take quite a bit of time—say, ten minutes at least. Finally, consider solving an algebraic equation like the one below:

$$x = \frac{-b \pm \sqrt{b^2 - 4ac}}{2a}$$

It, too, can be done with just hand calculation (as it was for many, many years), but, with the slide rule, you can solve it in a less than a minute, regardless of

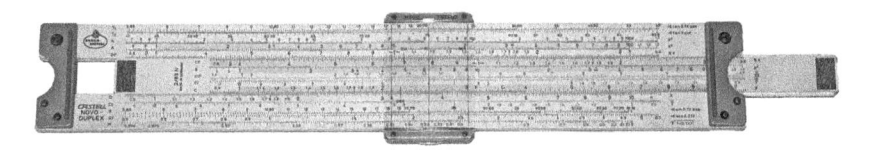

Figure 10.2 Log-log slide rule
(*Source:* http://commons.wikimedia.org/wiki/Category:Public_domain)

what values are given for *a*, *b*, and *c*. (You probably realize that just two of the values are required, as the third can be deduced from them—an easy conclusion with a slide rule). You get the idea. Slide rules were a quantum leap forward in ease of solution.

The slide rule was ingenious and manifestly useful. Moreover, it remained widely used in schools and industry (e.g., by engineers, mathematicians, physicists, statisticians, and almost anyone dealing with calculations) until finally being replaced by handheld calculators in the 1970s. Before then, slide rules were not quite as ubiquitous as cell phones are today, but certainly common.

A bit of history might be useful here. Shortly after the Scottish mathematician John Napier published a first concept of the logarithm in the early 1600s, calculating devices appeared. Some of these simple mechanical devices are considered to be early slide rules. An English minister and amateur mathematician named William Oughtred is credited with inventing the first of these. But his scale device did not use logs with the base that modern logs employ, and it was thus of limited utility. As soon as Oughtred's invention appeared, people realized that it was only modestly practical, but it still seemed a good idea. Soon, several other inventors offered their versions of a log-based calculating device. But these tweaks were only minor adjustments, and the small appliance remained impractical.

It was not until years later, in 1815, that Peter Roget (yes, of *Roget's Thesaurus*) made the modern "log-log" slide rule, which has several scales displaying the logarithm of the logarithm with a base 10, the modern base for this scale in most applications. The log scales slide past one another to align on the correct number. Roget's slide rule dramatically increased the device's usefulness, allowing calculations of all sorts.

These two simple mechanical devices—the stereoscope and the slide rule—were possible only because of the mathematics behind them, which itself stems from the work of the ingenious individuals we have met thus far. For quantification, both devices were great levelers in society, bringing together in a common experience the accomplished mathematician and the ordinary person. With these inventions, we see increased efforts to infuse numeracy into the daily lives of ordinary people, now operating on a wide scale. Quantification was advancing—not yet to the point of an internalized worldview, but certainly moving numeracy into the cognitive domain. We are thus trudging toward a transformation, surely and inexorably.

From the context of the post-Napoleonic era, and with these new inventions on the scene, we can see that quantification was decidedly moving into the experiential realm of ordinary people. No longer was a numerical achievement known solely within the confines of academia. Nor was it unusual for an ordinary individual to cite something numerical, like actuarial tables, tide movements, or phases of the moon. Mariners (finally, with chronometers) knew their position at sea (latitude and longitude) with relative accuracy. Every day, people were having more and more contact with things that are wholly quantitative. Despite the now-daily numerical experiences, however, people still held their personal identity in a qualitative realm. But, by bits and pieces, that too was changing.

****

Earlier, we saw that Laplace sought to bring quantification to Everyman, but his efforts were limited in scope and only modestly persuasive to society at large. However, a student of Laplace, Lambert Adolphe Jacques Quetelet, a mid-nineteenth-century French astronomer and mathematician, pursued the same intention but with even more zeal. In fact, today he is more famous for bringing quantification into the contemporary society than for making mathematical inventions. Quetelet expressed his point of view by saying, "I am less desirous to explain phenomena than to establish their existence" (Quetelet 1968, vii). He spoke this in the context of his "originalism" belief, giving God credit for all things that are then only "discovered" by humankind. This was a persistent view throughout the period.

Quetelet was a scholar of immense intellect, and one with a variety of interests. He was the first person to receive a doctoral degree (in 1819, at just twenty-two years of age) from the newly founded University of Ghent, in Franco-Belgium. While there, he studied two disparate subjects: mathematics and literature. He remained energetic throughout his life, writing dozens of books and more than three hundred scientific papers and even wrote a libretto for an opera.

Upon graduation, he started work as an astronomer in Brussels. There, he founded and directed the Royal Observatory and became active in academic circles. He was soon a public figure, something that he apparently enjoyed. As such, he stayed in close touch with one of his professors, Simon Laplace, who had a profound impact on his thinking and work. For all of his accomplishments, it is surprising his work generally included little mathematics, or only relatively simple ideas that were already well established. Regardless, his application of mathematics and probability to social science data was a significant contribution to quantification.

To Quetelet, promoting probability to ordinary people was like proselytizing for God, since to him, everything came from God. He advocated a philosophy wherein "constant causes produce constant effects" (quoted in Porter 1986, 55), meaning that living for God yields a godly and predictable, consistent outcome. Living apart from predictable behavior is a kind of deviant (read: not normal) behavior. For instance, if poverty leads someone to commit a crime, more poverty will yield the same result: more crime. He had allied ideas concerning "variable causes" (wherein change is the result of variation, such as in the four seasons) and "accidental causes" (for things that change randomly and act indifferently in any direction).

He saw life as almost literally fitting into a probability theorem, a sort of real-world application of Laplace's advanced version of the central limit theorem, something that captured his interest. Probability theory was Quetelet's route to understanding human behavior. To Quetelet, you have choice in a number of life's experiences, and the more your choices align in a particular direction—as, say, toward God or toward something else, such as criminal activity—the more likely that future choices will be of the same type, in the manner of "constant causes produce constant effects." According to his line of thought, this is like fitting life experiences to a normal curve. He was the first to hypothesize behaviors as causes for "errors" in a normal distribution. Remember, at the time, the normal distribution was still known as the "law of error," to reflect the mean and deviations about (above and below) it.

He believed that society can be understood—even "estimated"—by probability prediction. This is Quetelet's notion of predictable effects resulting from known causes, whether they be "constant," "variable," or "accidental." Hence, events that compose social order (such as marriages, births, deaths, and crime statistics) can be identified, measured, and plotted as a distribution, showing the features to be either normally distributed or skewed. He sought to learn as much as he could about such mathematically defined errors. In doing so, he believed he would understand much about how society functions and what gives it order. He wrote frequently about social order versus chaos.

He appreciated particularly that although social statistics are gathered individual by individual, separate peculiarities tend to be washed out when considered against those same choices or characteristics for the population. He said,

> The greater the number of individuals observed, the more do individual peculiarities, whether physical or moral, became effaced, and allow the general facts to predominate, by which society exists and is preserved. (Quetelet and Beamish 1839, 12)

But Quetelet also realized the value of the individual as a basic building block for both social order and the aggregate display of the bell-shaped curve—again, his idea of plotting data as a means to understand phenomena and hence preserve social order over chaos. From this understanding, he turned around the notion of looking at aggregated data and began to focus on the individual. From this perspective, he made the point that group statistics do not apply to the individual. This is the flip side of the individual being washed out ("effaced," in his words) when considered only as a small bit of data in the larger distribution of lots of numbers.

Whereas today the maxim of "group statistics do not apply to the individual" appears ordinary and matter of fact, Quetelet was the first to bring this perspective to public attention.

Quetelet's interest in the individual was made manifest in two particular and closely related ways: first, his focus on sociological phenomena, and, second, his desire to codify the average man. Both features came to dominate much of Quetelet's professional activities throughout his career and today constitute the areas for which he is best known. He had an intense interest in sociological phenomena (such as social demographics, health characteristics, and deviant behavior such as crime).

Prior to Quetelet, the focus of probability was usually on games of chance or on astronomy and geodesy. The extent to which social demographic data was captured and tracked or studied was certainly minor. Quetelet made it mainstream by working with zeal and from a particular philosophy of understanding society as a means to preserve social order over chaos. According to one historian of probability theory, "Adolphe Quetelet was among the few nineteenth-century statisticians who pursued a numerical social science of laws, not just of facts" (Porter 1986, 41).

Quetelet's second area of focus was describing *l'homme moyen* ("the average man"). He thought that, by describing the social behaviors of a lot of people, he could codify what was ordinary and what was not. He saw this as a problem of probability: estimating what was likely and what wasn't. His methodology was simple: gather statistics on many individuals and plot them as data in a normal distribution. He would thereby describe what was average and what was apart from average; namely, the mean of a distribution and its "errors." Recall, again, the term "standard deviation" was not used at the time. Rather, deviation scores were referred to as a "theory of errors." He was thus fitting data to a distribution as findings in a research problem and then drawing conclusions for the fit. This was a novel approach at the time.

His efforts to describe *l'homme moyen* became almost an obsession, for he sought data on all kinds of things, including physical features, sociological characteristics, mental capacities, and especially all manner of crime statistics. He collected data for literally hundreds of such variables. He called these the "facts of life." Per his professional predisposition, he viewed these facts of life in probability terms. Accordingly, as he collected data, he fit them to a normal distribution, noting especially the errors.

True to his academic disposition, he carefully noted his philosophy toward sociological phenomena and his methods and results for identifying and describing the average man. Quetelet was a prolific and detailed writer. His most famous work is *Sur l'homme et le développement de ses facultés*; this title would directly be translated as *On Man and the Development of His Faculties*, but somehow it came to be known as *A Treatise on Man and the Development of His Faculties* (Quetelet 1968). He wrote this work early in his career, originally publishing it in 1835. It was well received by his peers not only for its novel idea of characterizing the average man but also for its level of exacting detail in description.

Shortly after Quetelet published his work and saw its favorable acceptance, he wrote an essay in which he more fully laid out his beliefs for society generally—the constant effects notion introduced earlier. He incorporated this essay into a second edition of his main work and even added it to the original work's title, now is *Sur l'homme et le développement de ses facultés, ou Essai de physique sociale*. The translation of the last part of the expanded title is "Essays on Social Physics," which introduced a new direction for research and a new term to go along with it: "social physics." This new essay was popularly received, too.

Today, the complete work is viewed as one of the earliest publications on sociology, and it brings together several of Quetelet's ideas that he had previously considered independently: social order as defined by quantifying events into normal distributions; probability as a route to describing the average man; and Quetelet's belief that the science of probability should be applied to all sorts of social phenomenon.

Although generally popular, the work was not well received by everyone, particularly Auguste Comte, a contemporary of Quetelet. Comte was a French philosopher who also thought a lot about society and social order. Remember, the Napoleonic period—a time characterized by social upheaval and chaos—had just ended, and it was still uncertain whether the post-Napoleonic peace settling on the Continent would last. People had not yet reached a sense of tranquility in their social order, and personal homeostasis was unset.

Comte attempted to address this anxiety by offering a viewpoint in which science would pervade society, and order would result. He studied social phenomena empirically and eschewed metaphysical explanations of Quetelet and others. Science, he thought, was the route to understanding society, and, through science, social order would promulgate. This is the philosophy of positivism. (For reference, today "postmodernism" challenges the basic assumption of positivism.)

Comte described his way of thinking as "social physics." When he learned of Quetelet's *Essays on Social Physics* he was incensed and accused Quetelet of appropriating a term that he had invented. We do not know Quetelet's reaction, but we do know that his essay was widely read at the time and that Quetelet was a popular public speaker. Presumably, he refused to change the title. Although displeased, Comte apparently did not want to get into a row with such a public figure, so he changed his label for his notion to "sociology." This turned out to be a good decision by Comte, as he is now given credit for inventing the term, if not the science.

Early in his career, Quetelet had a notion of conducting a census of the Low Countries of the Continent. To him, this was a perfect application of his theory of social physics. He intended to describe rates for births and deaths and some other events, such as marriages, more accurately than was done before, and he would use the Laplace methodology of making probability ratios from a combination of observations in the manner of Tobias Mayer, which we saw earlier. The idea was to divide the region into a number of districts and then sample from each of them. However, since exact records of births and deaths were not uniformly kept, he was confronted with a problem of reliability. He realized that varied local record keeping could lead to bias in his counts. So, he devised a special application of Laplace's probability ratios in which he would use estimated numbers divided by reported numbers, making a ratio that he thought would smooth out many of the irregularities in the official records.

Unfortunately, a colleague—Baron de Keverberg of the Netherlands—pointed out to Quetelet that the heterogeneity in social structure among the districts went beyond just irregular records. For instance, the regions had differences in caring for the elderly, with longevity being affected. This fact made for dissimilarities in the birth and death rates. Also, he did not like the Laplace probability sampling idea. Apparently, the baron was doing more than just offering constructive criticism. He was influential in government funding, and he got the census monies stopped. In his defense, however, the government was rebuilding during this period, and money was in short supply for everything.

Quetelet, it seems, was not too upset, for he dropped the idea of a census, and so the inconsistencies between local records persisted.

But Quetelet, true to his active nature, simply redirected his professional energies. He continued his pursuit of describing the average man—*l'homme moyen*—almost obsessively counting everything in sight: rates of births and deaths; suicides; physical characteristics of conscripts; body dimensions of ordinary people; crime statistics; and on and on. Always, he fitted them in distributions and examined the errors, or deviations from the mean.

As Quetelet's ideal, the average man was a true expression of supreme goodness. He said, "If an individual at any epoch of society possessed all the qualities of the average man, he would represent all that is great, good, or beautiful" (quoted in Porter 1986, 103). In other words, the closer you are to the apex of the normal distribution (the mean), the more beautiful and good. Ultimately, for Quetelet, describing the average man was effectively giving structure to laws that govern chaos in society. He claimed to have come to this perspective while reading Aristotle, who also considered the ideal man, though not in the probability terms used by Quetelet.

Continuing his work in collecting "average" statistics, Quetelet began tracking obesity for ordinary people, a rather novel statistic at the time. At first, he simply measured body dimensions and plotted them on the normal distribution. But, with obesity in particular, he took a special interest and devised a standardized set of values, which he called the *Quetelet obesity index* (or QI). The index was almost immediately adopted by healthcare personnel, and it remains almost unchanged to today, although it is now more popularly called the "body mass index" (or BMI). Today, the BMI is employed as the international measurement of obesity. It is universally expressed in units of kilograms per square meter (from mass in kilograms and height in meters).

The World Health Organization has developed BMI charts showing cutoffs for various categories, including "Underweight," "Normal Weight," "Overweight," and "Obese." These charts are commonplace in healthcare facilities. There is even a reasonable chance that you have used them yourself—again, quantification makes an appearance in our daily lives! (Maybe, in this instance, one wishes the zeal toward quantification was more muted.)

As mentioned, Quetelet also had a special interest in tracking crime statistics. He not only examined whether a crime had been committed but also looked at the individual's behavior, such as if the accused showed up for the trial, and if the person did appear, the likelihood of a conviction. It seems he examined everything about crime for which data was available, such as crimes by knives, guns, poisons, drowning, or suffocation, and even death by hanging. He reasoned that, the more data was known about an individual, the greater was the probability of accurate prediction about the future behavior of that individual. He calculated that if the only piece of information available for an individual was that the person had been accused of a crime, there would be about a 61 percent chance of a conviction but that this number would change as additional information became known, such as whether the crime was against another person or against property.

The changing predictions followed his lifelong philosophy of "constant causes produce constant effects." Again, his interest was in determining the error in these group statistics for identifying significant deviations and, above all, preserving social order over chaos—quite a novel approach at the time.

He wrote of this theory in his thesis on probability, *Instructions populaires sur le calcul des probabilités* (Quetelet 1996; first published in 1825; translated as *Popular Instructions on the Calculation of the Probability* (Quetelet and Beamish 1839)). In that work, he advocated for broad application of his theory, famously saying,

> It seems to me that the calculation [*calcul*] of probabilities . . . ought . . . to serve as a foundation for the study of all the sciences, and particularly for that of observation. (Quetelet and Beamish 1839, ix)

For Quetelet, using probability to describe the average man was effectively giving structure to laws that govern civil society, as opposed to social chaos— this is the idea we saw earlier. It would seem he viewed order in society as an either-or proposition.

As time went on and Quetelet continued codifying the average man, his views on the topic evolved. No longer was it sufficient to merely record his facts of life to map an average or typical man. He began to portray the notion of the average man as an *ideal* to which all humankind should aspire. To Quetelet, to be at the mean in all facts of life was to be ideal. The mean became both the literal and the metaphorical apex for humankind. To whatever extent an individual deviated from the mean, he had fallen. According to Quetelet, for *l'homme moyen*, "virtue consists in a just state of equilibrium, and all our qualities, in their greatest deviation from the mean, produce only vices" (quoted in Porter 1986, 103).

In other words, man lived in a balanced state as represented by the mean in the facts of life, and deviations from that mean were the equivalent of defects. Quetelet had, apparently, a philosophy with no room for complexity of behavior.

One imagines he would have been sympathetic to views held by Jonathan Edwards and the Puritans in America (still popular there at the time), but, of course, this is conjecture, and there is no record whatsoever of Quetelet making any related comment.

Still, in one attempt to identify the ideal soldier, he plotted body dimensions for conscripts, including chest circumference. But he ran into an unexpected complication when he found that the conscripts' chest measurements varied considerably depending upon both the geographic area that a given individual was from and his ethnic heritage. Celts were decidedly different from Scots, for instance. Although the mean values for each group were not too dissimilar, the standard deviations were vastly disparate. He found that the largest deviations were for those of Scottish lineage. He reasoned that Scottish ancestry was actually an amalgamation of various heritages, including British, Welsh, Danish, and other Anglo-Saxon, leading to greater intrarace variability, whereas the Celts were more homogeneous in their descendance.

This discovery was more than just of novel interest to Quetelet. Following his concentration on the errors of the normal distribution, he wondered what amount of error would be significant, because, clearly, he would need different values for the conscripts of each heritage.

To determine what was a significant difference for each group, he tried a procedure that (true to his bent) used ratios in a probability of proportions. He contrasted the ratio of each group's chest statistic with that for everyone, regardless of ethnic heritage, and then compared them across the groups. Some readers will recognize from even this tiny bit of information that Quetelet had suggested a sort of rudimentary analysis of variance procedure, a statistical test that was not fully developed until much later. Regardless, this approach shows an ingenious component to Quetelet's efforts.

He used this information to suggest that he could predict height with uncanny accuracy. He measured conscripts' height and then, using the known deviation, he invented a rather clever system to extrapolate height measurements for a very large sample, which he construed to be a population of one million measurements. Then, he presented his data in an ingenious manner. Rather than plot heights in a histogram or on a distributional scale, he made a single, vertical bar and "filled" it with points, one for each height in his population. From tables of his height data, he made eighty equal divisions, noting from one million (his supposed population) the number of points that fell within each category. His scale, which he published in his book on social physics, is shown in Figure 10.3.

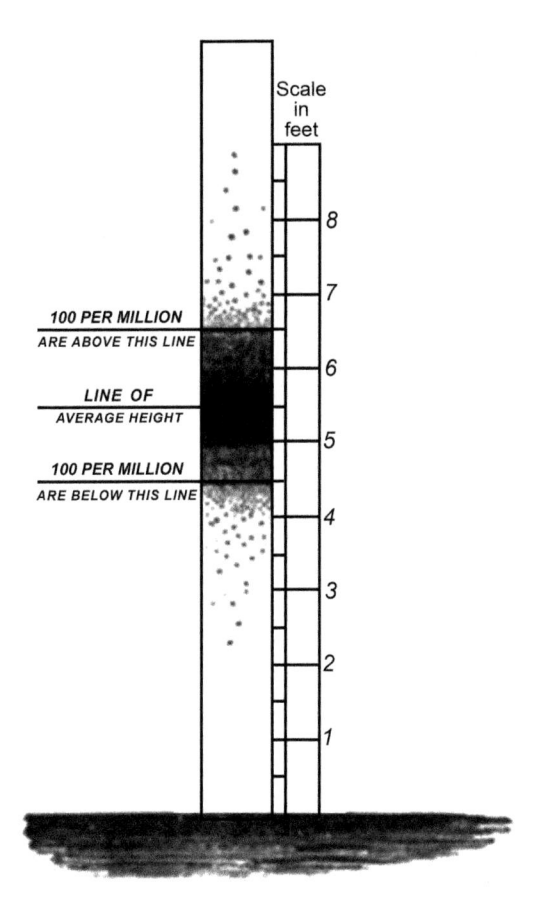

Figure 10.3 Quetelet's height measurements, distributed in categories with divisional lines
(*Source*: from A. Quetelet, *A Treatise on Man and the Development of His Faculties*)

From this, he developed a law of deviation in which any measurable event with enough similar measurements (i.e., ones that occur within the same range) would be governed by the same predictable deviation that he saw for the conscripts. He felt he could predict for a population the exact distribution for a large number of cases, and even how many would fall above and below any of his lines of division.

This work influenced Francis Galton greatly, who applied it to intelligence measurements in his seminal *Hereditary Genius*. In that work, Galton described Quetelet's system of distribution and imagined he could use it, too. I describe this more fully in Chapter 12.

And, as we see, Quetelet himself followed in the footsteps of his professor, mentor, friend, and primary influence, Laplace. He sought to apply quantification to social contexts and thereby bring it to a broad audience as they participated in everyday society.

He wanted to touch Everyman. While he was only modestly successful in this large effort, he did influence several important people from varying fields: Florence Nightingale, John Maynard Keynes, Francis Galton, Siméon Poisson, and others. Let us look at these influences individually.

Poisson was a mathematician who developed some highly technical papers on physics and invented an important probability distribution. We will meet him in Chapter 11.

Florence Nightingale, as is widely known, was a heroine of the Crimean War and, through her humanitarian work, she virtually invented the nursing profession. What is not as well known about her, however, is her interest in statistics and its application to social problems. She wrote scientific papers using probability to address such issues. Obviously, writing scientific papers was quite unusual for a woman at the time. She credited her activism to following Quetelet's notions on social physics, and we know that they corresponded.

Readers probably recognize that the English economist John Maynard Keynes is one of the most prominent economists of all time. Keynes freely acknowledged that his ideas on social physics were heavily influenced by Quetelet, particularly his acceptance of "constant causes" explanations for societal phenomenon. In 1921, Keynes wrote an important essay about social physics, titling it "The Application of Probability to Conduct" (reprinted in Newman 1956). In it, he sounds exactly like Quetelet. He says, "We might put it, therefore, that the probable is the hypothesis on which it is rational for us to act" (Keynes 1956, 1360). Indeed, upon reading the essay, one imagines that it could have easily been penned by Quetelet himself.

Keynes applied probability to problems in economics in his seminal work *The General Theory of Employment, Interest and Money* (Keynes (1936) 1973). It brought about the "Keynesian Revolution," attracting a worldwide following. He posited ideas on how to manage economies, particularly in times of severe unemployment, such as existed during the Depression, the time during which he wrote. The most lasting point of Keynesian economics is his theory on supply and demand. He said "supply creates its own demand" (Keynes (1936) 1973, 21), a line that started a generations-long discussion about its effects in practice.

The book remains one of the most influential ever written on economic theory. From its introduction in 1936 to at least the 1960s, it was required reading for students of economics, as well as practicing professionals. According to *TFE Times*, the book is No. 6 on the list of "10 Best Economics Books of All Time" (TFE Times 2017). (The no. 1 book is Adam Smith's *The Wealth of Nations*, first published in 1776.)

Later, we shall see in some detail that Quetelet's ideas were also picked up by the British statistician and sociologist Francis Galton.

****

In addition to these social influences, quantification was now also shaping other new fields, such as natural history, with the work of Georges-Louis Leclerc, Comte de Buffon, a French naturalist, mathematician, and cosmologist. Buffon wrote a single work during his entire career, but it is in thirty-six volumes, with at least two additional books published by a protégé after Buffon's death. This monumental work is *Histoire naturelle, générale et particulière* (*Natural History, General and Particular*) (Buffon 1780). The work describes and explains numerous events in the animal and mineral kingdoms. The work was extraordinary in the day for its novel content. It was popularly read across the Continent and translated into several languages, making it one of the most widely read books of the day. In fact, Buffon's popularity has been compared with that of Montesquieu, Rousseau, and Voltaire, making him among the most-read writers at that time.

Charles Darwin seems also to have been heavily influenced by Buffon, because many of his central ideas were ones that had been made earlier by Buffon. However, Darwin does not mention him at all. When asked about this, Darwin replied that he was not familiar with Buffon, (although, given the latter's popularity in natural writing, that is hard to believe).

Noteworthy to us is that Buffon engages probability in many of his explanations, something not done before. One particularly striking probability explanation has become somewhat famous because it was the first time someone used probability to address problems in nature. It is called "Buffon's needle problem." The problem he posed is this one:

> Suppose we have a floor made of parallel strips of wood, each the same width, and we drop a needle onto the floor. What is the probability that the needle will lie across a line between two strips? (Buffon 1780)

Figure 10.4 illustrates Buffon's needle problem.

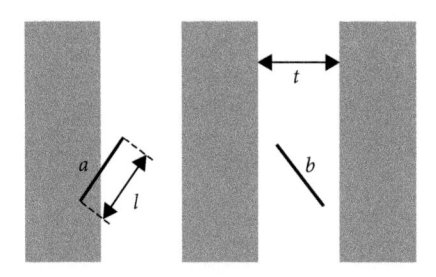

Figure 10.4 Illustration of Buffon's needle problem

(*Source:* http://en.m.wikipedia.org/wiki/Buffon%27s_needle)

The problem had vexed others for a long time, but Buffon solved it with geometric probability and integral calculus—the first to do so.

Most pertinent to our discussion, however, is his idea of bringing quantification to the natural sciences, another important first. Closer and closer it comes to the daily lives of ordinary people.

CHAPTER 11

Rare Events

Early on, I emphasized that, as much as anything else, the story of quantification is a journey of discovery into our inner being—who we are and how we came to a particular worldview. Quantification is our synonym for a transformed *Weltanschauung*: a comprehensive conception of the universe and of humankind's relation to it. It becomes our reality—our truth. Such a narrative is in contrast to a dry essay that just presents quantification as one fact building upon another.

In actuality, the story of quantification is complex and occasioned by dynamic events, best explained by a thorough and carefully reasoned narrative. No glib explanation or simple description can give it due. We know that the proximal cause for adopting a quantified outlook is the invention of methods for measuring uncertainty, as seen primarily through the science of probability theory. We see, too, that, during a relatively short period in history, there was a torrent of mathematical activity in this new science. And, significantly, contemporaneous events of history provided an encouraging context for such intellectual advancement. As I said in Chapter 1, these astounding mathematical inventions happened because they *could* happen. History gave impetus—a kind of tacit assent—to quantitative developments.

But the encouraging environment for quantification at this time in our story (roughly, the first half of the nineteenth century), was not uniform throughout the Western world or Eurasia. Historical and social events were strongly supportive of such intellectual productivity in Europe, Great Britain, and the Nordic regions, but less so in America, with even more pronounced deficits in tsarist Russia, and a particular paucity of widely spread scholarship in the Middle East and Islamic countries. The lack of progress in these regions was not due to universal causes but more to specific happenings within each area. Briefly, the reasons for the lag in each region are as follows.

In the United States, a stable government was well in place by this time, with the settling of federal authority by the case of *Marbury v. Madison* (US Court 1803). In it, the US Supreme Court ruled unanimously for its authority to declare a law unconstitutional, and Chief Justice John Marshall's words have been engraved on the wall of the Supreme Court building: "It is emphatically the province and duty of the judicial department to say what the law is." All thirteen of the original colonies were now states, and, between 1800 and 1850, an additional sixteen territories gained statehood, most of them on land acquired from Napoleon in the Louisiana Purchase of 1803.

Not all was encouraging in this new land, however. Tobacco and cotton farms, and the garish plantation homes of their owners, were springing up in the South, giving rise to a growing tension over slavery between northern and southern states, an issue intertwined with taxation. The brief War of 1812 with Great Britain (it actually occurred in the late summer of 1814) had been settled, and, although the true meaning of the conflict itself has been variously interpreted in American history, its consequences were lasting and important to the story of quantification (as we will see in Chapter 13). Out west, the Mexican–American War was building, with possession of several states at stake, including California, the grand prize.

Regardless of these troubling events, Americans, by and large, were a vibrant group of immigrants who were wildly optimistic about their young country. More than anything else, they were eager to get down to the business of building their nation. As individuals, they wanted to work hard and be self-sufficient, and they saw a direct link between that effort and earned rewards such as having a good family farm or small ranch or working productively as a local merchant.

Carrying forth this spirit of determination, the westward expansion began. We saw some of this enthusiasm for the "land of opportunity" earlier when examining the work of Laplace in Chapter 8. Notably, this time in American history was the era of "Manifest Destiny." As a slogan, Manifest Destiny is sometimes trivialized to suggest it is little more than a catchphrase for territorial expansion from the Midwest to the Pacific. However, such a modern-day reinterpretation of history is crude and uninformed because, in reality, the westward expansion movement was far more complex. For one, with enthusiastic government support, the pioneers lawlessly trampled on whoever possessed the land they wanted. One-sided treaties with native peoples were made and then often broken.

The idea of a westward expansion and Manifest Destiny was most enthusiastically endorsed by the Democrats, who used it as an excuse to ensure that the hoped-for new state of Texas would allow slavery. But not everyone agreed that permitting slavery in the new states was a good idea. Some early opponents included Abraham Lincoln, Ulysses S. Grant, and most Whigs, who later folded into the Republicans. As everyone knows, this issue was intertwined with national taxation and representation in national affairs in Washington, DC, leading some states to want out of the Union. These problems formed a portmanteau of reasons for the American Civil War, which did not start until 1861, when Confederates attacked Fort Sumter in South Carolina.

Despite these injustices and self-absorptions, much good came from the country's expansion west. The land the early settlers claimed was often unoccupied—for the most part, there were no defined borders to indicate ownership, other than "from here to there," as given sometimes to indicate a Native American territory; in other places, not even that broad indication existed. (An engaging and informative book by Stein (2008) describes how the states each came into their own.) With the westward population migration, warring and banditry between Native American tribes came to a halt. Towns and cities rose, and people made productive lives. Crops, orchards, and yields of every imaginable kind sprang forth. The railroad, too, played an important role in the western expansion. A new country—the likes of which the world had never seen—was being built.

Western exploration and expansion is exemplified by the historical painting *Westward the Course of Empire Takes Its Way* (also titled *Westward Ho*) by Emanuel Leutze. Painted in 1861, the 20 × 30-foot mural, shown in Figure 11.1, hangs in the US Capitol Building.

From these historical events, we can see that the American social and cultural milieu of the mid-nineteenth century (like the period before it) inhibited the spread of quantification. Ordinary folks were too preoccupied with building their new land for quantification to take hold. They were not psychologically ready for such a change. In this environment, there were few mathematical inventions emanating from third-level institutions across North America—despite America's many vibrant universities, including Yale, Harvard, the College of William and Mary, and the ten original great land-grant institutions west of the Mississippi River, including the Universities of Missouri, Nebraska, Iowa, and Wisconsin. Today, there are seventy-six land-grant universities, and the list is growing.

Hence, the sources of quantification were extant in Europe and Great Britain but not in America. That was yet to come.

Figure 11.1 *Westward the Course of Empire Takes Its Way* by Emanuel Leutze
(*Source*: http://commons.wikimedia.org/wiki/Category:Public_domain)

In Russia, the reasons for inhibiting quantifying developments were wholly different. During the mid-nineteenth century, Russian society was still semifeudal and lagged far behind developments on the Continent. The government of the Romanov family was weak and corrupt, losing war after war, such as the Crimean War and the Russo-Japanese War. The tsars were willing to sacrifice untold thousands of soldiers (some estimates run into the millions) just to keep the wars going as their means of hanging onto power. The education system at all levels was poor or nonexistent. Russians were not allowed to travel abroad, and few of them were aware of just how backward their country was in comparison to others.

Further, not only was the Russian leadership corrupt and uncaring about their subjects, they were woefully incompetent. In one famous illustration of their shortsighted myopia, the Russian tsars were so desperate for money they sold a huge plot of their territory—Alaska—to the United States for a mere $7 million, roughly two cents per acre. Most readers probably already know that the purchase was negotiated by William Seward, President Andrew Johnson's secretary of state, and was widely criticized in America, owing to Alaska's remote locale and its then-unrecognized value. It was derisively called "Seward's Folly." Soon, however, it was recognized that it was a valuable asset and that the purchase was a good one. Russia later made overtures to the US government to buy it back, but, by then, it was too late.

An important historical event in Russia is emblematic of why the country did not advance in quantification, or in many other arenas, for that matter: the Decembrist revolt of December 26, 1825. Prior to this date, Napoleon had wanted to expand the French empire eastward, but his incursions into Russia had been met with strong resistance. In fact, the Russian troops, despite being malnourished and lacking artillery, drove Napoleon all the way back to France. He lost more than 400,000 troops in the ill-fated campaign. While this was a military victory for the Russian homeland, the incursion had unexpected consequences; namely, many thousands of Russian soldiers saw other countries for the first time and suddenly realized just how far their own country lagged behind in development.

When they returned home, these soldiers learned that their supreme ruler, Tsar Alexander I, had died. The next in line for ascension to power was Alexander's son, Nicholas I. It was widely known among Russians that Nicholas was inept and corrupt. He was very unpopular. A group of officers and soldiers believed that the time was right for the military to take action. Not only did they want to overthrow the tsars, but they had a specific plan in mind. Here is an important part of the story that is less well known: the revolutionary leaders admired the notion of self-determination, especially for its role in America as embodied in the US Constitution. They thought it was a good model to follow and had plans to replicate its ideas in a new Russian constitution. They admiringly read Locke and Hume for their rationalism, and they knew of Jefferson's writings and even of Hamilton and Madison's *Federalist Papers*—all publications strongly influential to the spirit of a young America. They wanted to capture these ideals for Russia, too. They thought it would be a way forward—a way to end the norm of serfdom.

On that fateful December day, in what was, undoubtedly, a blizzardly cold St. Petersburg (the winter palace of the tsars), about 3,000 soldiers and officers assembled in Senate Square to make known their objections to Nicholas's ascension. According to reports, Nicholas directed his troops to surround them, amassing many times more soldiers than there were protesters. The protesters were instructed to leave, but they refused. Then, the order came down to shoot a few of them. The protesters fired back, and then, not surprisingly, gunfire was everywhere. Being vastly outnumbered, the protesters were quickly annihilated, but the leaders were kept alive for a trial. Upon conviction, some of the leaders were executed, and others were exiled to Siberia, to the village of Chita, where they served a life sentence of hard labor. Later, the American explorer George Kennan met them while surveying in Siberia for a planned telegraph line to Chita. Apparently, he admired them in many regards, saying "Among the exiles in Chita were some of the brightest, most cultivated, most sympathetic men and women that we had met in Eastern Siberia" (Kennan 1891, 336). Figure 11.2 depicts a public trial for some of the Decembrist leaders.

Figure 11.2 Decembrist revolt leaders at a public trial in St. Petersburg, 1826
(*Source*: from http://russianculture.files.wordpress.com/2010/11/x1-z5_094.jpg)

The Decembrist revolt failed that day, but it did embolden the Russian people to express dissatisfaction with the Romanovs and tsarist rule. From that day forward, unanimity in purpose grew among the Russian people as they began to flex their own brand of freedom. It took quite a while, but their sentiment eventually led to the Russian Revolution in 1917.

As you can see, the realities of life in Russia at the time left little room for intellectual pursuits such as advancements in mathematics or probability. It explains why the movement to quantification was slow there.

The reasons for negligible scholarly achievements in the Middle East and Islamic countries during the early nineteenth century were different still, yet obvious, including (1) perpetual warring among countries, tribes, and cultural and ethnic factions, and (2) a feeble academic environment at all levels. In this part of the world, war had been the norm for hundreds of years, stretching across generations. Particularly destructive were conflicts between the Ottomans (who ruled over Turkey and surrounding lands) and the Iranian Safavids (who claimed to have descended from the Islamic prophet Muhammad, although scholars doubt this). The holy war for Constantinople—from the Crusades of about 1095 to 1291 CE and lasting until the city finally fell in 1453, signaling an end of the Byzantium Empire—epitomized the clash of Islam and the West. This period in history foretells the current jihad against the West. The causes of such constant conflict are deep-seated and complex, and certainly beyond our scope to explore. (For an intimate history see Crowley 2005.) With regard to our purpose, however, this context made for a challenging environment for academics and intellectuals to contribute to worldwide scholarship. As a result, virtually no original, significant advances in statistics or probability theory emanated from this region of the world, despite it having had some mathematical innovations earlier in history.

The lack of consistent schooling in the Middle East at this time, as you can imagine, inhibited intellectual advancement and the movement to quantification by the populace as a whole. Despite being ancient in terms of culture and civilization, in the nineteenth century, these states had only primitive public education systems. Even a full century later in the twentieth century, most still had no laws for compulsory schooling, and many towns and villages had not built any schools at all, as reported in the first international survey on the topic, in the 1948 United Nations *Universal Declaration of Human Rights* (UDHR 1948). Even when schools were available, they were only weakly focused on teaching literacy and numeracy, giving more emphasis to instruction on local customs and religions.

At this time, the region had almost no third-level educational system at all, and the few institutions extant in the mid-nineteenth century—the American University of Beirut and Cairo University (still two of the best universities in the region)—were not on par with those on the Continent or in America, although these and other institutions have improved since then.

With colonization by European countries beginning at this time, some schooling did come, although it was mostly restricted to children of the colonizers and a few ruling elites. More commonly, though, children of these groups were sent to Europe for formal education. The vast majority of local children remained without proper instruction in reading or elementary arithmetic.

Deplorably, this deficiency continues today, although most of these countries have built schools and have finally enacted laws for compulsory education. Still, doing even this basic government service has taken more than a century after virtually all other countries. But troubles persist. A 2016 report by the United Nations Children's Fund, or UNICEF)—updated annually as *The State of the World's Children*—reports that, in 2016, almost 50 percent of children (more than 24 million) in Syria, Iraq, Iran, Yemen, Libya, and Sudan were denied education because of ongoing wars and governmental inadequacies. More than one-quarter of school buildings have serious war-related damage. They label it "a vicious cycle of disadvantage" (UNICEF 2016). It is beyond disgraceful that these governments do not provide such a basic service for all the children under their jurisdiction and for whom they are responsible.

Napoleon's invasion of Egypt at the turn of the century did bring new markings of scholarship, then by the founding of Egyptology to study the pyramids and other historical ruins and artifacts. As a side benefit, Napoleon's incursion slowed the recurrent looting of historical sites.

This brief description of events in the Middle East and Islamic countries during the nineteenth and early twentieth centuries conveys only a hint of the history and culture of these people and lands. However, it does provide enough context to understand some of the reasons for the paucity of scholarly accomplishment during this period and to understand why it would be slow to integrate a modern, quantification mindset among its people.

From the above, we see that, in the middle years of the nineteenth century, the journey toward quantification continued, but unevenly: heavily so on the Continent, in Great Britain, and in the Nordic countries, but more gradually in America; then, quite a bit more slowly in Russia, and more slowly still in the Middle East and Islamic countries.

Still, quantification continued, despite its uneven pace.

****

Unexpectedly, a brilliant English engineer named Charles Babbage also contributed to bringing a mindset of quantification to ordinary people. He was considered even to be a polymath: one who is extremely gifted in many areas. But, unlike most of the others in our story, Babbage was not a statistician, nor did he make technical advances to probability theory, despite his keen interest in the field. He was a gifted mathematician, however. It is surprising he is not more widely known, especially as he held the most prestigious position in all academia: the Lucasian Professor of Mathematics at Cambridge University, from 1828 to 1839. Recall, this is also "Newton's chair," which was occupied in the late twentieth century by Stephen Hawking.

Still, Babbage's role in quantification is direct and important. He is credited with inventing a machine that is ubiquitous in modern society—one that you very likely own: the digital, programmable computer. Many historians of computers consider Babbage to be the true "father of the computer" (Halacy 1970).

On learning about Babbage, one imagines that, like today's rediscovery of Nikola Tesla (who discovered alternating current, thereby making electricity practical to use, and is the namesake of the electric car company), interest in Babbage will revive at some point. Both men were brilliant engineers, and their work is characterized by detailed plans and drawings for inventions that were either never built or only realized in practical application many years after their deaths.

And one imagines fame will also come to another remarkable individual, Ada Lovelace, whose full name was Augusta Ada Byron. She was the daughter of famed poet Lord Byron, although her parents separated shortly after her birth, and she never knew the famous poet. Ada Lovelace was an English mathematician and an associate of Charles Babbage. She wrote an early computer code for Babbage's machine and accordingly, she has been called the first computer programmer.

Babbage was especially facile with logarithms, and he used them extensively in his early study of mechanics. But he was frustrated by a widely recognized problem of the day. Commonly used mathematics tables were replete with errors. Remember, at the time, the values in tables were calculated by hand and then recorded in columns on a piece of paper or typeset. A man of exactitude, he deliberately set out to correct the situation.

Babbage was a friend of Laplace and admired Gauss greatly, both of whom were also of astounding intellect. One imagines that if either of these mathematics geniuses were to address the problem of inaccuracies in math tables, they would look to invent an algorithm or perhaps several of them, each to address a different table. Babbage took a different route—he sought to make a mechanical computing device that allowed a formula for input and then would efficiently produce accurate values to use in any mathematics table.

True to his engineering training, he made plans and drawings to specify his ideas. He revised them continually, each time providing more and more detail and refinements. In the end, Babbage made 500 large design drawings, 1,000 sheets of mechanical notation, and 7,000 sheets of related handwritten notes. In them, he specified a large-scale calculating machine, which he called a "Difference Engine." Later, he renamed it "Difference Engine No. 1," to accommodate his planned refinements.

Difference Engine No. 1 was a large and complex contraption, measuring more than 6 feet in diameter and more than 15 feet tall, with a second section that extended its length by an additional 25 feet. It had thousands of parts—hundreds of them movable! He had to invent and fabricate many of the separate pieces on his own, requiring him to also invent purpose-built machine tools. Babbage's Difference Engine No. 1 iterated number counting via a series of wheels, each of which progressed to turn the number wheel next to it. He described his contraption as one that was "eating its own tail" (Curley 2010).

Babbage's machine accomplished his intention of producing accurate numbers for mathematics tables, and he received wide acclaim for inventing it. He was particularly admired in academic circles.

He built several of the machine's parts along the way, often displaying them in his own house, where notables of the day, including Charles Darwin and Charles Dickens, would stop by to marvel. Even just looking at its pieces caused a feeling of wonderment, as they fit together like a complex puzzle. When he activated them by turning a hand crank, levers fell in a cascade of rolling tumblers, one onto the next. There were so many cylinders, each turning onto the next, that one could watch them fall like dominos and listen to a very satisfying sound of click after click. It must have been fun to see.

Soon after inventing his calculating machine, while on a trip to Italy, Babbage learned that he had been nominated to be Lucasian Professor of Mathematics at Trinity College, Cambridge. At first, he did not want to accept the post so he could continue to work on his Difference Engine. His friends, however, prevailed upon him, and he did accept the post after all. In the meantime, he settled in London and married, and, eventually, he and his wife produced eight children. Sadly, only three of them survived to adulthood.

During this time, he and Quetelet met, but, apparently, they had a somewhat tense relationship. Quetelet was a founding member of the Statistical Society of London, which was one of the first professional organizations devoted exclusively to studying and advancing statistics. Later, a group of statisticians included Babbage to evolve the organization into the Royal Statistical Society, but he soon lost interest in the group when no one was available to keep up its administration. A planned invitation to Quetelet never came about.

Babbage continued thinking about computing machines and eventually planned Difference Engine No. 2. This one would be different from his No. 1 engine. Smaller and much more sophisticated, it could handle all sorts of computing tasks beyond just calculating numbers for tables. He envisioned Difference Engine No. 2 to be not just a better calculating device, but a programmable, analytic computing machine—an unheard-of idea. Significantly, in addition to handling numeric data, it could accommodate symbols such as letters.

Babbage's analytical machine would have two parts, corresponding to modern computers: a central processing unit (or CPU) and a memory storage unit. Being programmable, it would read input from a set of punch cards that contained instructions for the machine's operations, as well as original data. Some readers may already know that punch cards were used as input to computers until relatively recently, around the 1970s or so, when computer language advanced to the point wherein an operator (now, "programmer") could type machine code directly.

I remember using punch cards with mainframe computers in the 1970s and '80s. I even keep a stack of them from the old days, but now they are just interesting as history and no longer useful. It was also in the 1970s that ingenious youngsters like Steve Wozniak, working in his parents' garage with his high-school friend Steve Jobs, made Apple-1, a very small computer that was easy to use. They called it a "personal computer." Soon, they were joined by another exceptionally bright young man, Harvard dropout Bill Gates, who was a whiz at programming and marketing. He made an operating system for the personal computer that he licensed to IBM for installation on their computers (foregoing their offer to buy it outright) for a small payment for each one they sold. He used that money to found Microsoft, an early software company that started with just seven employees. Today, of course, Apple and Microsoft are two of the largest— and richest—companies in the world.

Back in the 1830s, however, true to his meticulous nature, Babbage set about making plans and drawings for Difference Engine No. 2. As before, he made hundreds of drawings and notes that described each piece in detail.

There is yet another part to the story of Babbage's progress on Difference Engine No. 2, and this one does not relate to the machine's technical aspects. His work was funded by the government, and officials were becoming skeptical of his ever producing anything useful. Also, he had trouble locating a suitable building to accommodate the planned monstrosity, despite its being smaller than the No. 1. The building needed to be very long, well lit, and dust-free. Eventually, the government lost confidence in Babbage's work and, in 1842, refused to give him more money. With this unhappy development, he simply lacked the resources to build Difference Engine No. 2. Expressing frustration, he wrote, "The drawings and parts of the Engine are at length in a place of safety—I am almost worn out with disgust and annoyance at the whole affair" (Babbage 1834).

Babbage never built his Difference Engine No. 2 in full, but, in 2002, more than 150 years after he was forced to abandon his unfunded project, a group of Cambridge University students, evidently Babbage devotees, did build it. They used only the materials available to Babbage and faithfully followed his plans and drawings. When it was completed, they discovered that Babbage's computer worked exactly as he predicted, which highlights his amazing accomplishment. Their creation is at the Science Museum in London, and a duplicate resides in California at the Computer History Museum in Silicon Valley, near Stanford University.

To see and hear this machine work is marvelously entertaining. It is truly a wonder, not only for its purpose but in the beauty of its mechanical engineering. A picture of it is given in Figure 11.3. Remember, this machine is more than 8 feet tall and 20 feet wide. Turning the hand crank requires a fair amount of force; one must stand beside the machine, which is elevated on a wooden platform. To our good fortune, a YouTube video is available for just this purpose at the referenced site (see Computer History Museum and Microsoft Research 2012). What a beautiful and graceful machine.

To think, Babbage's Difference Engines—first, the behemoth No. 1 and then his plans for No. 2—were one of the first versions of a modern computer, which has spun into modern evolutions like the smartwatch and the mobile phone. Today's "Difference Engine" weighs only a few ounces and fits on your wrist or in your pocket, yet has exponentially more computing power and storage capacity than did Babbage's inventions. But realize they are still just programmable analytic devices with two elemental parts: a processor and a memory unit—all springing from the imagination of Charles Babbage, who worked nearly two centuries ago.

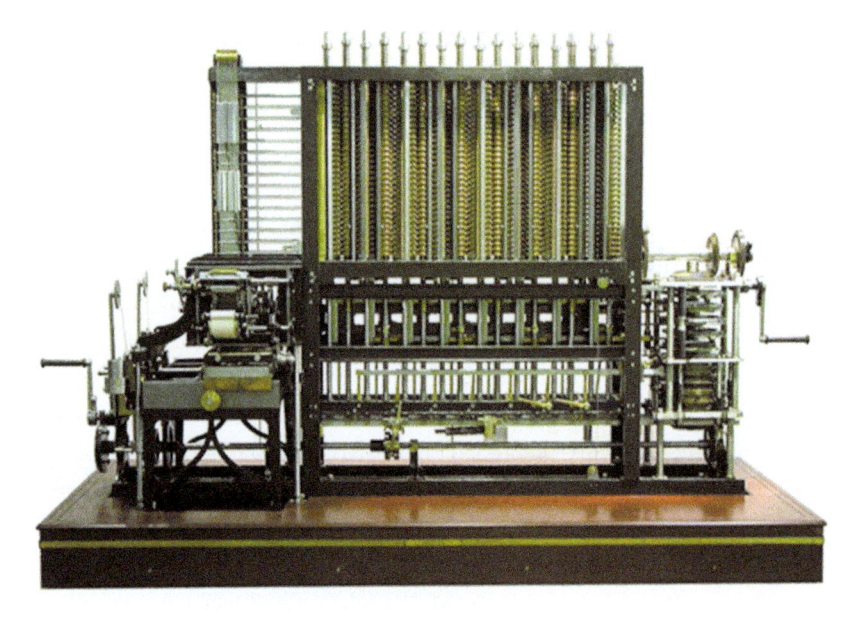

Figure 11.3 Babbage's Difference Engine No. 2

(*Source:* http://commons.wikimedia.org/wiki/Category:Public_domain)

The following slight digression is an end note for those who may claim that the first true computer was ENIAC (Electronic Numerical Integrator and Computer), built in 1943 by John Mauchly and J. Presper Eckert. Their machine was the size of a small house (at 1,800 square feet), and was built with 17,500 vacuum tubes, 7,200 diodes, and miles of wire. ENIAC was different from Babbage's mechanical Difference Engines in that it was electrical, using circuits to open and close electronic pathways for various calculations. While this was a significant innovation, it also made the machine a monstrosity to use. Whenever a new program was to be run, ENIAC required that the wires first be rerouted, a daunting task that took many hours for even a slight change. But, elementally, ENIAC drew its design from Babbage's ideas for the computer's two parts: a central processor and a memory unit.

****

Earlier, we saw that both Laplace and Quetelet employed probability in research on crime and related social behaviors. Only a few years later, Siméon Denis Poisson, a pure mathematician and physicist, continued their work, but followed Laplace's methodology over that used by Quetelet. He advanced their work by developing a particular kind of distribution useful in a specialized circumstance when there are few occurrences over an extended time period, which could be used for, say, showing the likelihood of severe earthquakes over a period of years or forecasting the number of cars that might drive through a red light at a particular intersection. These events occur, certainly, but only infrequently.

With the kinds of probability distributions we have seen thus far—binomial and Gauss's normal distribution (the bell curve)—it is not easy to display these occasional circumstances. When illustrated in these ways, the distributions plot to a figure that is either almost flat or so heavily skewed that it is not easy to follow. The "Poisson's distribution" fits these rare-case scenarios, and, by it, one can readily see the likelihood of occurrence for an infrequent event. In a moment, I'll explain the Poisson distribution more fully and illustrate it with some examples, but, first, a bit about Poisson himself.

Poisson grew up near Paris in a family of modest means. His father was in the military, although he later deserted due to the poor healthcare at the time, but he saw to it that his son received a good primary education. Siméon was a very bright youngster, although not of genius level, and he was especially hard-working at his school subjects. He showed a special aptitude for mathematics. When he entered École Polytechnique in Paris, his professors (one of whom was Legendre) noticed his mathematical ability and allowed him special liberty to study the mathematics subjects of his choice. While only eighteen years old, he wrote two lengthy and notable papers, both on the calculus of finite numbers and differential equations. Legendre and Professor S. F. Lacroix (another of Poisson's instructors) recommended that the papers be published in an important academic outlet of the day called *Mémoires présentés par divers savants étrangers à l'Académie* (Memoirs Presented by Various Foreign Scholars at the Academy). For a young person only two years into university, this was quite an accomplishment.

Poisson demonstrated a laudable work ethic through his life, as he eventually published many books and more than 300 technical papers, some of them so long that they were considered treatises. However, most of these publications did not offer original inventions; rather, they typically extended the work of others or provided numerous examples of applying a given theory or idea. The noted historian of statistics Stephen Stigler characterized Poisson's work as follows:

> His work in these areas (mechanics and the theory of heat) was solid but not of a class with his predecessors, Lagrange, Laplace, and Fourier. In all of this work, his role was that of a competent and insightful extender rather than that of a bold originator, and the same is true of his work in probability and its applications. (Stigler 1986, 182)

Stigler's rather harsh assessment of Poisson is not to suggest that he was unsuccessful in his work; rather, just that he was not particularly original. Upon graduating from École Polytechnique, and with the enthusiastic support of his mentor and, by then, friend Legendre, Poisson was appointed to the faculty, filling the shoes of a vaunted professor, Jean-Baptiste Joseph Fourier. Fourier was renowned in his field at the time and is still today, and his academic chair was a prestigious one.

I digress here to mention Jean-Baptiste Fourier, even if only briefly, to convey the significance of Poisson being named his replacement. Fourier worked in theoretical physics, where he made numerous important contributions. He is best known for his Fourier transform, something which Lord Kelvin called "a mathematical poem." This transform deals with heat transfers and vibrations, and, in it, Fourier invented a way to express how heat decomposes as a function of time into the frequencies that make it up, in the same manner in which a musical cord is heard as the frequencies of its constituent notes. The Fourier transform is so widely used in modern physics that it is a virtual necessity in the field. If Fourier had invented only this, it would still show his stature. His replacement (soon to be Poisson) would have to be someone who could carry forward the position's reputation.

Here is one more interesting note on Fourier: he also invented a way to assess how radiation degrades Earth's (and other planets') atmosphere, creating what he first termed "greenhouse gases." Even superficially, we recognize that the greenhouse effect is a part of explaining how a planet warms. He made this discovery in 1798 while accompanying Napoleon on a trip to Egypt as a scientific adviser. While there, he demonstrated how Earth is warming year by year due to greenhouses gases. This finding of global warming—made long before carbon or fossil fuels were commercially exploited and when Earth's population was less than one billion—is regularly cited in today's climate-change debate.

For our story, Poisson, evidently, did fill Fourier's shoes as a professor: he gained a reputation as an excellent teacher and quickly worked through the academic ranks to full professor in 1806, at just twenty-six years old. Further, Poisson must have been an able administrator, too, because, for the next twenty-five years, he was also responsible for making academic appointments at his university, as well as for developing the curriculum. Here again, we see that he worked ceaselessly to carry out his all of his pursuits as professor and busy university administrator, as well as remaining a researcher and prodigious writer. Obviously, Poisson was a man of enormous energy and drive. Reportedly, he said, "*La vie c'est le travail*" ("Life is work"; Wikisource 2018).

Having gained recognition early in his career, Poisson was admitted into the prestigious Académie Royale des Sciences de l'Institut de France. There, he enjoyed a certain status. He was by then friends with several other famous members, including Laplace, Legendre, and the chemist Antoine-François Fourcroy. As a scholarly group, they held fast opinions on contemporary academic issues.

One topic of the day—the theory of light—was being increasingly scrutinized by scientists. Poisson and his Académie colleagues believed in the particle theory of light as opposed to the wave theory, a notion that was gaining acceptance in scientific circles outside of Poisson's group. In 1818, Poisson thought he would put the debate to rest once and for all. He suggested that the Académie hold a contest in which they would offer a prize for the most persuasive essay proving one theory over the other. Poisson's own entry argued the particle theory. He wrote a paper that did not use much mathematics or present scientific results; instead, his thesis relied on intuition. His premise was that an elemental notion of the wave theory was false. He pridefully disseminated his paper widely and was ready to receive the prize and the attendant recognition. However, it did not go well for Poisson. The judging committee was headed by the mathematician Dominique-François-Jean Arago, who conducted his own experiments—with accompanying calculations—to prove that the wave theory notions were correct after all and that Poisson's rejection of the theory was itself faulty. Poisson was humiliated, and his reputation suffered. This stain followed him for the rest of his career. Meanwhile, Arago, who was also connected to political circles, went on to become prime minister of France in 1848.

Before exploring Poisson's work, I pause to point out how to properly pronounce his name, since it is so often mispronounced, except of course by persons who are fluent in standard French. His name is included in his most famous contribution to measurement: the Poisson distribution. Sometimes Poisson is mispronounced as "poison." (It is decidedly not the "Poison" distribution!) The correct pronunciation of his name is "*pwa-son*," the final syllable rhyming with French "*bon*." Many readers will know that *poisson* is the French word for "fish," or a fish course in a meal. I imagine he would like to have had his name pronounced correctly, especially by future generations of mathematicians.

Poisson's magnum opus was *Recherches sur la probabilité des jugements en matière criminelle et en matière civile* (*Research on the Probability of Judgments on Matters Criminal and Civil*), published originally in 1837 (Poisson and Sheĭnin 2013). In it, he extended the work of Laplace, particularly his work on jury behavior in criminal trials. One particularly vexing circumstance routinely arose for researchers working in this area: namely, how to display the relative infrequency of crime incidents for an entire population. When considered as a frequency by a binomial distribution (i.e., convicted of criminal activity: yes or no), the overwhelming indicator was no, making frequency plots for these infrequent events highly skewed and not very useful. Poisson found a way to consider such seldom-occurring events.

In statistics and probability, a seldom-occurring event is called a "rare event." This is a circumstance that happens only once in a long while among many occurrences, such as experiencing an earthquake among many days without earthquakes, a home run among many times when batters do not hit it over the fence, or failure of a manufactured part among the many times that it performs properly. As you can imagine, plotting rare events and figuring their probability is extremely useful to research and planning. Some examples of real-world questions will help to give the idea of what is conveyed in a Poisson probability. Consider these questions:

- A traffic planner may ask: "What is the likelihood that five cars will drive through a red light at a particular intersection in a day? What about ten cars?"

- A health researcher wants to know: "What is the chance of discovering at least one coliform bacterium in a 10 cc sample of river water?"
- A hospital administrator may want to know: "What are the odds in the hospital of having five births during the hours from midnight to 6 a.m.?"
- A charitable foundation officer, making plans for next year, may ask: "What is the likelihood of reducing malaria by 10 percent over the next year in a particular country if our monetary contributions increase by 10 percent?"
- A manufacturing company executive may ask: "What is the probability of having five parts fail out of every thousand produced in a month?"
- A real estate developer may want to know: "What are the chances of selling six new houses every month in a given neighborhood?"
- A game officer may want to know: "What is the likelihood of the moose population in a given national forest doubling every five years?"

Notice several features of these scenarios: (1) they are expressed as a probability, (2) they occur relatively seldom among the many times they do not occur, and (3) the probability is within a given time frame. Time is the most common variable used in Poisson probabilities, but another scalable unit—such as a liquid measure like ounces or the pressure for gases—could be used.

Poisson considered such events as binomials: they either occur (yes) or they don't (no), like predicting heads in a coin toss or whether a car drives through a red light. But he realized that such frequency counts (as is done in the binomial) for rare events would mean counting only a few for yes and many for no, resulting in a distribution that is so lopsided when plotted as to be unhelpful. Poisson looked at them at a much more granular level and focused on making a probability for observing an exact number (as small as zero, and typically fewer than ten) of rare events. In other words, a *Poisson probability* is the likelihood of observing a given number of, say, "yeses" for a rare event among many occurrences. Such a probability is a Poisson probability. Hence, a Poisson probability is much like a frequency count of a binomial, but it is expressed as a probability. There are technical differences between the binomial and the Poisson probability, but they need not concern us here.

Poisson probabilities are often plotted for several values, as in Figure 11.4, to create a Poisson distribution. Seeing the curves of a Poisson distribution also helps us to understand this special kind of probability. The figure gives three scenarios, each a Poisson distribution and shown by a different line. The horizontal axis is specified as a unit interval, which is usually time. In this figure, the

Poisson distribution

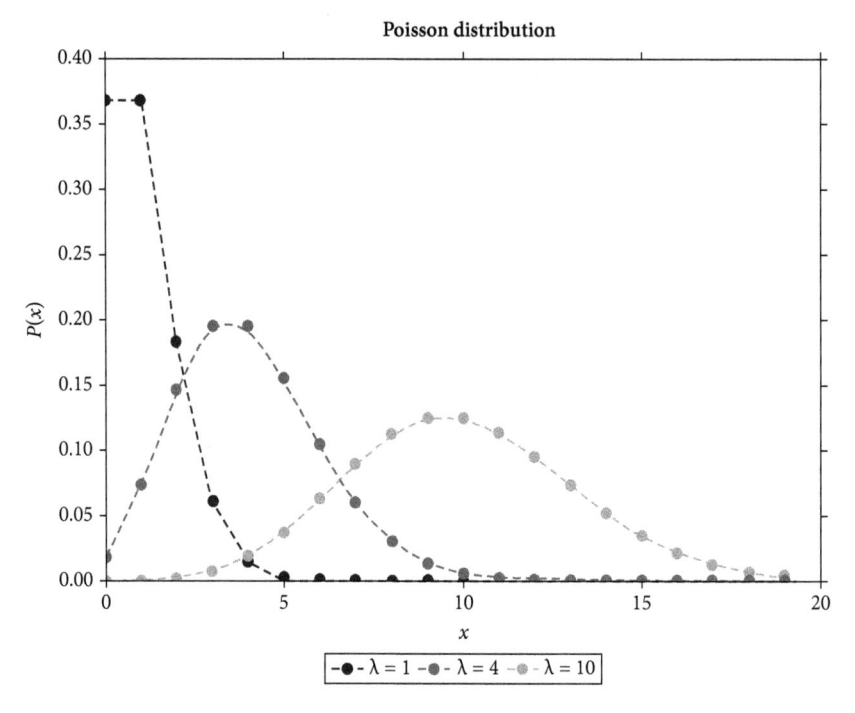

Figure 11.4 Illustrative Poisson distributions

numbers range from 0 to 20, which could be minutes, hours, or years. Let us continue, using the previous example of forecasting the number of cars that might drive through a red light at a particular intersection. For this, we use the unit of hours, so the baseline is zero to twenty hours.

The vertical axis (along the left side), is the probability. Theoretically, since probability goes from 0 percent to 100 percent, the scale would range from zero to one hundred. But, as is typical, the display is truncated to save space (given that, in this example, about 37 percent is the highest percentage, the scale is truncated just above it, at 40 percent). The researcher stipulates the number of occurrences for which a probability is determined. In this example, three circumstances are specified in the figure (1, 4, and 10), meaning that we want to know the likelihood of the event (a car driving through the red light) occurring just once, four times, or ten times. Then, the three Poisson distributions are calculated. Each curve represents the probability (which is read off the vertical scale) within the time frame (read as hours on the horizontal scale). For example, suppose we wish to discover the probability of one car running a red light within a five-hour period (see the left curve). Reading off the vertical scale, we see there is a less than 1 percent likelihood of this event. However, within a given five-hour interval, there a roughly 20 percent likelihood of seeing four cars run the red light (see the middle curve). For seeing ten cars run the red light in a ten-hour interval, there is about a 13 percent chance (see the right curve).

With this understanding, then, we can readily imagine that Poisson distributions are a special type of odds ratio. But they are not calculated in the same manner as the odds ratios of conditional probabilities that we saw earlier. Nevertheless, calculating Poisson probabilities and plotting the Poisson distribution are not technically difficult, although this does take us afield of our purpose. The interested reader may find any of number of Internet sites and textbooks that work through the computation. Also, obviously, there are statistical assumptions being made (about the independence of each time interval, the proportionality of disjoint time intervals, etc.), but they do not concern us here.

Poisson invented his probability calculation and attendant distribution early in his career, publishing the idea in *Recherches sur la probabilité*, his important work introduced earlier. But as obviously useful as the idea seems, it was not immediately picked up by mathematicians or other researchers of his day. This is surprising, as quantification as a concept was rapidly advancing at this time, and Poisson was a noted public figure. It was not until sixty years later, in 1898, that a Russian economist named Ladislaus von Bortkiewicz grew interested in the Poisson distribution and included an example using deaths by horse kicks in the Prussian cavalry in his first statistics book, which had the peculiar title *Das Gesetz der kleinem Zahlen* (*The Law of Small Numbers*; Bortkiewicz 1898).

His example was certainly meant to be serious in intent, but it was also novel and captured the attention of many individuals, giving wide exposure to the Poisson distribution. Over the years, this example has become famous for illustrating the Poisson distribution—it is referred to as "Poisson's horse kick data" or similar.

Bortkiewicz collected the rather unusual statistic of annual deaths in the Prussian cavalry by horse kick. He gathered the values for a twenty-year period (from 1875 to 1894) for ten army corps, yielding 200 total observations. He compared the actual number of deaths (122 out of 200 points of data) to that predicted by Poisson probability, thereby demonstrating the accuracy of Poisson's method. His results are shown in Table 11.1.

In the table, note there are four columns: number of deaths, Poisson probability, expected number for 200 corps, and actual number for 200 corps. The rows show values for the number of deaths in a year: 0 to 5. The first row is information about 0 deaths per year in the 200 total observations: the Poisson probability of 0.543 yields the Poisson estimate of 108.67 years (out of 200)

Table 11.1 *Annual deaths from horse kicks in the Prussian army (1875–94)*

Number of Deaths	Poisson Probability	Expected Number for 200 Corps	Actual Number for 200 Corps
0	0.543	108.67	109
1	0.331	66.29	65
2	0.101	20.22	22
3	0.021	4.11	3
4	0.003	0.63	1
5	0.000	0.08	0

without any deaths, perfectly accurate to the observed 109 years with 0 deaths. For an estimate of 1 death per year, the probability estimate is 0.331, yielding 66.29 years when there was only a single death. This estimate is compared with the actual 65 years in which just one death occurred. Again, the Poisson probability estimate is very close, as it is for each of the other rows of probabilities. He thereby demonstrated the accuracy of Poisson probabilities. For whatever reason, Bortkiewicz's illustration almost immediately became well known and led many others to adopt Poisson probabilities as a useful methodology.

A popular party game today uses Poisson probabilities in its solution, although it is doubtful that any of the guests realize it. The point of the game is to guess the likelihood that any two players have the same birthday. The guests simply state their birthdate, and the fun is to discover whether two share the same date. It turns out that, with only about two dozen guests, there is a nearly 50 percent chance that at least two share a birthdate, although, obviously, an individual's birthday could fall on any of 365 days. The probability calculations are relatively simple but a bit tedious to explain; they can be found on many Internet sites, including one sponsored by the University of Massachusetts (see University of Massachusetts 2017).

****

Like Laplace, Poisson believed in the universal applicability of probability, as he stated many times in *Recherches sur la probabilité*. As such, he was fascinated with how numbers operate and explored not only the binomial theorem but also other related number theorems. Recall from Chapter 4 that we explored three foundational number theorems: the binomial theorem, the law of large numbers, and the central limit theorem.

In fact, Poisson gave a name to Bernoulli's theorem (originally *Theorema aureum*), calling it "the law of large numbers," the name we use today. He extended it to the circumstance wherein there is not always a fifty–fifty chance for each of two outcomes and there can be an unequal success probability across a number of trials. This was called "Poisson's law of large numbers." It is not nearly as well known as Bernoulli's law because the instances in which it arises are fewer.

But Poisson did introduce a notion that is common: namely, that of a random variable. As we saw earlier in Chapter 9, in probability theory, a "random variable" is one that must assume a discrete number, such as 0, 1, 2, or 577. Some examples of "discrete random variables" are the number of children in a given family, and the number of minutes someone waits in a line at a grocery store. It is random because the outcome is completely unknown: choose another family and the number of children may or may not be different, or take another trip to the store and the time you wait in line may be completely different. Another type of random variable is the "continuous random variable," which is a number along a scale, such as height or a test score.

In exploring the central limit theorem, Poisson provided a calculus proof to it that had not been given by Laplace when he invented the theorem some years earlier. Laplace had devised the theorem but had relied upon intuition and common logic as his rationale. The final proof for this foundational notion, however, did not come until later when Baron Augustin-Louis Cauchy, a French mathematician and physicist and contemporary of Poisson, proffered a rigorous calculus proof for the central limit theorem. Later still, in the twentieth century, a group from Russia, called the St. Petersburg Mathematics School, validated Cauchy's work, which is now accepted as definitive.

Cauchy made a related mathematical invention, too, that is a demanding application of Poisson's probability distribution and is itself a continuous probability distribution, although its application is primarily in physics rather than in probability and statistics. This mathematical framework is called the "Cauchy distribution," but, in physics, it is also known as the "Cauchy–Lorentz distribution" or sometimes just the "Lorentz distribution," after the mathematician Hendrik Lorentz.

All of this rigorous work has special meaning to the overall flow of quantitative thinking, too, because it signals a period in the development of probability theory in which precise, rigorous, and thorough proofs are required before developments would be widely accepted. By this new demanding standard of the mid-nineteenth century, probability theory was given credence as a rigorous and independent discipline within mathematics. As one historian of mathematics notes,

> The mid-nineteenth century saw a new era in mathematics with the introduction of rigor in analysis, spearheaded mainly by Cauchy and then followed by many others. Among other things, rigor meant that every mathematical concept used had to be explicitly defined in terms of known concepts and that all assumptions in a mathematical proof had to be made explicit. Rigor also meant that mathematical ideas were based on precise and exact concepts, rather than on intuition. (Gorroochurn 2016a, 139)

For our story, from the early to mid-nineteenth century social and cultural events described in the opening pages of this chapter to the closing notes on the demands for rigor and exactitude in probability, people's thinking has been moving ever closer to quantification. In routine areas of life for ordinary folks (that is, nonmathematicians and nonacademics), there is an induction of quantification.

CHAPTER 12

Regression to the Mean

By the mid-nineteenth century, quantifying events and experiences were everywhere for nearly everyone. Statistical and probability information popped up in many parts of daily life, including accurate oceanic navigation, market-based economic activity, and governmental functions, such as city planning and having a sophisticated military organization. Questions of astronomy (i.e., learning the orbits of celestial bodies) as well as those of geodesy (i.e., estimating the size of the earth's circumference and determining the distance between towns) were informed by reasonably reliable probability calculations. This flood of numeracy meant that people were beginning to interact with a quantitative environment with almost routine frequency. Still, however, the frequent contact with quantifying events did not, by itself, cause people to transform their worldview—but this momentous paradigm shift was just around the corner.

Two related watershed events started the transformation to a quantifying worldview for Everyman. This evolution in worldview was qualitatively different from exposing people to more and more quantitative events, although that had a lot to do with it. Instead, it was the start of a reformed thinking: a new way of seeing the world and one's place in it. Indeed, our very sense of "self" was transformed. These two momentous events were (1) developments in transportation that made it relatively easy for large numbers of people to travel farther and farther distances from their hometown and (2) the tremendous economic expansion of the time, which led shortly to a full-blown industrial revolution, first in England and then later in America.

Transportation for masses of people was advanced at this time by building suitable road systems to replace muddy horse and cart paths. And, at almost the same time these roads were being built, public passage made a second leap forward with the introduction of powered passenger trains, made possible by the invention of the steam engine. (Rudimentary "trains" did exist before the steam engine was invented, but they were mere carts pulled by horses or oxen, and they could only haul a few people for a short distance.) Finally, good roads and trains were themselves fuel for an economic expansion of huge proportions, unlike anything before in history. With the ability to easily travel outside their hometown, people saw a larger world and thought more and more about their place in it.

In related progress, people's work, their occupations and labors, underwent a revolution of its own. For the first time, the economy was seen as a cohesive whole rather than the separate enterprises of individuals going about their daily labors. Now, people were part of a larger economic system comprising everyone. Suddenly, there was "the economy," a phrase meant to suggest collective efforts, regardless of how disparate the elements. As such, it became almost reified, taking on a life of its own—a distinct colossus, if you will. For the first time, scientists and governmental researchers began to systematically study the economy with regularity and purpose, and they did so by employing the new methods of probability and statistics.

Further, the economy had rapidly become integral to social and cultural developments, affecting nearly all public aspects of an individual's sphere of activity. Now, the separate aspects of an individual's labors were simply subsumed into the "economic engine" of society. The economy was the steam-powered engine of society itself. The colossus was growing so fast that people could scarcely absorb its happenings into their own lives. It directed their interactions with their environment and gave rise to people's thinking about it as never before.

At first, the economic expansion began modestly, with distinct but still important events, such as establishing stock markets, and a few novel and practical inventions, such as the washing machine. But then—like the energy released from a steam-powered whistle multiplied a millionfold—it took off. The Industrial Revolution began. Some consider its beginning to be marked by the invention of the steam engine and its greatest impact to have been driving steam-powered railways (Figure 12.1). We will look at this period a bit more—specifically, how it affected our route to quantification—in Chapter 13.

The Industrial Revolution was more than a plethora of utilitarian inventions. It was consequential in that it began a process of quantitative infusion—a new outlook on the world seeping into the very pores of cognition.

Figure 12.1 Engraving of *The Stockton and Darlington Railway*, opening day on September 27, 1825

(*Source:* http://commons.wikimedia.org/wiki/Category:Public_domain)

All of these developments had their roots in statistics and probability, as we have seen. Recall a few of the links between advances in society and developments in probability: Newton and Laplace invented a useful calculus that allowed calculations of all kinds, such as determining the weight load for bridges, among countless other applications. Legendre and Gauss independently invented the method of least squares, which was used for determining relationships between geodesic points for accurate oceanic navigation. Then, Gauss offered his much more useful probability density function, which is applicable in research. On a very practical note, Gauss invented the heliotrope, a machine useful for triangulation to determine distances between towns (we saw it in Chapter 9). And, as we learned, Quetelet began to describe with statistical precision *l'homme moyen* ("the average man"). Babbage invented calculating devices (Difference Engines No. 1 and No. 2 and his analytic computing machine—truly the first computer with a CPU for processing information and a storage area to hold the information). This paved the way for mechanical devices of all sorts, including the steam engine, one of the most important inventions of the nineteenth century.

The important takeaway is that advances in society were not reached independently of mathematical advances; rather, they happened *because* of these mathematical accomplishments.

The fact that these inventions rapidly became a part of everyday life for nearly all people means that individuals were interacting with them routinely and without any unusual reaction. To an ever-increasing degree, people saw their lives as defined by numbers. Viewing ordinary events as wholly random and unpredictable was a fading perspective. The sense of a quantified worldview had not yet fully set in, but it was filtering into the public's mindset, rapidly and inexorably.

Hence, our story of how we came to have a quantified worldview and how that fundamentally changed us is not yet complete, but we are on well on our way. For the first time since we began this journey with Isaac Newton and Dr. Johnson, the worldview is truly changing.

******

Most readers are aware that Darwin's 1859 book *On the Origin of Species* was revolutionary in thought and upset social and cultural mores of the mid-eighteenth century across the globe. In the United States—at the time a country of strong Puritan prudence, as we have already discussed—the book's impact was perhaps most widely realized through the 1960 Stanley Kramer film *Inherit the Wind*, a fictionalized Hollywood account of a popular play of the same name. In the movie, the actor Spencer Tracy portrays the defense lawyer for the high-school biology teacher John T. Scopes, who (in real life) was convicted in 1925 of breaking a Tennessee state law by teaching Darwin's theory of evolution, which states that humankind evolved from lower animals—most directly from the primates. The incident was infamously called the "Scopes Monkey Trial," and Tracy's portrayal of Scopes' lawyer, Clarence Darrow (completed with tussled white hair and a rumpled shirt), is iconic for his impassioned defense speeches.

An opening note to both the play and the movie cautions the audience that this recounting of the trial is done with "literary license" and is not intended to be a "historical documentation" of what really happened. (In fact, later historians and many others have identified the movie's many inaccuracies, exaggerations, and omissions.) But this cautionary note to audiences was largely ignored. In the public's mind, the image was set of a man persecuted for teaching the then-modern theory of evolution. It got people arguing about religion's role in forming one's view of life, something unusual at the time. I clearly remember these intense discussions because, at the time the movie came out, I was still in high school, and I remember some of them taking place during my high-school biology class!

Darwin's thesis was indeed well known throughout the United States and beyond, and Darwin is a central player in the story of quantification. Readers will recall that, in Chapter 3, I discussed how observation has played an inventive role in scientific inquiry and that it is the starting point of the journey to quantification. I cited Darwin reporting his findings as an example of this kind of observation. Today, his observational methods are loosely classed as "naturalistic inquiry."

Darwin justified his theory solely by intuition, relying upon observation as his only scientific tool. In his investigations, he did not employ an empirical methodology, although such was well developed by his time, nor did he engage in any mathematical reasoning, despite probability theory being available to him. He simply observed a natural phenomenon and drew his conclusions.

Modern scholarship is growing increasingly suspicious of Darwin's authenticity. He is questioned on three fronts. First, he is faulted for his logic on evolutionary development. One critic meticulously identified numerous technical shortcomings in Darwin's work on evolutionary biology and wrote that he was "skeptical of claims for the ability of random mutation and natural selection to account for the complexity of life" (Gould 2002, 144). A second criticism of Darwin is that his claim of having originated the theory of evolution is easily refuted because, even at the time of his writings, there were numerous documented and well-known forerunners (see Thompson 1981). And the third, and perhaps most serious charge of all, is that Darwin liberally plagiarized several earlier writers on evolution, especially his own grandfather, Erasmus, from whom he "may have borrowed" (Darwin's own words) many sentences and even whole paragraphs (Bergman 2002). Much of this lack of scholarship was discovered and widely discussed during Darwin's lifetime, and his scientific contemporaries were exasperated by the fact that his writings included almost no references or attributions.

Still, with wider audiences outside of academic circles, Darwin was popular, and he made much of his livelihood from public speaking about the theory of evolution and his travels to exotic lands. It is almost impossible to overstate his influence for the next one hundred years or more on communal thinking about the origins of humankind. In fact, he is one of the most broadly influential persons of modern time.

An interesting aside is that Darwin's fame as inventor of the theory of evolution follows a storied pattern of false attribution, wherein scientific discoveries are often credited to a person who is not the actual inventor or discoverer. This is sometimes called "Stigler's law of eponymy," after the University of Chicago statistics professor Stephen Stigler (1980), who cites many examples, including Halley's comet, Hubble's law, and even the Pythagorean theorem.

Now, in our story, Darwin again serves as an illustration, but this time showing the broadening impact of quantitative thinking on the discipline of sociology. Darwin is following (by nearly a hundred years) the introduction of probability theory into sociology by Tobias Mayer, as well as Quetelet's and Comte's "social physics." Moving quantification into social disciplines is indeed an important part of the quantification story, and, despite his obviously weak scholarship, Darwin's influence is undeniable in bringing to public awareness the notion of employing scientific inquiry in sociology. By his monumental influence, quantification was moved from being restricted to the physical sciences into the social sciences of human traits and behaviors. The early work of Laplace and Quetelet set the stage for its introduction, but the enormous influence of Darwin made the move complete. Hence, obliquely but importantly, quantification was advanced by Darwin.

This is also the point in our story at which "truth" comes in. We see at this time that the facts of creation are being questioned with the new theory of evolution. But the material significance of such questioning is not in learning new information about humankind's evolutionary development. Rather, its real meaning lies in the stirrings in our own minds for what constitutes reality as ontological truth. That is, people now began asking, "What is truth?"

With this step in the measuring of uncertainty—that is, quantification—society was beginning to redefine itself. This was a momentous turn, for sure. The worldview of individuals (and collectively, the populace) was thus inclined to being transformed at this time.

****

Darwin also plays into our story in yet another way: through his contact with a childhood friend who was also a relative, his half-cousin Francis Galton. Darwin and Galton shared the same grandfather, Erasmus Darwin. The son of Erasmus's first wife was Charles Darwin's father. Galton's mother (Violetta Darwin) was the daughter of Erasmus Darwin by his second wife. Hence, Charles Darwin and Francis Galton were blood relatives. And, as is frequently observed in their photographs, they were rather alike in physical appearance, too.

While Sir Francis Galton was not as influential across a broad population as was Darwin, he was decidedly more consequential in quantitative developments. Certainly, he was the superior scholar and more careful researcher; he was far more productive than his relative.

The scope and quantity of Galton's accomplishments are almost unbelievable. In brief, he worked primarily in research on intelligence and heritability, where he left a legacy of monumental achievement. Not only did he develop several theories about intelligence, most of which were later verified, he was also the first to employ probability and statistics in their exploration. He was one of the first to examine races and cultures in legitimate scientific investigation, giving rise to ethnography. He also founded psychometrics (the science of measuring mental abilities) as a distinct field of study.

In many of his investigations, he invented new and novel methodologies. For example, he discovered "reversion to the mean," evolving it into "regression to the mean." This work led to his inventing "co-relations" as a predecessor to the notion of correlation. He is also responsible for a plethora of meaningful terms in statistics and probability theory, including *percentile* and various offshoots such as *quartile*, *decile*, and *ogive*. He even worked on fingerprints as a new tool for personal identification, which would be useful in criminology and elsewhere.

He had a feverishly inventive imagination. And, obviously, Galton was an incredibly hard worker, who (you may imagine) would echo Poisson's comment that "life is work."

He documented his work in more than 340 scholarly publications, a simply astounding rate of production. His best-known work is *Hereditary Genius: An Inquiry into Its Laws and Consequences* (Galton 1869). This is the first scholarly study of genius and greatness; it includes Galton's thoughts on several of his aforementioned studies. The book came relatively early in his career, and the ideas are often immaturely developed, but he would more fully develop many of his ideas in later publications. Several of his papers are of seminal importance to particular topics.

Surprisingly, despite his skill in mathematician, he was not interested in complex mathematics; rather, he focused on practical applications of statistics and probability and employed only simple mathematics in his publications. He almost never included the higher-level mathematics (such as calculus) so often used by his predecessors.

Galton was born in 1822 to a wealthy family living near Birmingham, England. He showed signs of high intellectual ability at an early age, and, by age six, he reportedly knew most of the *Iliad* and the *Odyssey* by heart. At school, he was allowed to devise his own curriculum, deciding for himself what books to read and what subjects to study—an unusual freedom at the time. His father wanted him to pursue medicine and become a doctor. During his first program at Birmingham General Hospital and then King's College, London, he did study medicine, but at Cambridge he grew interested in mathematics and dropped medicine to focus on mathematics and probability theory.

While Galton was still a young man, his father died, leaving him a substantial inheritance. With financial means at his disposal, he began to travel. True to his nature, he was not content to merely sightsee; he used this time to do original work in anthropology and meteorology.

One early trip was to the Middle East and then to southern Africa, where he studied the Bushmen of the Kalahari Desert, which is the largest desert in the world, covering parts of present-day South Africa and neighboring Botswana and Namibia. His study of the Bushmen is an example of early ethnographic research. In particular, he made varying measurements of their physical characteristics and plotted these in a manner similar to the one Quetelet used.

The Bushmen consist of a loose collection of tribes who live simple, peaceful lives in an incredibly hostile environment, following ancient customs and habits. For example, they use broken ostrich egg shells to store rainwater for drinking and are known for going long periods without any water at all. They are nomadic, ranging across the Kalahari region, have no writing, and speak a remarkably complex language made up mostly of distinct clicking sounds that outsiders seem largely incapable of copying.

Their life and customs have been documented in films and books. Readers may recall the movie *The Gods Must Be Crazy*, a 1980 South African comedy about the Bushmen, one of whom has a funny adventure when he sees an empty Coke bottle fall from the sky (actually, it fell from a plane that was obscured from view by the clouds). More serious study was given by Sir Laurens van der Post—an Afrikaner author, war hero, political adviser to British heads of government, humanitarian, conservationist, and later member of Virginia Woolf's Bloomsbury Group—in *The Lost World of the Kalahari* (Van der Post 1958). The book is a wonderful read that takes you on his intense expedition into what was then an almost unexplored region of southern Africa to meet the fascinating Bushmen. The book is—not incidentally—a first-rate work of sociology; and, I might add, exceptionally well written.

Galton's trips were made in a time ripe for explorers (although the Golden Age of the Explorers was still fifty years off), and his adventures, along with Darwin's parallel journeys, caught the public's imagination. Both men gave lectures that were popularly attended. Galton included ethnographic scholarship in his talks, bringing maps and elaborate descriptions of customs and cultures of the native peoples he encountered.

True to his personality and professional interests, he documented these observations in several books, one of which proffered advice to would-be explorers and had the wonderfully interesting title of *The Art of Travel: Or,*

Shifts and Contrivances Available in Wild Countries (Galton 1855). As his travel experiences increased, Galton added information to the book and published subsequent editions. The fifth edition was the most famous and is still in print. Some topics in the table of contents include "Fords and bridges," "Game, other means of capturing," and the curious headings "Savages, management of" and "Hostilities." Figure 12.2 shows a woodcut from the cover which depicts Galton fording a river by hanging onto a horse's tail. Another African explorer, Samuel Baker, took exception to Galton's method with this entrancing critique:

> In that very charming and useful book by Mr. Francis Galton, *The Art of Travel*, advice is given for crossing a deep river by holding to the tail of the swimming horse. In this I cannot agree; the safety of the man is much endangered by the heels of the horse, and his security depends upon the length of the animal's tail. In rivers abounding in crocodiles, which generally follow an animal before they seize, the man hanging on to the tail of the horse is a most alluring bait, and he would certainly be taken, should one of these horrible monsters be attracted to the party. (quoted in Galton Institute 2017)

For his adventurous scientific exploits, Galton was awarded a gold medal from the Royal Geographical Society, an honor that piqued Darwin because he was not similarly recognized. At the time, the Royal Geographical Society had

Figure 12.2 Woodcut on cover of Galton's *The Art of Travel*
(*Source*: from, F. Galton, *The Art of Travel: Or, Shifts and Contrivances Available in Wild Countries*)

only recently been formalized, being the outgrowth of a dining club in London where members held informal dinner debates to share experiences of their journeys. Like Galton, most of the group's members were learned men with scholarly achievements, and they emphasized that their adventures were not merely travels to far-off lands but an opportunity to learn about new cultures and geographies. This was quite novel and forward-looking at the time, and it gave a beginning to ethnographic research. The Royal Geographical Society remains one of the most famous learned societies for advancing scientific adventures. We'll encounter this noteworthy society again in Chapter 15 when I discuss Shackleton and the Golden Age of the Explorers.

Galton's interest in ethnographic research led to another curious development of his: fingerprinting and the use of fingerprints for positive identification. He was not the first to suggest looking at fingerprints in police work, but he was an early researcher on the topic. He was among the first to address fingerprinting as a serious field of study. He collected many thousands of them, and, as original work, he engaged probability theory to demonstrate the uniqueness of the individual parts of a fingerprint, including devising a method for deciphering a blurred fingerprint. This groundbreaking work provided law enforcement with a rational basis for their use in identifying individuals.

True to his detailed nature, Galton cataloged his more than 5,000 fingerprint samples by their minute parts, which led him to develop a classification system. His system remains the basis of how fingerprints are cataloged today. His publications on the topic are among the first ever written, and on the cover of one of his works is a picture of his own fingerprints and palm prints.

In his academic study of fingerprints, Galton employed probability in a field where no one else had previously thought to do so. Quantification was indeed advancing in unexpected ways.

****

I have already mentioned Galton's first major work, *Hereditary Genius*, and now we look at this work more carefully. It is one of the most renowned books of the nineteenth century. The ideas that Galton put forth in it—about intelligence and about heredity and its possible effect on human behavior—are hugely influential still today. Nearly all work in the field of behavioral genetics and substantial inroads into today's research on many forms of cancer directly stem from it. His innumerable contributions to this early study of intelligence are carefully recounted in a 2002 book by the Galton scholar Arthur Jensen (himself a man of astounding intellect—on par with his subject). Nor is Jensen the only one to extol Galton's many accomplishments. His friend and early biographer, Karl Pearson (whom we'll see again in Chapter 13), describes Galton and his work in almost adoring terms. Clearly, Galton was an extraordinary individual.

Galton was particularly interested in studying variation in social phenomena and human characteristics, and he intended to further the work of Quetelet, especially in sociology. He thought that variability in human features (both physical and cognitive) could be best explored from the perspective of the normal distribution. Remember, back then it was still called the "law of error," and this phrasing figured prominently in Galton's thinking. In a departure from the ideas of Laplace and Quetelet, who used the mean value for certain human characteristics because it represented what was average or "normal." Galton thought that the key to understanding human behavior lay in studying deviations from the mean.

To do this, he first needed to map a given human physical characteristic, like height or eye color, or mental ability, like a certain amount of short-term memory, on a normal distribution. He began his investigations by considering whether the ability was present in a particular individual. Recall, this is a binomial: yes or no. So, Galton wanted to examine a distribution of this binomial for many individuals. Of course, he studied the pioneering work of Gauss, who showed that the binomial expansion, by the central limit theorem, eventuates to a normal distribution. But Galton wanted to see it in actuality. Recall that Gauss's proof was done theoretically with calculus and only existed on paper. Galton wanted to demonstrate it practically.

We saw earlier in Chapter 7, in following the work of Quetelet for specifying a predictable distribution, that Galton thought Quetelet was onto something important. One famous example of this in *Hereditary Genius* is his application of it to measuring intelligence. Galton glowingly described Quetelet's system with its measurements of the heights of conscripts and reasoned that it must also apply to intelligence and IQ measurements. He said,

> Now, if this be the case with stature, then it will be true as regards every other physical feature—as circumference of head, size of brain, weight of grey matter, number of brain fibres, &c. ; and thence, by a step on which no physiologist will hesitate, as regards mental capacity. This is what I am driving at; that analogy clearly shows there must be a fairly constant average mental capacity in the inhabitants of the British Isles, and that the deviations from that average—upwards towards genius, and downwards towards stupidity—must follow the law that governs deviations from all true averages. (Galton 1869, 31–2)

This was quite a remarkable statement, even for the time—and one wholly unsupported by empirical evidence for this variable. So, for his experimentation, he invented a "machine" (as it was thusly called, though it looked more like a board game) named "Galton's bean machine," although in most of his writings Galton called it a "quincunx." As a noun, "quincunx" is an arrangement of five things in a square or rectangle, with one at each corner and one in the middle. The next two figures (Figures 12.3 and 12.4) illustrate Galton's bean machine. The first, Figure 12.3, shows the original drawings of several versions of his machine, although he probably built only the first version.

Far beyond just a clever invention, the notion of making a working model to test his ideas is still considered revolutionary today. The respected statistics historian Stigler claims, "Galton's quincunx was a brilliant conception" and, further, that "it is simply one of the greatest mental experiments in the history of science" (Stigler 1989, 74).

The second figure (Figure 12.4) shows a working model of Galton's bean machine and so makes it easier to understand how it works. In the quincunx, one drops balls or beans at the top of the board, one at a time. A ball rolls through a narrow chute and into a field of randomly placed pins. As in a modern-day pinball machine, the ball bounces back and forth randomly among the pins until it exits at the bottom, falling into one or another channel (there are

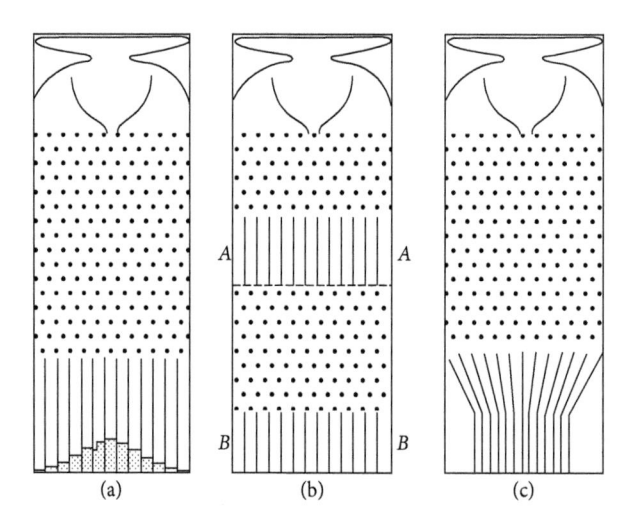

Figure 12.3 Galton's bean machine: (a) the quincunx, (b) the double quincunx, (c) the convergence quincunx

(*Source:* from F. Galton, *Hereditary Genius: An Inquiry into its Laws and Consequences*)

Figure 12.4 Working model of Galton's bean machine, "the quincunx"
(*Source*: http://commons.wikimedia.org/wiki/Category:Public_domain)

about fifteen channels). Any single ball could roll into any channel, thus demonstrating the random nature of an independent trial in an experiment. The several balls dropped into the machine are like many coin tosses or many independent trials of any experiment. In the same manner as Bayes's early experiments with billiard balls rolling to the left or right of a diagram (see Chapter 6, Figure 6.1), the balls in Galton's machine represent a probability distribution of k successes in n trials.

As we can imagine from viewing the working model (Figure 12.4), after a sufficient number of balls are rolled (trials), the quantities in all the columns assume a Gaussian distribution (the bell-shaped curve). If one were to repeat the whole process (say, the next day), the same bell-shaped distribution would be observed, true to the dictates of the central limit theorem. Hence, whereas Bayes, and then more formally Gauss, provided a calculus proof for the central limit theorem, Galton arranged a physical demonstration of it. Very clever!

Now, with validating evidence in hand that the binomial does indeed represent the normal distribution, as Gauss proved mathematically, Galton devised a novel way to use this information. Remember, he was not interested in identifying the mean of a distribution but in studying its variability. Hence, he turned around the customary interpretation, explaining, "we may *work backwards* [italics in original], and, from the relative frequency of occurrence of various magnitudes, derive a knowledge of the true relative value of those magnitudes, expressed in units of probable error" (Galton 1875, 37-8). In other words, the normal curve is usually interpreted as yielding data on the most common or prevalent characteristic, as shown in the mean. Galton was interested in using the normal curve to identify the so-called errors, or deviations from the mean. He wanted to express those deviations in terms of their distance from the mean (their *magnitude*).

To do this, he defined an upper limit for a curve, something he called an "ogive." An *ogive* is a curved shape that arcs to a point, like the diagonal arch in Gothic architecture or the part of a bullet that is tapered near its point. When plotting a human characteristic (e.g., short-term memory), he labeled it on the horizontal axis (x-axis) in terms of probable error (i.e., "How likely is it to be above or below the mean?") and on the vertical axis (y-axis) as percentage in the population, from 1 to 99, or the probability from 0 to 1. When the goal is to measure a particular characteristic for an individual, by Galton's arrangement, one could read it off the ogive's curve the percentile for the individual. For example, if the characteristic of interest is height, then knowing that an individual is a certain height, their percentile for height could be determined by just reading the percentile value off the ogive. By this, Galton invented percentiles.

Early in his research, Galton also—and significantly—applied this idea to examining mental abilities rather than just physical features. (At the time, all mental abilities were simply lumped together as "intelligence"). He focused his work on exploring the variable of intelligence in people by taking mental measures with all manner of purpose-built, simple contrivances (such as looking down a row of playing cards for assessing depth perception). He set up an "Anthropometric Laboratory" where he collected many thousands of samples from people. Soon, he began to sort them by crossmatching categories, such as country of origin, sex, race, ethnic heritage, and more. He made all kinds of comparisons between categories, highlighting especially the differences. He noted that one sex or race or country of origin would be higher or lower than another on the variable of intelligence. He was not the first to make these classifications or comparisons, but he was the most well-known researcher to do so and was probably the first to engage probability theory, such as percentiles, for the comparisons. In the temper of that time, such classifying and labeling of individuals was neither uncommon nor considered illegitimate.

In 1884, Queen Victoria proudly supported a special and highly publicized conference, the International Health Exhibition, designed to show the monarch's concern for the health and general welfare of her subjects. At it, Galton set up his Anthropometric Laboratory, where individuals from all walks of life could come and be tested for intelligence and other physical characteristics, such as depth perception and hearing. Four million people attended the exhibition, and his display was immensely popular. Figure 12.5 shows a photograph of Galton's Anthropometric Laboratory at the exhibition. Afterwards, he rebuilt

Figure 12.5 Galton's Anthropometric Laboratory

(Source: from Karl Pearson, *The Life, Letters and Labours of Francis Galton*)

his laboratory in Cambridge, where he tested more than 17,000 individuals during the 1880s and 1890s, charging them fourpence each for the first set of meaurements, and threepence thereafter for any subsequent measurements (Galton 1885).

Then, Galton carried the idea even further. He now wanted to explore how intelligence transfers from one generation to the next: heritability. Following the contemporaneous experiments of the Austrian monk Gregor Mendel, who was experimenting with the genetic tracing of generations of pea plants, Galton turned to looking at heredity as the sole and direct cause of whatever human ability he mapped, such as depth perception or short-term memory.

Most famously (or, infamously), Galton's mapping of intelligence focused on the differences between his categories of classification. From there, it was only a small step to suppose that humankind could be "improved" by selective mating of humans with the most desirable heritable characteristics. This was his groundbreaking work in founding "eugenics"—a term he probably invented—and the eugenics movement. Initially, Galton meant the term to describe his so-called errors in his observations of intelligence, but its meaning quickly evolved into suggesting improvement of intelligence by selective breeding, sort of like Darwin's notion for natural selection, only here done not in nature but with humans.

The idea was not controversial at the time, and Galton was certainly not alone in his belief. Others who strongly supported eugenics included Teddy Roosevelt (who wrote a grisly letter about its use to improve humankind), Winston Churchill, George Bernard Shaw, Francis Crick (winner of the 1962 Nobel Prize for discovering DNA), Helen Keller (who, although blind and deaf, was an enthusiastic supporter and called for physicians to be "juries for defective babies"), and Alexander Graham Bell (who served as president of the Second International Congress of Eugenics and chairman of the board of scientific advisers to the Eugenics Record Office). And these leaders had many followers in the movement.

Moreover, the notion for selective breeding of humans was hardly original to Galton. Scholars have traced its origins to ancient Greece, as well as several other countries. Even Plato suggested selective breeding of humans as early as 400 BCE. Still, Galton was eugenics' most prominent advocate during the nineteenth century, and his name is still associated with it today. As we saw from the list of notables above, the eugenics movement was widely accepted across the globe and remained popular long after Galton. It was not until between the world wars, when the Nazis in Europe adopted the idea as rationale for their reign of terror, that the eugenics movement lost advocates and the term assumed a negative connotation. As Nazi atrocities became known, people everywhere abandoned the idea. As example, the journal *Eugenics Quarterly* became *Social Biology* and changed its focus entirely to reflect a more informed scholarship.

One Galton scholar informs us that there were two sides to the man, each expressed in a different period of his life. For most of his life and career, Galton was a man of "extraordinary curiosity, inventiveness, and investigative zeal," but his later years were marred by his enthusiasm for eugenics (Jensen 2002, 145). Jensen argues that we should recognize Galton's tremendous achievements in the first period but reject the flawed perspective of the second.

Today, the very idea of eugenics is (correctly) considered racist. Regardless of his early advocacy for the eugenics movement, Galton's work in heritability has led to modern-day studies in genetic biology. Of course, today's research is not racist, despite its focus on the role of genes in our very composition and functioning. Such work is important in medicine (e.g., much of cancer research is in genetic biology) and elsewhere—work carried on by thousands of scientists, from every background and belief, worldwide.

When offering explanations for his findings, Galton used the phrase "nature and nurture." He did not invent the alliterative expression, but he did bring it into common use, and it remains associated with him. Even today, "nature versus nurture" is a common expression.

For our interest in how quantification advanced to the general populace, Galton was decidedly important in broadening probability studies. For the most part, he sought to bring a broader view of the normal distribution to social interests and to apply probability to new areas of inquiry, such as intelligence and heritability. He believed that statistics could have a broad impact across nearly all societal studies. Later, such studies assumed the name "psychology."

He loved the discipline, charmingly saying:

> It is difficult to understand why statisticians commonly limit their inquiries to Averages, and do not revel in more comprehensive views. Their souls seem as dull to the charm of variety as that of the native of one of our flat English counties, whose retrospect of Switzerland was that, if its mountains could be thrown into its lakes, two nuisances would be got rid of at once. An Average is but a solitary fact, whereas if a single other fact be added to it, an entire Normal Scheme, which nearly corresponds to the observed one, starts potentially into existence. (Galton 1907, 62–3)

This quote is now rather famous, for both its wit and its insight.

****

Finally—for the point of advancing the story of quantification, especially as it moves to a broader audience—I will briefly explain some of Galton's achievements which were merely mentioned earlier in the chapter. These include (1) his discovery of regression to the mean, (2) his invention of statistical correlation, (3) the founding of psychometrics (although some prefer to credit Sir Ronald Fisher with this—we meet him in Chapter 13), and (4) his formation of the construct of intelligence.

To study the psychological phenomena that fascinated him, Galton moved his Anthropometric Laboratory from its temporary locale at the International Health Exhibition in London to a more permanent facility. He intentionally patterned the rooms with a layout like that of hard-science laboratories. Over the next few years, he brought many thousands of people to his laboratory, where he took all kinds of physical and psychological measurements. Among Londoners, it was a popular spot, and being invited there was quite the novel experience.

Galton's discovery of regression to the mean came from studies he did in this Anthropometric Laboratory. Working from the samples of measured physical characteristics that he collected and then plotted in various ways, Galton noted a phenomenon that, at first blush, seems ordinary but actually is fraught with statistical complication. He observed that parents usually produce children who are closer in height to the average than they are. In an important 1886 paper titled "Family Likeness in Stature," he wrote:

> It appears from these experiments that offspring did not tend to resemble their parent seeds in size, but to be always more mediocre than they—to be smaller than the parents, if the parents were large; to be larger than the parents, if the parents were very small... The experiments showed further that the mean filial regression toward mediocrity was directly proportional to the parental deviation from it. (Galton 1886, 42)

Stated more clearly, Galton found that adult children's height tends to be somewhat closer to a mean height than to their parents' heights: Tall parents often have children somewhat shorter than themselves, and vice versa for short parents. Remember, this was in a day when parents and their offspring likely had roughly the same nutrition throughout their lives. He called the phenomenon "reversion to the mean." He repeated this point in several others of his publications.

In a paper that became one of his most famous—"Typical Laws of Heredity" (Galton 1877)—he saw this phenomenon as a mathematical problem, and he devised a simple numerical adjustment to the differing heights of the sexes (for both fathers and mothers as well as for sons and daughters) to standardize them and thereby allow for meaningful comparisons. He then plotted these standardized heights and observed the reversion to the mean phenomenon. He concluded that there are two components to height (as summarized in Gorroochurn 2016a, 199):

1. A first component that is unusually extreme and is expected to remain extreme
2. A second component that is not extreme and is expected to remain near the center of the distribution

Galton then went further in his reasoning, this time supplementing it with his familiarity with the central limit theorem. He realized that the phenomenon of characteristics following generations was extant in other variables beyond height. In fact, the same data trend could be observed in most variables, so long as the variable conforms to some statistical assumptions. At this point, he concluded that reversion was not only symmetric but also not limited to inherited characteristics.

His "reversion" was actually an application of the bivariate normal distribution. With this revelation, he changed the name of the phenomenon to *regression to the mean*, which is the term we use today. Figure 12.6 depicts Galton's reversion bivariate normal distribution plot of median heights of adult children to parents.

From these mechanisms, he realized that his observations were not limited to the few cases he studied. His sample comprised 205 families, with parents and adult children. Rather, he knew they could be extrapolated to all families whose genetic characteristics were similar to his sample. In addition, he noted that the height relationship did not always follow directly from one generation to the next but sometimes skipped a generation. That is, one could observe the phenomenon better when looking across generations, as from grandparents to adult grandchildren. With this realization, Galton invented the genetic construct of "atavism"—the reappearance in an individual of characteristics of some remote ancestor that have been absent in intervening generations.

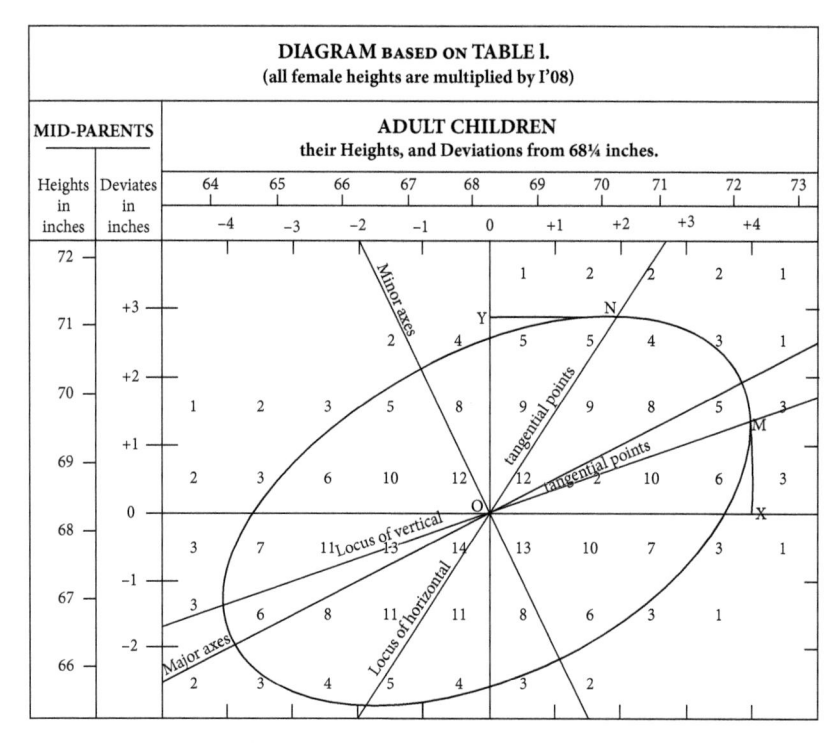

Figure 12.6 Galton's "reversion" bivariate normal distribution plot of median heights of adult children to parents

(Source: from F. Galton (1886). Regression toward mediocrity in hereditary stature. *The Journal of the Anthropological Institute of Great Britain and Ireland*, 15, Plate X after 248)

Looking deeper into his bivariate normal distribution plot of median heights of adult children to parents, he continued his reasoning, and discovered a "co-relation." In what is now a well-known quote, he said,

> Two variable organs are said to be co-related when the variation of the one is accompanied on the average by more or less variation on the other, and in the same direction. (Galton 1889a, quoted in Gorroochurn 2016a, 206)

Today, such relationships between phenomena (or "variables," in the language of researchers) seem obvious, but, in Galton's time, this had not been systematically studied. He noted that the term "co-relation" was close in sound and implication to concepts in heritability, so he changed it to "correlation," as we call it today. Galton's statistical exploration was thus a direct evolutionary line from "reversion" to "regression" and then from "co-relation" to "correlation."

To be clear to those with a technical research perspective, Galton discovered the concept of a correlational relationship and not a true statistical correlation with concomitant mathematical support. The mathematical coefficient of correlation (r) had not yet been invented, although we will see in Chapter 13 that this important piece was soon provided by his friend and colleague Karl Pearson. Important, too, is the fact that Galton almost immediately realized that his correlational relationship does not, ipso facto, imply causation. We know this everywhere today as "correlation is not causation."

He reached his conclusion mostly by intuition, providing only the simplest of mathematical underpinnings. Nonetheless, Galton's intuition was spot on, because later researchers confirmed his conclusions by supplying his missing mathematics. His omission of sophisticated mathematics to his discoveries is surprising, because he was trained at Cambridge in mathematics and was certainly capable of handling its complexities.

His understanding of the fact of correlational relationships without causation is what gives depth to his discovery. And, although this is seen by today's researchers as obvious, it was not so wholeheartedly accepted at first. At the time, Galton's notion of correlation was taken by careless researchers as causal and applied almost everywhere: to anthropometric measurements, crime, poverty, alcoholism, and more. Some were all too eager to also apply it to Darwin's weak observations, hoping to supply a semblance of proof for Darwin's theory that he himself did not.

Galton warned against this misinterpretation of his work on correlation and was supported by another influential contemporary evolutionary biologist and founder of biometry, Raphael Weldon. Weldon was a strong admirer of Galton's work and vehemently criticized the sloppy research of others who misapplied it. And, in a critique of Darwin, Weldon said that Darwin missed an opportunity, because "the problem of animal evolution is essentially a statistical problem" and not one of simple observation (quoted in Pearson 1938, 26). He argued that a correlational relationship alone would not substitute for solid empirical support.

Galton apparently conceived of the correlation concept in a sudden realization—an "aha moment." He charmingly recounts this instant in his memoir:

> As these lines are being written, the circumstances under which I first clearly grasped the important generalization that the laws of heredity were solely concerned with deviations expressed in statistical units, are vividly recalled to my memory. It was in the grounds of Naworth Castle, where an invitation had been

given to ramble freely. A temporary shower drove me to seek refuge in a reddish recess in the rock by the side of the pathway. There the idea flashed across me, and I forgot everything else for a moment in my great delight. (Galton 1908, 300)

Some historians note that as an older man, Galton's memory may have been in error (or perhaps the lapse is in Fisher's biography of Galton, when he retells the story), as a careful review of the evidence suggests that he actually came to his discovery somewhat earlier, in a different place (Stigler 1989, 73–4). Still, the tale's essentials are not in dispute: Galton invented correlation, and he did so in a moment of revelation.

****

Combining his invention of correlation with his prior work on intelligence, Galton came to one of his most lasting influences: a singular focus on the accuracy of the instruments he was using to make his mental appraisals. Recall from Chapter 3 that, even as early as Galileo, scientists were attentive to the accuracy of their data. Galton was one of the first, however, to apply probability to systematically study errors in observations in social physics. He thus developed a whole new discipline: that of "psychometrics." In fact, Galton is often credited with being the "father of psychometrics," although this honor is variously attributed to others, too.

Psychometrics is the science of measuring unobservable individual characteristics, such as achievements, attitudes, opinions, and the like. Today, appraisal of most mental attributes is referred to as studying "latent traits," although often psychometricians and psychologists make further labels for talents and other human attributes that are clearly cognitive but can also be behaviorally exhibited, such as playing a piano.

In Chapter 16, we explore psychometrics itself as an independent field of research; however, for introduction, the following notes are important. In practice, psychometrics is mostly focused on the construction and analysis of measurement instruments, especially as they concern validity and reliability in collecting data on latent traits. Today, the terms "validity" and "reliability" are taken to mean the appropriate and meaningful interpretation and use of scores yielded by a measure. They are not characteristics inherent in an instrument itself. In other words, a test instrument, independent of context, is neither valid nor reliable; rather, these properties are imbued in the decisions made on the basis of the yielded scores. Technically, validity is a sophisticated notion with independent properties, and reliability is subsumed under it, indicating a degree of mathematically verifiable consistency (see Osterlind 2010). Addressing psychometric questions typically involves sophisticated statistics that overlap with probability theory.

We see that, overall, Galton was someone who appreciated statistics not only for its utility but also for its elegance. As he said in his impactful work *Natural Inheritance*,

> I find [statistics] full of beauty and interest. Whenever they are not brutalised, but delicately handled by the higher methods, and are warily interpreted, their power of dealing with complicated phenomena is extraordinary. (Galton 1889b, 62)

The fact that Galton moved research on intelligence into realms of empiricism and that such work was then being published in America is early evidence of quantification spreading to the New World. Developments in probability and the movement to quantification had finally moved across the pond.

CHAPTER 13

Interrelated and Correlated

Galton's protégé and biographer, Karl Pearson, was one of the first to chronicle the history of statistics. In doing so, he made the following observation: "It is impossible to understand a man's work unless you understand something of the environment. And his environment means the state of affairs, social and political, of his own age" (Pearson and Pearson 1978, 360). This statement is much to the point of this book, too, for throughout it, I have tried to provide a meaningful context for the mathematical achievements that led to our quantified worldview.

Since opening our story in Chapter 1, I have maintained that the astounding technical, quantifying inventions, discoveries, and advancements we have witnessed were made because they *could* happen, in deference to a period in history whose events encouraged scholarship. Pearson made his remark in a book whose full title gives even more detail to this notion: *The History of Statistics in the 17th and 18th Centuries against the Changing Background of Intellectual, Scientific, and Religious Thought* (1978).

In parallel fashion to Pearson's recounting of the history of statistics in earlier centuries by also chronicling contemporaneous historical events ("background of intellectual, scientific, and religious thought"), we now view the phenomenon of quantification in the nineteenth century through the context of the events of that period. We saw at the end of Chapter 12 that, with Galton's invention of psychometrics, quantification was finally spreading to America. Accordingly, we now explore the what, how, and why of this expansion of quantification by looking first at the societal events in that country.

The Error of Truth. Steven J. Osterlind. Oxford University Press (2019). © Steven J. Osterlind 2019.
DOI: 10.1093/oso/9780198831600.001.0001

Two events—even beyond the Western expansion described in Chapter 12—define this century in America: the Industrial Revolution and the Civil War. Because societal influences are not cleanly bounded for either of them (despite the Civil War's battlefield dates of April 12, 1861, to May 9, 1865, by proclamation), it is not possible to precisely tie a particular advance in the quantitative fields of mathematics or probability to a specific happening in either era. Nonetheless, there is enormous convergence of technical developments in probability and statistics with these historical periods, and this is what I describe herein.

Historians often cite the invention of the cotton gin as instrumental to the beginnings of both the Industrial Revolution and the Civil War. This machine was invented by Eli Whitney in the late eighteenth century and finally patented (after a number of delays) in 1808. Whitney's invention bears on the Industrial Revolution in two important ways. The first is its impact on the economy of the Antebellum South of the United States (covering the period from the late eighteenth century to 1861, the start of the Civil War), and second is the machine itself, for its innovation of having interchangeable parts.

Prior to Whitney's machine, cotton could only be processed in small amounts. Before it can be spun into cloth, the seeds must be removed from the cotton bolls. When done manually, the task is slow and tedious to the extreme. Even more important, it is punishing on a worker's hands and back. The cotton gin transformed the production of cotton by doing the job mechanically, and that made all the difference. Separating seeds from bolls by a machine was many times faster than before and was not so physically grueling on the operator. Thus, with the cotton gin, huge quantities of the crop could be processed efficiently. From the Antebellum South, the soft, clean bolls were sent north to awaiting textile mills where the cotton was spun into cloth and garments made.

At first blush, this may appear to be a wonderful advance (and mechanically, it was), but there is another step in producing cotton that the cotton gin did not change: picking it in the fields. At the time of Whitney's invention, slavery was legal in Southern states, and land owners relied on enslaved Africans for plentiful, cheap labor. With vastly increased crop production, Southern economies soared. Some powerful landowners built plantations (some of which encompassed many thousands of acres) and built for themselves gigantic, ostentatious houses. But the backbreaking work of picking the cotton continued as before, only now with many more slaves.

At the same time, several states were seeking to join the Union, and whether they should allow slavery was hotly debated as an issue of state's rights. Initially, the public disagreement was less focused on the moral issue of owning slaves and more on whether to count them as citizens who should be taxed (with the levy to be paid by their owners). Soon, however, in addition to states' rights, the argument broadened in scope to include the ethics of slave ownership. As the northern states demanded full taxation (and because slavery featured prominent in the South's position), the South saw secession as the answer. As readers likely know, the period included the notorious "Missouri compromise" and a number of important court cases (e.g., *Dred Scott v. Sandford*).

There are, of course, libraries full of books and videos telling this involved story more completely. The classic work is Shelby Foote's three-volume masterpiece, *The Civil War: A Narrative* (Foote 1986). Foote's wonderful work is also the basis for much of Ken Burns's engaging video series *The Civil War* (Burns et al. 2004). Experiencing both Foote's book and Burns's video is time well spent (Foote is interviewed extensively in the video series).

We can see, then, that the cotton gin was instrumental in contributing to events that eventually led to the American Civil War. Nearly everyone was either personally or very closely affected by the experiences of this era in American history. In terms of our current inquiry, the period slowed progress toward quantification in the same manner that most Europeans had been constrained in thought by the Napoleonic era—in both times, people were fully occupied with the strains of daily existence. But, obviously, any period in history as complex as this had mixed influences, and some of them were supportive of the path to quantification, although clearly these countervailing influences were not nearly so strong as to negate the effects of the war on the public's mindset. Nonetheless, two simple examples will illustrate this point: the 1860 United States Census and, from sports, the even earlier formation of the National Baseball League (in 1857), initially the National Association of Base Ball Players (NABBP).

The United States Census of 1860 was particularly important because the issue of states' rights was front and center in public debate. Due to the period's severe tensions, the census was abbreviated, but it still tabulated a 35 percent increase in population over the previous census in 1850, from 23 million to nearly 34 million. Relevant to the heated public debate, this number included more than three million individuals listed as slaves. Since each family unit was personally contacted for the census, it brought counting and the notion of aggregating data into direct contact with millions of ordinary folks. Figure 13.1 shows a page from that census.

Another fact of the day that brought numeracy to a vast population—and thereby fostered quantification—is seen in the game of baseball. Immediately upon the official close of the Civil War in 1865, baseball was already gaining a following: NABBP league games had been played since 1857. It happily and

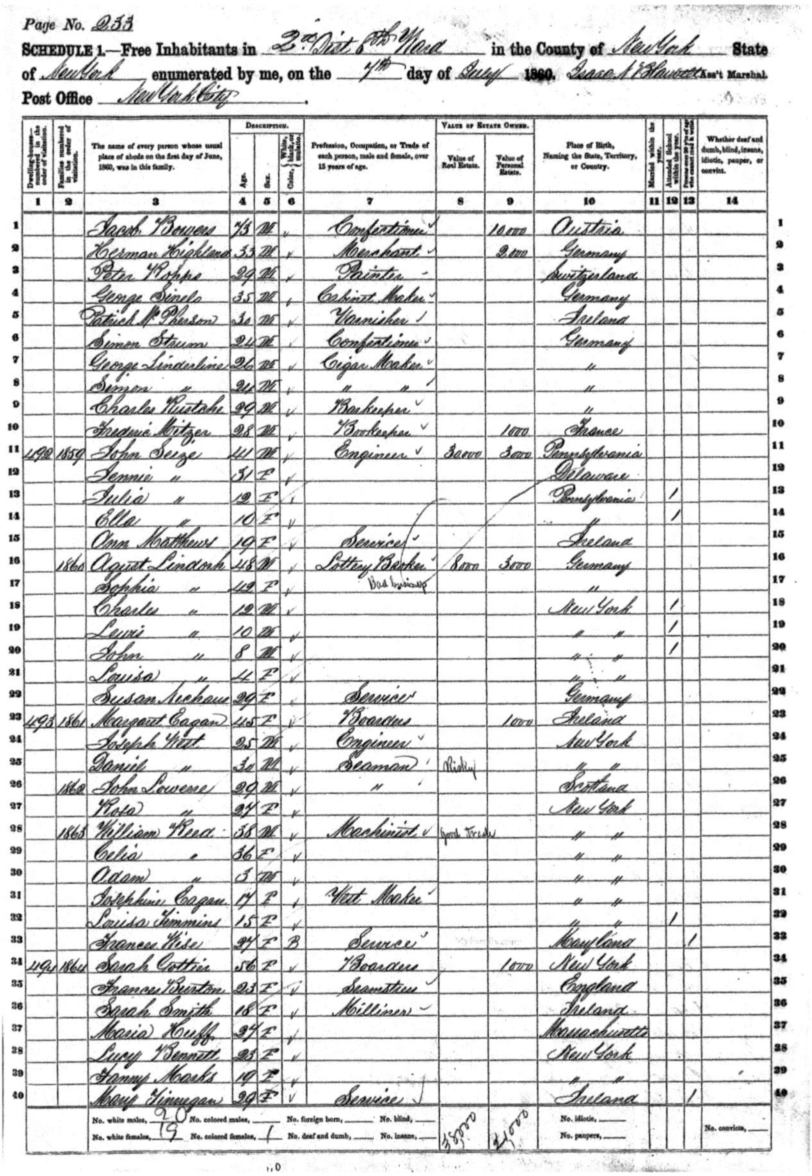

Page No. 253

SCHEDULE 1.—Free Inhabitants in 2nd Dist 8th Ward in the County of New York State of New York enumerated by me, on the 7th day of July 1860. Isaac N Blauvelt Ass't Marshal.

Post Office New York City.

	Dwelling-houses—numbered in the order of visitation.	Families numbered in the order of visitation.	The name of every person whose usual place of abode on the first day of June, 1860, was in this family.	Description. Age.	Sex.	Color, {White, black, or mulatto.	Profession, Occupation, or Trade of each person, male and female, over 15 years of age.	Value of Estate Owned. Value of Real Estate.	Value of Personal Estate.	Place of Birth, Naming the State, Territory, or Country.	Married within the year.	Attended School within the year.	Persons over 20 y'rs of age who cannot read & write.	Whether deaf and dumb, blind, insane, idiotic, pauper, or convict.
	1	2	3	4	5	6	7	8	9	10	11	12	13	14
1			Jacob Bowers	73	M	✓	Confectioner		10000	Austria				
2			Herman Highland	33	M	✓	Merchant		2000	Germany				
3			Peter Kopps	29	M	✓	Painter			Switzerland				
4			George Sinels	35	M	✓	Cabinet Maker			Germany				
5			Patrick McPherson	30	M	✓	Varnisher			Ireland				
6			Simon Sturm	24	M	✓	Confectioner			Germany				
7			George Linderline	26	M	✓	Cigar Maker			"				
8			Simon "	24	M	✓	" "			"				
9			Charles Kustahe	29	M	✓	Barkeeper			"				
10			Frederic Mitzer	28	M	✓	Bookkeeper		1000	France				
11	492	1859	John Seeze	41	M	✓	Engineer	30000	3000	Pennsylvania				
12			Jennie "	31	F	✓				Delaware				
13			Julia "	12	F	✓				Pennsylvania		/		
14			Ella "	10	F	✓				"		/		
15			Ann Matthews	19	F	✓	Service			Ireland				
16		1860	August Lindorh	48	M	✓	Lottery Broker	8000	3000	Germany				
17			Sophia "	42	F	✓	Bad business			"				
18			Charles "	12	M	✓				New York		/		
19			Lewis "	10	M	✓				" "		/		
20			John "	8	M	✓				" "		/		
21			Louisa "	4	F	✓				" "				
22			Susan Nechaus	29	F	✓	Service			Germany				
23	493	1861	Margaret Eagan	45	F	✓	Boarders		1000	Ireland				
24			Joseph West	25	M	✓	Engineer			New York				
25			Daniel "	30	M	✓	Seaman	Risky		" "				
26		1862	John Lowene	29	M	✓	"			Scotland				
27			Rosa "	27	F	✓				New York				
28		1863	William Reed	38	M	✓	Machinist	Good Trade		" "				
29			Celia "	36	F	✓				" "				
30			Adam "	3	M	✓				" "				
31			Josephine Eagan	17	F	✓	Vest Maker			" "				
32			Louisa Timmins	15	F	✓				" "		/		
33			Frances Wise	27	F	B	Service			Maryland			/	
34	494	1864	Sarah Cottier	56	F	✓	Boarders		1000	New York				
35			Frances Burton	23	F	✓	Seamstress			England				
36			Sarah Smith	18	F	✓	Milliner			Ireland				
37			Maria Huff	27	F	✓				Massachusetts				
38			Lucy Bennett	23	F	✓				New York				
39			Fanny Marks	19	F	✓				" "				
40			Mary Finnegan	29	F	✓	Service			Ireland			/	
			No. white males, 20 No. white females, 19	No. colored males, ___ No. colored females, 1			No. foreign born, ___ No. deaf and dumb, ___ No. blind, ___ No. insane, ___	38000	21000	No. idiotic, ___ No. paupers, ___				No. convicts, ___

Figure 13.1 Page from the 1860 *United States Census*

(*Source*: http://commons.wikimedia.org/wiki/Category:Public_domain)

easily became the national pastime. Its popularity at that time has never been equaled by any other sport.

A student of the game and one of the best writers to capture its deep meaning to Americans, George Will, claims that, in post-Civil War America, baseball was a calming influence, helping to heal the wounds of a divided nation (see Will 1990, 295). He makes this observation in one of the best books on the sport ever written: *Men at Work: The Craft of Baseball*, although it is now a bit dated (Will 1990). As is widely realized, since the first league baseball game was played, the sport has had an obsession with keeping game statistics on everything imaginable, from runs, hits, and errors to (seemingly) what occurred during which phase of the moon. Even today, there are sponsored contests that draw a huge following, matching contestants on their knowledge of arcane baseball statistics. Also, here is an interesting, relevant side note: the official site of the sport states that the game's inventor, Abner Doubleday, made up the initial set of rules in 1839 and later went on to become a Civil War hero.

Also in this era, and despite the end to Civil War fighting, the so-called Indian Wars continued, forcibly (and shamefully) relocating Native Americans to reservations and sometimes even breaking agreements in order to force them onto less desirable lands. One battle of the Great Sioux War of 1876—infamous in song and legend as "Custer's Last Stand"—took place over two days (on June 25 and 26) along the Little Bighorn River on the Crow Indian Reservation in southeastern Montana Territory. A 36-year-old Army general with personal ambition, George Custer, disobeyed his orders to hold fast to his locale on the edge of the prairie and instead led his troops onto native lands, with the intention of forcing the people there onto a reservation. The famous eponymous battle ensued. When the action ended, all of General Custer's 600 troops lay dead in the prairie grass, leaving a single horse named Comanche as the only Army survivor. About 180 Crow warriors had been killed, too. This was a sad and shameful day in American history.

Will puts this event and the game of baseball in a perspective that is both enlightening and unexpected. He said, "The day Custer lost at the Little Bighorn, the Chicago White Sox beat the Cincinnati Red Legs, 3–2" (Will 1990, 293). While Custer's loss is well known, there is less awareness that many in his regiment were recruited from baseball teams, with them often saying they were eager to get back to playing the game as soon as the battle ended. This was a remarkable juxtaposition of events.

More than any other sport in America, baseball, with its immersion in game statistics, is emblematic of quantification, since its invention to today.

****

Eli Whitney's impact on the Industrial Revolution and the Civil War era extended beyond simply the cotton gin, as revolutionary as that invention was. He also designed an "in-line process" for making muskets in quantity by manufacturing and assembling them in pieces at stations. For this process, Whitney's workers were stationed in a line where each one would add a specific part to the gun and pass it on to the next worker. Prior to this manufacturing innovation, guns were made one at a time by a single worker, a process which took a lot longer. Using Whitney's manufacturing method, these long guns became cheap to produce. Due to this innovation, Whitney is considered the "father of mass production."

Ever the clever inventor, Whitney also engineered the guns to have interchangeable parts, another important first. The advantage of interchangeable parts is that the guns could be repaired in the field by infantrymen. Thanks to his rapid in-line process of mass production and the feature of interchangeable parts, Whitney's muskets were clearly attractive to large buyers. Whitney demonstrated the steps in his mass-production process before a session of the US Congress, where he assembled a gun from its separable parts in just a few minutes. At the conclusion of his presentation, he was immediately awarded a huge contract for many thousands of muskets.

Some historians have noted the supreme irony that this one man had so much to do with both sides of the Civil War: first, by advancing the South's economy through his cotton gin and, second, by producing vast quantities of guns for the North's military war effort, helping it win.

Plainly, the Industrial Revolution on both sides of the Atlantic was one of the most momentous periods in all of human history. Nearly everyone in the Western world was directly affected by it, many to a life-altering degree. Often, historians note that even more significant than the inventions and innovations of the period was the fact that the Industrial Revolution brought forth tremendous and sustained progress in the standard of living for most people. Folks of meager means and only ordinary opportunity were better off almost overnight, or certainly within the span of one generation, living lives of much more convenience. They had better healthcare, superior education, and, for a time anyway, there was less direct impact on daily life from national wars and conflicts.

For context, remember, this improvement is judged by the living standards then, sans "presentism"—one need only check Dickens to be reminded of daily

Figure 13.2 Workers at machine works company during the Industrial Revolution
(*Source:* http://commons.wikimedia.org/wiki/Category:Public_domain)

hardships in working conditions brought on by the Industrial Revolution, too. Figure 13.2 illustrates some of the infamous circumstances wherein men and women—often while still in their youth—toiled in tedious and frequently dangerous conditions.

Pertinent adjectives to describe the Industrial Revolution spring forth in such volume that the words themselves could fill a treatise: imaginative, inventive, hard-working, creative, visionary, inspired, resourceful, ingenious—and I must mention two obvious descriptors: industrious and revolutionary. There is considerable debate among historians and sociologists about what years encompass this period and which inventions and innovations to include in its discussion. I follow several historians who cite the invention of the steam-powered railway as the Industrial Revolution's start. By this, it began in England in the early years of the nineteenth century when queues were long, with folks waiting their turn to ride the new Stockton and Darlington Railway in 1825 (see Chapter 12, Figure 12.1). Some folks climbed on board just for the novelty of this newfound freedom, but more were discovering it to be a means to expand their businesses and other enterprises.

Ocean liners, too, were being powered by steam, and transatlantic travel was becoming more practicable—although, of course, beyond the reach of most people. Inventions, discoveries, and innovations of the same ingenuity as steam-powered engines were happening so fast that people could scarcely keep track, much less absorb their impact. Nonetheless, everyone knew about the newly harnessed power source and other inventions, and it set forth a spirit of "I can do that, too." Worldwide, there was a new sense of possibility. In America especially, the production of steel at huge steel plants (mostly located in the Northeast) and the drilling of oil in vast oil fields in Texas, Oklahoma, and throughout the West, came to dominate industry. The textile industry, spurred on by the invention of the cotton spinning machines, was flourishing, first in Great Britain but soon all along America's East Coast.

These formative years in American history are chronicled in a marvelously informative book titled *What Hath God Wrought: The Transformation of America, 1815–1848* (Howe 2007). (Note: this is Volume 5 in the comprehensive series *The Oxford History of the United States* and, in 2008, won the Pulitzer Prize for History.)

As one might expect, numeracy and quantification were integral to much of this period. Suddenly, nearly every new thing was described in numerical terms: how big, how high, and how powerful. Technical words like "watts," "voltage," and "force" came into ordinary parlance. I imagine that, before such terms became the most economical and accurate way to describe the next new invention or innovation, they were not so popularly used. Now, though, people were progressively realizing that a numerical point of view was growing in influence on their daily lives—quantification was advancing.

Even more fundamentally, the sense of self seemed primed by the times to transform so as to keep apace of the quantifying events. Indeed, the relationship between mathematical inventions and the times in which they were done is almost symbiotic: people changed their lives and way of thinking *because* of the times—and simultaneously, many inventions and innovations that constituted the Industrial Revolution were themselves possible only because of the new mathematical discoveries.

Or, from the perspective of quantification, the people who pioneered these mathematical and probability ideas gave possibility to the inventions of the Industrial Revolution in the first place, which then fostered for us a transformed view of the world and our place in it. The concept that mathematical advances occurred because they *could*, which I have mentioned so many times before, is clearly evidenced here.

As everyone knows, during this period, not all inventions were mechanical; a plethora of technological inventions sprang up, too. Some were life altering in their impact. One such invention can be traced to an exact date: on March 26, 1876, newspaper headlines printed in large type the news that Alexander

Graham Bell had intoned into a cone-shaped microphone, with his voice crackling, these unassuming but now-famous words to his assistant in the next room: "Mr. Watson, come here. I want to see you." He had invented the telephone. With it, life for nearly everyone—rich and poor—was changed.

Yet another influential piece of machinery that was important to the Industrial Revolution was the sewing machine. Although various machines used for sewing had been around for a long time, the modern sewing machine really took off for popular use around 1851, when Isaac Merritt Singer added a foot pedal and applied for a patent. However, it seems a rival named Howe had already invented such a machine—he won a legal judgment against Singer, wherein Singer would pay Howe a royalty of $1.10 for each sewing machine sold. Singer's machine was wildly popular, as ordinary folks could now sew and repair their own clothes in a fraction of the time it took to do by hand. Singer also copied Whitney's idea of having interchangeable parts, making the machine simple to repair, sometimes even at home. His sewing machine came to symbolize modernity and played an important role in the Industrial Revolution.

Isaac Singer was an ambitious entrepreneur. After working on his sewing machine, he devised an innovative sales strategy: he added a sales force of nattily dressed individuals who went door to door selling the sewing machine to middle-class homes. And, to make the purchase more attractive, Singer (together with a lawyer named Edward Clark) devised a "hire-purchase arrangement" by which people would immediately take possession of the machine and then make payments over time—the first installment plan. These innovations were quickly adopted by others and became central to the progress of the times.

Following Karl Pearson's observation (given in the opening lines of this chapter) that an awareness of the times is a prerequisite to understanding your work, we are now prepared to discuss the next significant quantitative development: Pearson's own coefficient of correlation.

****

In 1892, Karl Pearson published a book that changed the world of scientific research: *The Grammar of Science* (Pearson (1892) 2004). For a generation, it was virtually required reading for anyone engaged in scholarly work, almost regardless of the discipline. Einstein recommended it as the first book to read among his friends at his book discussion group *Akademie Olympia* ("Olympia Academy"). The title is itself telling of its comprehensive scope and unconventional approach. In it, Pearson describes the laws of modern science, both stating how they were applied (in his time) and then setting a standard for researchers that previously had not been well developed.

On the title page, he gives his book a motto that has become famous to scientists: "Statistics is the Grammar of Science." He elaborates the motto's meaning by emphasizing to scientists the importance of employing creative imagination rather than mere fact gathering, while at the same time remaining objective and dispassionate in following a careful methodology. In the preface (in quotes taken from its second edition, the best known today), he says,

> There are periods in the growth of science when it is well to turn our attention from its imposing superstructure and to carefully examine its foundations. The present book is primarily intended as a criticism of the fundamental concepts of modern science, and as such finds its justification in the motto [Statistics is the Grammar of Science] placed upon its title-page. (Pearson 1900b, ix)

As is apparent, Pearson intends no less than to upend scientific research, placing empiricism at its center while simultaneously encouraging scholars to exercise care for objectivity and dispassion in their work. Elaborating this point, he adds:

> The classification of facts and the formation of absolute judgments upon the basis of this classification—judgments independent of the idiosyncrasies of the individual mind—essentially sum up the aim and method of modern science. The scientific man has above all things to strive at self-elimination in his judgments, to provide an argument which is as true for each individual mind as for his own. (Pearson 1900b, 6)

And, in summary,

> It [science] claims that the whole range of phenomena, mental as well as physical—the entire universe—is its field. It asserts that the scientific method is the sole gateway to the whole region of knowledge. (Pearson 1900b, 24)

Further, and in an innovative fashion, he made an assertion that shocked the world of researchers, declaring "all science is description and not explanation" (Pearson 1900b, vii). Clearly such words were sorely needed at the time, for as we saw in Chapter 12, when Galton warned that his description of a correlational relationship did not ipso facto mean causation, researchers were often careless about interpreting their findings.

Pearson's approach to scholarship and definition of a philosophy of science has shaped much of scientific research from the time of the book's publication to today. Surprisingly (and regrettably), though, apart from intellectual circles, his book is scarcely known. Today, although most scholars are attentive to protocols for methodology in their work, it is indeed unfortunate that few of them are aware of the origin of their rules.

From even this brief introduction to *The Grammar of Science*, it is easily seen that the work is both philosophy of science and technical treatise. As a work of philosophy, Pearson opens the book with thoughts that suggest a quantified worldview—one that would "profoundly modify our theory of life"—is needed to appreciate "so revolutionary a change" of recent scientific advancements. He begins the book:

> Within the past 40 years, so revolutionary a change has taken place in our appreciation of the essential facts in the growth of human society that it has become necessary not only to rewrite history, but to profoundly modify our theory of life and gradually, but nonetheless certainly, to adapt our conduct to the novel theory. (Pearson 1900b, 1)

In this exposition, Pearson sees the world of science divided in two: the philosophical and the technical—but he resolutely believes they should be unified. To this end, while imploring the careful researcher to follow rules of scientific discipline, he coaches future researchers to be concerned with their humanity as well: "Directly or indirectly, the individual citizen has to find some reply to the innumerable social and educational problems of the day" (Pearson 1900b, 5). Here, he goes beyond Laplace, Quetelet, or even Galton, who each sought to bring sociology into their expertise. To Pearson, scientific research is qualitatively different; it is but an avenue to social consciousness—they are one and the same. Quantitative work and social ethics are "much the same character," in that they identify and define ordinary people going about daily life:

> Geometry might almost be termed a branch of statistics, and the definition of the circle has much the same character as that of Quetelet's *l'homme moyen* [average man]. (Pearson 1900b, 12, note 1)

Having established form and context for scientific research, Pearson then turns his book to the technical side of his work. In this section, filling nearly half of the book, he covers a wealth of technical topics. His thoughts include a critique and extension of Newton's original laws of motion and delve into conjecture on the space–time continuum. Perhaps most famously of all, he writes early notes on the physics of relativity, anticipating Einstein's theory of relativity (notes which Einstein acknowledges as his inspiration; we'll examine them briefly in Chapter 16). His chapter headings give a flavor of this section of the book. They include "Space and Time," "The Geometry of Motion," "Matter," "The Laws of

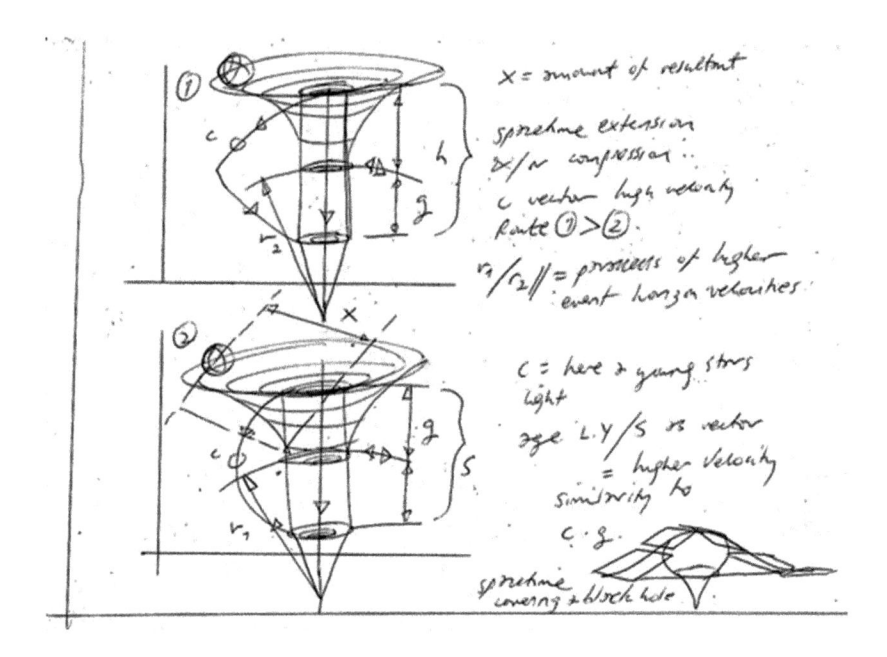

Figure 13.3 Karl Pearson's notes for *The Grammar of Science*.
(*Source*: from http://benabb.files.wordpress.com/2010/05/img_0081.jpg, accessed August 21, 2018.)

Motion," and "Modern Physical Ideas." A drawing of his notes for the book is displayed in Figure 13.3.

For persons with a bent for more exploration of Pearson's concepts, I recommend it. It is wonderful reading, with nugget after nugget of insight, all composed in an engaging style (see Pearson 1900b). Clearly, *The Grammar of Science* is a brilliant work of scholarship, one with many firsts. By his book and its clever motto, Pearson is sometimes credited with having invented mathematical statistics as a distinct discipline, although just as often that title is given to others.

To our purpose, the author's words reinforce quantification more directly than any scholarly text we have seen thus far. Hence, it moves quantification significantly forward.

****

In an especially important spark of genius, Karl Pearson developed a mathematical coefficient that made the Galton's correlation truly statistical, now a concept with a derived proof and an empirical interpretation. Pearson did this by borrowing an idea from physics—a particular expression of a numerical scale called "moments"—and applying it to a normal distribution. "Moment" has a special definition in physics in a sine wave and some other mathematical applications, but, for our purpose, we may think of them as standard deviations: each moment represents one standard deviation, plus or minus.

Pearson's addition of a mathematical structure to Galton's correlation has given rise to a formal name for the statistic: the "Pearson product–moment correlation coefficient." The name itself is more thoroughly descriptive than is common for statistical terms: it is honorary to Pearson, expressive of the relationship's quantitative structure, and broadly hints at its calculation by multiplying the moments (i.e., standard deviations) of a distribution. The final word, "coefficient," is essential in its proper interpretation, for it points directly to the degree of association or influence between the included variables.

Pearson used the letter r to denote his coefficient (mathematics readers will realize that the Greek letter ρ is used in some technical contexts when the statistic is applied to a population of values and not just the sample statistics). Familiar examples of r are displayed in Figure 13.4.

Pearson laid out his coefficient in his famous book and expounded upon it much further in a number of follow-up papers (e.g., Pearson 1905, 1907). In these, he extended his account to include relevant considerations such as issues dealing with the size of the sample, realizing what happens when the variables are themselves skewed, or the circumstance of accommodating more than two variables.

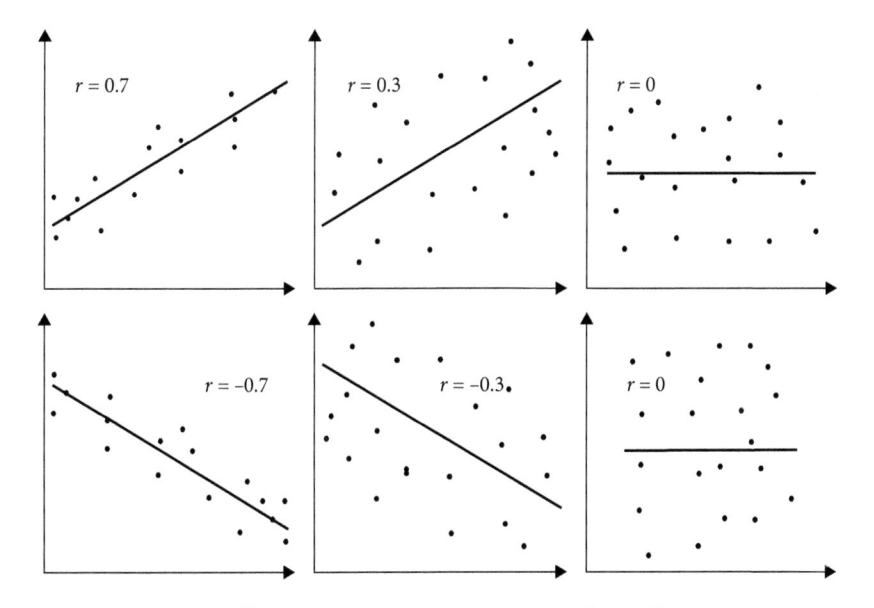

Figure 13.4 Pearson product moment correlation coefficient

He then built upon these concepts to introduce many other statistical terms, such as "skewness," "kurtosis," and "goodness of fit." (This last term, particularly, is notable, and I will describe it momentarily.) He was the first to use the term "standard deviation," and he formally introduced the "contingency table," now a routine way of comparing frequency counts between categories. For example, a 2 × 2 contingency table has four cells with a dichotomy for each of two variables, as when, say, biological sex (male and female) is compared with incidents of a heart attack (yes and no). By themselves, these ideas of Pearson's have led many others to develop still more procedures and statistical tests, some of which I'll address in later chapters.

Pearson's legacy is all the more remarkable when you consider that he was not formally trained in statistics, although he did have a mathematical education. Knowing what we know now about him, it is not surprising that he regularly read some of the great philosophers, such as Berkeley, Locke, Kant, and Spinoza. Although he did not write philosophy as such, it is obvious this education influenced his work in statistics and probability. We already saw this in his *Grammar*, and, throughout his mathematical writings, he routinely cited philosophical ideas.

****

Given that Pearson was a protégé of Galton and an admirer of Darwin, it is not surprising that almost everything Pearson did stemmed directly from these two predecessors. Most famously, he followed them in biometrical studies, especially as they were used in personal identification, such as with physical characteristics. From early in his career, he had been interested in their findings on heredity, and, accordingly, he produced several works on eugenics, looking at such things as immigration and criminal background as variables in sundry heritability studies.

Although Galton and Pearson were friends and each admired the other's work, they did not produce much scholarship together. But, near the end of Galton's life, Pearson wanted to initiate an academic journal devoted exclusively to advancing theoretical statistics. So he, along with Weldon (mentioned earlier as one who cautioned against misinterpreting correlational relationships), asked Galton to join the effort. Galton happily agreed, although his role was mostly honorary. Together, in 1901, these three academics founded *Biometrika*. Since its establishment, *Biometrika* has been one of the leading journals of mathematical scholarship, near the very top even today. The articles present only the most serious and noteworthy ideas, and to have work published in the journal is a feather in any academician's cap. For its nearly 120 years of existence, it has been published by Oxford University Press.

Pearson originally called the journal *Biometrica* (spelled with a *c*) to reflect biometric studies, but another colleague, Francis Edgeworth, changed the journal's name to be spelled with a *k* to indicate a broader scope of interest.

Edgeworth was a notable scholar in economics, and he introduced the so-called Edgeworth box, a graphical representation of how economic resources are distributed. Today, in virtually all courses and texts on introductory economics, the Edgeworth box is standard fare for study. Edgeworth himself was an interesting fellow. Evidently, he was born into a wealthy Anglo-Irish family—his education was entirely by private tutors, and he never attended school. Later, however, he attended Trinity College, Dublin, the prestigious Irish university.

Returning to Pearson, one work in particular showed the influence of both Galton and Darwin. At the time of its publication, it received immediate attention from his peers and has had lasting impact, even to today. The two-volume work is titled *The Chances of Death and Other Studies in Evolution* (Pearson 1897). Its opening pages of the second volume are most unusual. While proposing to discuss his studies involving lifespan and evolution, he recites medieval notions of death and reproduces several woodcuts from the fifteenth-century artwork called *The Dance of Death*, by Hans Holbein the Younger. These strange woodcuts show skeletons in contorted positions and caricatures of persons with emaciated bodies. Pearson discourses on these macabre woodcuts as though he is reviewing previous literature to set the stage for his own work. Bizarre! Fortunately, he moves past this to more conventional scholarship wherein he presents several ideas on heritability.

The book's influence arises from one idea in particular that has become quite famous. He introduces the notion of the "chi-square goodness of fit test." It is commonly referred to as "Pearson's chi-square." This simple but effective statistical test is routine in research today. There is scarcely a beginning statistics textbook or other introductory material that does not mention Pearson's chi-square statistic and his goodness of fit. In fact, to the degree that an early student of statistics or a beginning researcher may be aware of the name Pearson, it is probably thanks to this statistical test.

Pearson, following his earlier specification of significance levels for a correlation, set out to devise a way to determine whether a distribution of sampled data was close to or far from a normal distribution. The idea is perhaps easiest seen through example: suppose you collect data on the incidence and severity of crime in a neighborhood and want to determine whether the levels observed from that one sample are close to what could be expected if the same data was collected across all neighborhoods (considered the population). To state this in statistical terms, as a researcher would, the question becomes "Is the sample data similar to the population data?" Pearson's chi-square goodness of fit test applies to such binomial data. To take another example, suppose you flipped a coin many times to see whether the observed results followed a hypothetical binomial distribution for the population. Recall from earlier discussion the three theorems of numbers that are relevant here: the binomial, the law of large numbers, and the central limit theory. Pearson's test specifies significance levels for determining the closeness of sample data to these theorems; namely, goodness of fit. This is one of the first such statistical tests of difference.

To create his statistical test, Pearson needed to devise a way to gauge the distance of observed data from the theoretical population. His method for doing this was quite clever and insightful with regard to the three theorems, while at the same time remaining simple and direct. He set out to exactly specify a population for binomial values, meaning he needed enough data for it to be considered not just a very large sample but the population of values itself (which, of course, is very, very large—even theoretically infinite).

To collect such data, he used many samples of coin flips. In fact, he needed enough sample coin flips to aggregate to what could be taken for a population of all possible coin flips. Initially, he flipped a coin 2,400 times. Apparently, Pearson tired at this task, because he asked a student of his (!) to continue collecting data until he had an additional 8,178 samples. He then added even more data from outcomes of Monte Carlo roulette tables. Finally, with sufficient data in hand, he compared given samples to his "population data" and looked at the errors (he first used the term "deviations," and then "standard deviations").

He amassed the data in a set of response matrices. From these, he devised significance levels to identify the relative closeness of any given sample to the population. In other words, if the three basic theorems of numbers are correct, what is the probability that any sample data fit them? From this was derived Pearson's chi-square goodness of fit test. Figure 13.5 shows Pearson's actual distributions.

In 1900, only a few years after mentioning his method in *The Chances of Death*, Pearson expounded on it in an article that is now famous and is considered to be the formal introduction of his goodness of fit test. The article certainly has one of the longest titles in history, one that nearly explains the whole thing: "On the Criterion that a Given System of Deviations from the Probable in the Case of a Correlated System of Variables Is Such that It Can Be Reasonably

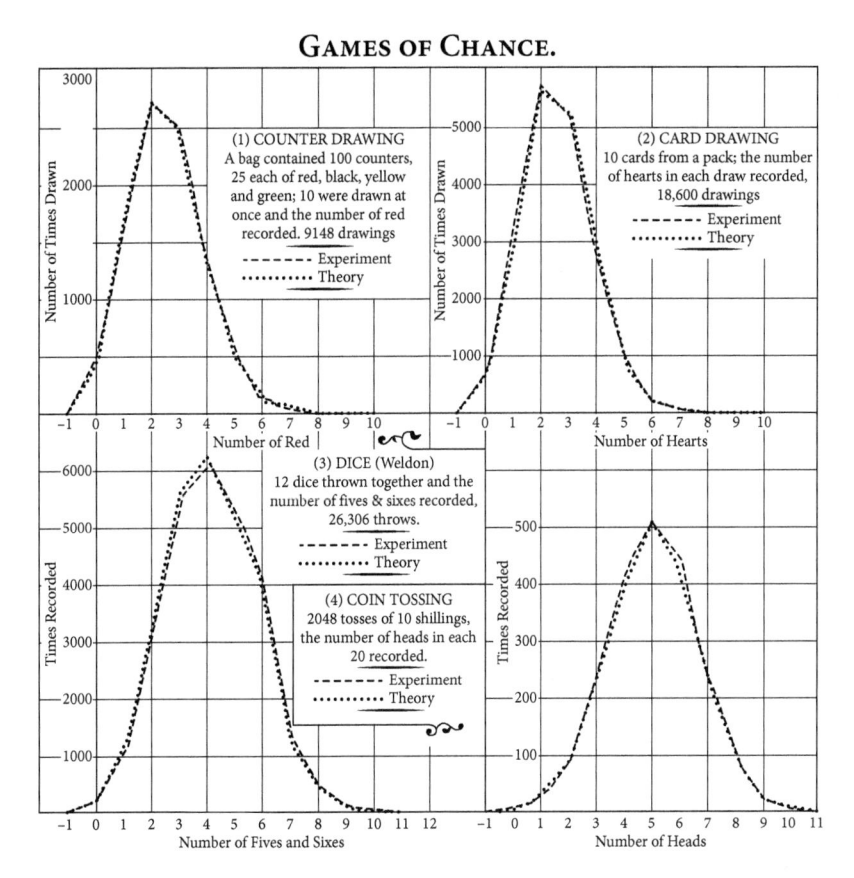

Figure 13.5 Pearson's distributions for assessing normality assumptions

(Source: from K. Pearson, *The Chances of Death and Other Studies in Evolution*)

Supposed to Have Arisen from Random Sampling" (Pearson 1900a). It is a bit surprising that the journal's editor would allow such a drawn-out title for an article. Maybe Pearson's academic stature held the same sway over the editor as it had earlier over his dutiful, coin-flipping student. Regardless, Pearson's chi-square goodness of fit test is one of the most popularly used inferential statistics, then and now.

****

As we have seen several times before, however, discoveries and inventions, both physical pieces of machinery and technical advances in mathematics, rarely happen cleanly. Nearly all have direct antecedents or have been discovered or invented elsewhere. Such is the case with the chi-square statistic. In 1876, a few years before Pearson's publication, a German professor of mathematics at the RWTH Aachen University (a leading technical university) and innovator in geodesy named Friedrich Helmert, published a nearly identical (and slightly more sophisticated) procedure for comparing a sample's distribution to that of a normal distribution. Hence, it seems that Helmert may actually have invented the chi-square first.

Even at the time, Helmert's invention did not go unnoticed, as several contemporaneous statistics textbooks describe Helmert's chi-square distribution. The context seems to have been that Helmert's work, and that of surrounding texts, was neither translated out of old German nor widely distributed. It seems, then, that both Pearson and Helmert can lay legitimate claim to invention of the chi-square statistic. This is another instance of simultaneous and independent progress in probability and statistics.

To our modern benefit, Helmert, a fine mathematician, has not been forgotten. Together with our mathematics giant, Gauss, Helmert published a textbook on least squares that enjoyed wide acclaim. Moreover, today Helmert is remembered for a particular kind of statistical contrast. In statistical contrasts, a variable with several categories—called "levels"—is considered. For example, suppose "income" is the variable of interest and there are, say, five levels, from lowest to highest. Several combinations are possible for comparing the levels. For instance, each income category (i.e., level) can be contrasted with some reference, which is typically either the highest or the lowest income category.

With the "Helmert contrast," each level of the income variable is compared with the mean of the subsequent levels. That is, the lowest income level is compared with the mean of the higher four levels, the second lowest income level is compared with the mean of the higher three levels, and so forth. In a "reverse Helmert contrast," each level of the variable is compared with the mean of the previous levels. Helmert contrasts are used in many fields. One interesting application today is when diagnosing the severity of brain damage: a given brain scan is contrasted to reference brain scans for various levels. Helmert contrasts are also routine in economics and census evaluations.

CHAPTER 14

Discrepancy to Variability

By this time in our story of quantification, the last half of the nineteenth century, statistics and probability theory were both relatively established sciences, yet still young enough that important inventions still awaited, particularly developments for gauging the veracity of research hypotheses by inference and in the new field of psychometrics for cognitive assessment. Overall, quantifying experiences had become so common, though, that ordinary people were beginning to use them in daily life: in business decisions, in government reports on all manner of topics, in insurance, in budgeting and the broader economy, and elsewhere. More ordinary people than ever before were gaining an overt awareness regarding both prediction and the assessment of probability. These notions were seeping into everyday conversation and decision-making. In short, people were beginning to *act* on the new knowledge surrounding them.

The next advances to quantification occurred in research, hypothesis testing, and the statistics for determining significance in group differences. These are the topics we explore in this chapter and the ones after. Also, our focus shifts to America, as that is the locale for many of these developments.

During this period, mathematics saw the introduction of new topics that trended the field toward increasing vagueness and concepts that had only an opaque philosophical rationale and almost no calculus proof, even though having a clear philosophical rational and a solid calculus proof had previously been the criteria accepted by traditionalists as necessary for adopting new ideas. These new developments included imaginary numbers (numbers that cannot be readily expressed with a real value, written by mathematicians as $i^2 = -1$), non-Euclidean geometry (remember the "orthodrome problem" discussed earlier as an example), and symbolic algebra, requiring exceedingly complex, often

The Error of Truth. Steven J. Osterlind. Oxford University Press (2019). © Steven J. Osterlind 2019.
DOI: 10.1093/oso/9780198831600.001.0001

hierarchical algorithms. Don't be concerned if you are not familiar with these arcane forms of mathematics—few folks are. Our point is to appreciate the turbulence of the times in mathematical developments—something that upset traditionalists.

On person who was especially perturbed by the philosophical wanderings of mathematics was a mathematics professor at Christ's Church, Oxford, named Reverend Charles Lutwidge Dodgson. Dodgson was a decided traditionalist on these matters, and, although naturally shy (he spoke with a noticeable stutter—he called it his "hesitation"), he talked often about his disagreement with the new direction for mathematics, particularly as it affected Euclidian geometry, his teaching subject. He felt so strongly about the issue that he wrote two important books (actually three, but his second book just continued the story begun in his first) to try and influence others back to the traditional point of view. By their titles and cover art, these books could not have seemed further apart: one had the appearance of technical treatise with a plain cover and an academic title, whereas the first edition of the other depicted a young girl on its cover, and its title suggested a typical Victorian-era children's book. Upon reading them, however, it is apparent that both books are about mathematics and argue the same point: namely, in favor of a traditionalist view in mathematics while highlighting a perceived lack of coherence in the new mathematics of imaginary numbers, symbolic algebra, and non-Euclidean geometry.

The latter of these books, which was the first book Dodgson wrote, is one you will certainly recognize: he published it under the pen name Lewis Carroll and titled it *Alice's Adventures in Wonderland*. He followed this work with its sequel, *Through the Looking-Glass*. Far more than a fanciful children's story, *Alice in Wonderland* (its shortened, popular name) is a tale of a young girl forced to experience ordinary events (like a tea party) in an absurd world. There are "mad hatters" and clocks that only go to six o'clock and no further. Wormholes lead nowhere, and strange characters work endlessly to no avail. In the sequel, Alice falls through a mirror into an alternative world where further mind-bending escapades await.

As most readers of these books quickly realize, this child's tale is not meant only for children; rather, it is filled with symbolism and metaphor. Knowing that these elements actually refer to the direction of the new mathematics makes their interpretation more intelligible. In fact, *Alice* is a difficult book for young children to read because they are unlikely to be aware of the symbols, and Carroll (Dodgson) often sets up convoluted scenes, describing them with formidable vocabulary. Nonetheless, the storyline is wonderfully fanciful, and the artwork is famous. (In fact, Dodgson drew the first illustrations himself, but then commissioned John Tenniel to draw the versions we see most often today.)

The Alice fictions did reach children, however, in their reinterpretation by Hollywood. Hollywood has made at least three films of the tales, making the scenes less baffling and using a toned-down vocabulary. Most folks hold that Disney's 1951 movie is still far and away the best version.

The story's mathematics is unveiled in an explanatory article titled "Alice's Adventures in Algebra: Wonderland Solved," which offers an almost academic discourse. For example:

> The madness of Wonderland, I believe, reflects Dodgson's views on the dangers of this new symbolic algebra. Alice has moved from a rational world to a land where even numbers behave erratically. In the hallway, she tried to remember her multiplication tables, but they had slipped out of the base-10 number system we are used to. (Bayley 2009)

And, as a mathematician who has studied the book wrote in "The Hidden Math behind *Alice in Wonderland*,"

> Perhaps the most obvious example [of its math] is the Cheshire Cat, which disappears, leaving only its grin, an obvious reference—critical in Dodgson's case—to increasing abstraction in the discipline. Most of the story was based on situations and buildings in Oxford and at Christ Church. For example, the "Rabbit Hole" down which Alice descends to begin her adventure symbolized the actual stairs in the back of the college's main hall. (Devlin 2010)

About a dozen years after *Alice's Adventures in Wonderland* appeared, Dodgson published under his own name a more conventional academic text; it, too, carried the same traditionalist message. This book is titled *Euclid and His Modern Rivals* (Dodgson 1879). It is a textbook that explicates Euclid's famous *Elements of Geometry* (which Dodgson referred to as his "Manual").

Euclid, as readers likely know, was an ancient Greek mathematician, and a foundational figure in modern mathematics, especially geometry. To a significant degree, Euclid influenced most of the mathematicians we have seen throughout this story, from Newton, to Gauss, to Einstein. Almost singlehandedly, he conceived the entire field of geometry and devised a set of proofs and theorems leading to number theory, which itself is much of the basis for calculus. (Recall, Gauss said, "Mathematics is the queen of the sciences and number theory is the queen of mathematics.") Across the globe, students of mathematics still study Euclidian geometry.

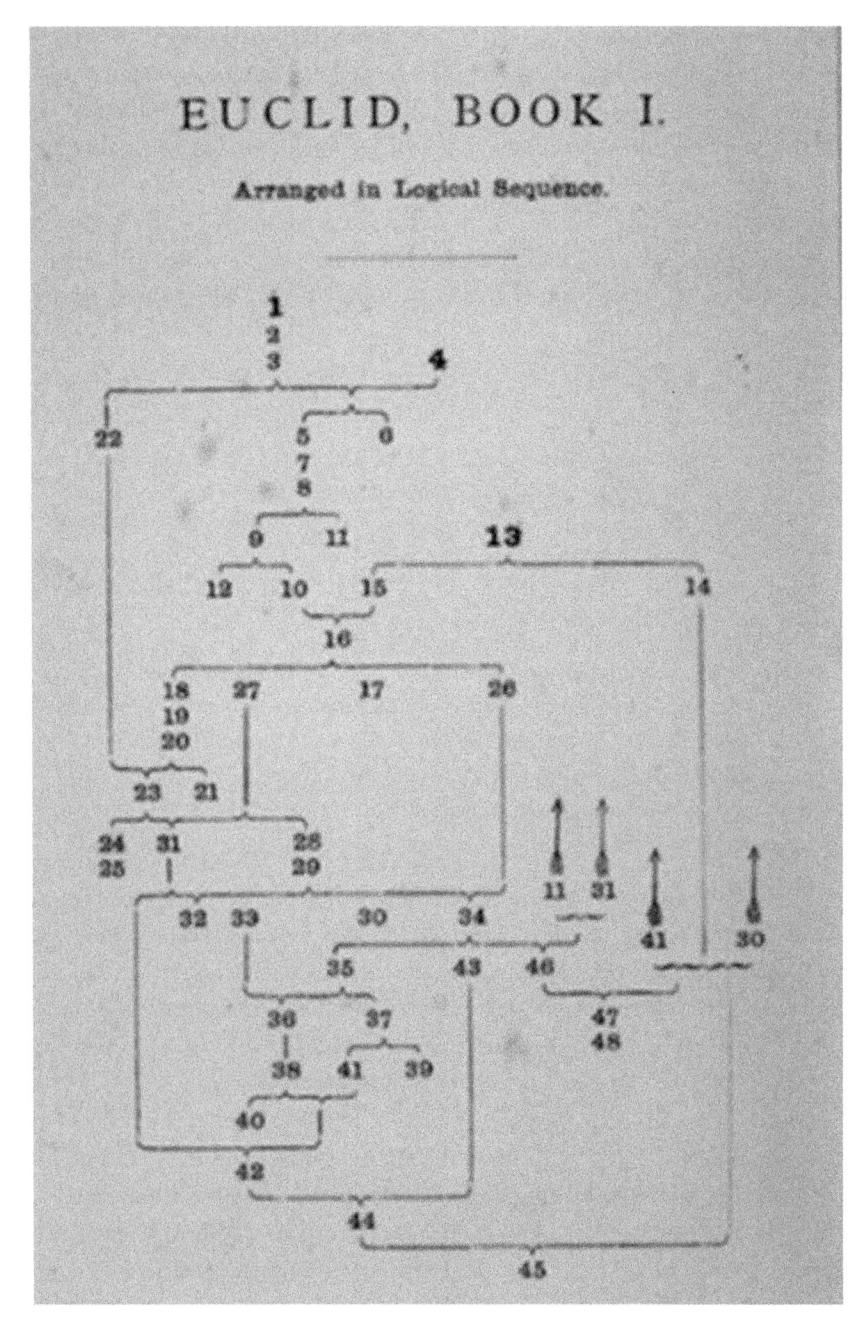

Figure 14.1 Frontispiece from Dodgson's *Euclid and His Modern Rivals*
(*Source*: from C. Dodgson, *Euclid and His Modern Rivals*)

Dodgson, for all the years of his teaching, followed a two-centuries-long practice of using Euclid's *Elements* as the text for his geometry classes. He believed it to be the standard for teaching geometry. *Elements* was, after all, a masterpiece of scholarship. But, despite his adherence to Euclid's scholarship, he disagreed with its organization.

Even more upsetting to Dodgson, however, was that he believed the new trend toward indistinctness in mathematics was encouraging many of his colleagues to forego Euclid's text in favor of newer ones. Dodgson saw this as a mistake and intended his book to influence them to come back to geometry's traditional moorings. Its very title (*Euclid and His Modern Rivals*) suggests his academic argument. In the preface, he asserts that he wrote it for three reasons: (1) he wanted to use just a single book in his class rather than teach from sections of many texts as he had previously done; (2) he believed he could use all of Euclid's "Manual," (i.e., *Elements*); and (3) after exhaustively searching, he could not find any better book. In this third point, Dodgson shows his disdain for the new, alternative geometry textbooks.

To address his disagreement with Euclid's organization of the book, he came up with his own outline. Figure 14.1 shows Dodgson's preferred sequencing of Euclid's theorems. Tellingly, he labels his reorganization of Euclid's theorems as being "Arranged in Logical Sequence"!

Personally, Dodgson was educated at home; early on, he showed a strength in mathematics, and he read prodigiously. His health was generally frail (throughout his life, he suffered from migraine headaches, as well as his stutter), and, although he tried sports, he was humiliated by his ineptitude. It seems he generally kept to himself, and he never married. But, certainly, he knew Christ's Church well, as many scenes from *Alice* and his other books are caricatures of actual places on campus. It was the only college with which he had a formal affiliation. He studied there as a student, and, upon graduation, he was hired as an instructor of mathematics. He eventually rose to full professor and held that position for the remainder of his career. It is reported that he was an engaging instructor and popular with students and colleagues alike. He played chess and made up word puzzles that typically had some twist of logic. This penchant is seen throughout *Alice's Adventures in Wonderland*. At least one book (though I suspect there are others) is devoted to presenting the puzzles of his literary works.

****

A brief editorial note about sequencing events in our story is needed at this point. As readers recognize, I have generally introduced the people and events that contributed to quantification in rough chronological order. To continue that order, I would now introduce the early inventors and developers of psychological testing (IQ and achievement tests, mostly), including Alfred Binet, James Cattell, Théodore Simon, Lewis Terman, L. L. Thurstone, and David Wechsler. These individuals were important in bringing quantification to a broad populace through their work in this specialized field. However, their efforts spread over a relatively long period, stretching from the mid-nineteenth century to the mid-twentieth. Also, in this age, others were working in pure mathematics, statistics, and probability theory. It seems a bit disjointed to slavishly adhere to a strict chronological order, because that would mean jumping back and forth between those involved in psychological testing and the pure mathematicians (including probability theorists). Hence, I will postpone describing the originators of psychological testing, keeping them together in a later chapter (Chapter 16) devoted to that topic.

****

Many readers are undoubtedly aware that a source of national pride for the Irish is its Guinness beer, brewed in Dublin on the same site since 1759 (with a 9,000-year lease at St. James's Gate!). For many, drinking a pint (or so) is as much a statement of cultural identity as refreshment; and all know the long-running campaign "My Goodness, My Guinness" with the famous Guinness toucan. Posters and knickknacks of every imaginable variety promote the dark brew. Most know, too, that Guinness has a distinctive taste and when first poured (there is a protocol to even this act), the bubbles rise, leaving a thick foam head, rather than fall, as for most other beers.

More to our point, the Guinness empire has historically been a leader in introducing research in manufacturing processes and that is how the brewery plays into the story of quantification. In 1901, the first Guinness research laboratory was established to examine water quality, hops, and more. A promising young man named William Gosset worked in the laboratory. Gosset was asked by his superiors to address a specific problem: determining the optimal ratio of hard to soft resins in the hops used by the brewery to flavor the beer. Hops are the flower of the plant *Humulus lupulus*, containing both hard and soft resins. Their ratio is what determines a beer's degree of bitterness, a necessary part of its flavor.

The brewery acquired hops from many sources, and Gosset realized that, in order to answer his primary question, he must first determine the consistency of the ratio across the many batches purchased. It is reported that, in one batch of eleven samples, there was an average 8.1 percent soft resins while, in another batch, from fourteen samples, the average was 8.4 percent soft resins. He certainly knew of the central limit theorem and that he could integrate the data into a cumulative frequency distribution. But, as we saw earlier, applying the theorem requires quite a few samples (considered to be thirty at minimum). He simply did not have enough samples to apply the theorem here. So, he formalized the problem as one of determining consistency across relatively few samples.

Gosset conceptualized the problem as one of determining by how much the values in his small sample differed from the normal distribution. Cleverly, he made his own version of the standard normal distribution (we saw the standard normal distribution in Chapter 9) by simulating many samples and specifying that it had a defined mean and standard deviation of 0 and 1, respectively; he called this the "z-distribution." He took just two samples of his hops and compared them to his theoretically derived z-distribution; he then repeated this several times. To his surprise, he discovered that, in 80 percent of these comparisons, he was within 0.05 degrees of the true number (the term "degree" is a reference to the analytical geometry underpinning the statistical fitting of samples to a given distribution). In other words, most of the time, his comparison of these small samples to his standard normal distribution was quite close, although not exact. He repeated the whole thing with three samples and found that, now, 88 percent of the time he was within 0.05 degrees of the true number. When he used four samples, he was even closer. He repeated this two-sample comparison several times—enough to make a new distribution, which he called the "t-distribution."

The t-distribution is more descriptive of small samples because of its slightly smushed shape: it looks like a bell curve that has been punched in on the sides so that the middle is higher and the tails are longer and flatter. By using a 0.05-degree criterion to test the difference between the t-distribution and the z-distribution, Gosset had invented a new approximation method that is useful when only a few samples can be had: the "t-test."

To apply his t-test to varying research conditions, Gosset described three variations, including (1) a "one-sample t-test," wherein the mean of a single group is gauged against a known mean, typically 0 from the z-distribution; (2) an "independent samples t-test" which compares the means for two groups; and (3) a "paired sample t-test" that compares means from the same group at different times, say, one week apart. These three types cover most common research scenarios.

Gosset wrote his findings in a report for the brewery, titling it "The Application of the 'Law of Errors' to the Work of the Brewery." His Guinness superiors were thrilled with his work, for he had successfully addressed the problem originally posed to him, far beyond their original expectations. Gosset was a bright individual, and, although fairly new in his career as a researcher, he realized he was on to something. We know now, more than one hundred years on, he was indeed "on to something" because his t-test and t-distribution are standard fare in statistical analyses today. Their significance is captured by a historian of the period, who declared, "Very few achievements in statistics have been as momentous as Student's [Gosset's] discovery of the t-distribution" (Gorroochurn 2016a, 348).

Still, Gosset wanted to be sure of what he had invented. He needed to have someone verify his method. Even more generically, he also sought to devise a mathematical model for his method but was unsure of how exactly to do it. To this end, he requested a consultation with Karl Pearson of University College, London, who had a well-known mathematics laboratory. The consultation became an extended study: Gosset spent a year there. During this time, Pearson was taken by the young man's interest in applying probability to address practical problems, and they developed a friendship that lasted throughout their lives, despite Pearson being thirty-three years the senior. We saw in Chapter 13 that, earlier, Pearson had had a parallel friendship with Galton, who was thirty-five years *his* senior. The friends were, in seniority, Galton, Pearson, and the younger Gosset.

Back in Dublin after his experience working with Pearson, Gosset wanted to publish his findings on his small-sample distribution approximation. He contacted his friend and mentor Pearson, who was editor of *Biometrika*, the premier statistical journal of the day (it remains so today). Pearson agreed to publish Gosset's work, and it appeared in a now-famous article titled "The Probable Error of a Mean" (Student (William S. Gosset) 1908, 1). The paper was a touchstone in the development of statistical tests using a criterion. Ronald Fisher, another famous statistician whom we will meet momentarily, notes this in his praise of Gossett for giving impetus to all studies of distributions: "The study of the exact distributions of statistics commences in 1908 with 'Student's' paper 'The Probable Error of a Mean'" (Fisher 1925, 23). And, of course, distributional statistics is fundamental to research analysis today.

But, in a well-known quirk of circumstance, the article appeared not under Gosset's name but under the pseudonym "Student." Many readers (and students of statistics) will recognize Gosset's work as "Student's t-distribution," and its interpretation as "Student's t-test." Figure 14.2 shows Page 1 of this famous article.

VOLUME VI   MARCH, 1908   No. 1

BIOMETRIKA.

THE PROBABLE ERROR OF A MEAN.

By STUDENT.

Introduction.

ANY experiment may be regarded as forming an individual of a "population" of experiments which might be performed under the same conditions. A series of experiments is a sample drawn from this population.

Now any series of experiments is only of value in so far as it enables us to form a judgment as to the statistical constants of the population to which the experiments belong. In a great number of cases the question finally turns on the value of a mean, either directly, or as the mean difference between the two quantities.

If the number of experiments be very large, we may have precise information as to the value of the mean, but if our sample be small, we have two sources of uncertainty:—(1) owing to the "error of random sampling" the mean of our series of experiments deviates more or less widely from the mean of the population, and (2) the sample is not sufficiently large to determine what is the law of distribution of individuals. It is usual, however, to assume a normal distribution, because, in a very large number of cases, this gives an approximation so close that a small sample will give no real information as to the manner in which the population deviates from normality: since some law of distribution must be assumed it is better to work with a curve whose area and ordinates are tabled, and whose properties are well known. This assumption is accordingly made in the present paper, so that its conclusions are not strictly applicable to populations known not to be normally distributed; yet it appears probable that the deviation from normality must be very extreme to lead to serious error. We are concerned here solely with the first of these two sources of uncertainty.

Figure 14.2 Portion of Page 1 from Gosset's *Biometrika* paper published as "STUDENT"
(Source: Permission from Oxford University Press.)

Several reasons are speculated for why Gosset published under the pseudonym "Student." None are fully corroborated, but all are fun to imagine. In one, the Guinness brewery wanted to hide from the competition that it was relying upon scientific approaches for decisions over using what was thought to be the wisdom of the senses garnered from generations of brewers. Another less folksy reason is that Karl Pearson made up the name and imposed it upon Gosset because, as founder and editor of the most prestigious academic journal of the day, he did not want to have his journal associated with a pub's stock-in-trade.

And yet, what better way to discuss quantification than over a pint of Guinness!

****

Gosset was a friendly, gregarious individual. As well as his friendship with Pearson, he had another professional acquaintance, this time with someone a bit nearer his own age, Ronald Fisher. Together, they worked on developing the *t*-distribution, eventually leading to its mathematical proof. Most of their contact was by letters written over an eight-year period in the 1920s; at least eighty of these letters survive for reading today (Box 1981).

The two men communicated by letter, rather than meeting in person, because they lived in Ireland and England, respectively, in a time when the politics of Irish independence dominated much of public discussion and governmental energy. Although not a long journey, travel between the countries was limited and not common beyond intragovernmental contacts. The Irish Nationalist Charles Parnell had, only a few years earlier, as an influential member of the British House of Commons, laid the groundwork for Irish independence. This was on the heels of the Potato Famine in the 1840s, when one million people starved to death, and nearly two million more fled to the United States or, if less well heeled, to Australia. All the while, the British sat idly by, ruling in Dublin just 100 kilometers (60 miles) to the west and doing almost nothing to help. The problems festered and finally led to the 1916 "Easter Rising" against British rule in Dublin, where vastly outgunned Irish protesters were shot in front of the Post Office (a noted civic building in the central part of the city). Soon, events turned revolutionary with the guerrilla-style Anglo-Irish War and its adherents in the Irish Republican Army (more commonly known by its acronym IRA), leading to the decades-long conflict known as the "Troubles."

In 1921, a ceasefire was signed, and the country was divided into two parts, the Republic of Ireland and Northern Ireland, after the ineffectual Éamon de Valera's inept negotiations with Great Britain for home rule. Unfortunately, this did not end the Troubles. The popular first chairman of the Provisional Government of the Irish Free State, Michael Collins, was assassinated just seven months into his term, and there was dissatisfaction about having the homeland divided. The Troubles continue to this day, in an ebb and flow, although now mostly isolated to Londonderry and Northern Ireland. Belfast remains a divided city.

Against this social and political backdrop, Gosset and Fisher maintained their friendship and collaborated on statistical developments.

Fisher was a British statistician of prodigious accomplishments. To many individuals, he is considered the "father of modern statistics." This title is certainly true for the inferential statistics employed in so much of today's research. A historian of early statistics opens a chapter on Fisher with this accolade: "'Monumental' and 'revolutionary' are words that come to mind when attempting to assess Fisher's impact on mathematical statistics" (Gorroochurn 2016a, 371). Another statistician gives him this praise: "Even scientists need their heroes, and R. A. Fisher was certainly the hero of twentieth-century statistics. His ideas dominated and transformed our field to an extent a Caesar or an Alexander might have envied" (Efron 1998, 95). You can see, with Fisher, we are meeting someone important in quantification.

Further, readers may already know, too, that Fisher is the inventor of the famous "lady tasting tea" problem, which I will describe momentarily. (It sets the stage for modern hypothesis testing by statistical inference.) First, however, I give some background and context for Fisher's accomplishments, although his achievements are so numerous I cannot describe them all. I will describe only some of Fisher's work as it contributes to our story of quantification. Readers with a special interest in the history of statistics may wish to explore this titan's work further, but I offer this bit of warning on reading Fisher's original writing: his composition is endlessly awkward. I refer you to a good biography or summary of his work, such as that by Gorroochurn (2016a).

The relationship between Gosset and Fisher began when Fisher was still a student. He admired Gosset's work and thought he had something to offer by suggesting a technical modification to better account for the "degrees of freedom" used with his t-distribution. "Degrees of freedom" is an important statistical concept that is best understood by working with it, even if only a little. It allows for estimating the alignment (called "model fit") of observed data (i.e., the sample) with a hypothesized distribution, most often the normal distribution. The underlying geometry that is necessary to fully understand this concept need not concern us here; rather, what is important is that it is needed for the proper interpretation of research hypotheses because the number of subjects comprising a sample varies from one study to the next, and the concept of "degrees of freedom" allows for this variation.

For our purposes, Fisher's modification with degrees of freedom strengthens Gosset's t-tests. Further, Fisher's work in this area also led him to his pivotal statistical invention of the F-distribution and the F ratio.

His work on Gosset's distributions led Fisher to an acquaintance with Pearson, which unfortunately degenerated into a personal feud. I mention this because it may have inhibited Fisher from attaining even greater heights, and it shows that these notables were still human. While yet a student, Fisher sought to publish his work in Pearson's journal *Biometrika*, the same journal where Gosset (as "Student") published his important work, "The Probable Error of a Mean." Pearson did publish young Fisher's paper, which set Fisher's reputation as a rising statistician. But, two years later, Pearson published an article criticizing Fisher's work. Fisher was quite unhappy because he realized Pearson's comments would be widely read among his circle, and he believed that the point of criticism was incorrect. It turned out that Fisher had got it right in the first place. He wrote a paper defending his original work and demanded that Pearson publish it. Pearson did so, but not as a separate piece; rather, he included it only as an appendix to another paper. This miffed Fisher greatly. Then, two others of Fisher's papers were rejected by Pearson—much to Fisher's growing anger.

Despite their personal tension, it seems that Pearson appreciated Fisher's abilities: he offered him a very good job, that of Chief Statistician at the prestigious Galton Laboratory. Fisher, who apparently had not yet got over his bad feelings, refused. In fact, he vowed to never again publish in *Biometrika*, and, although, over the years, he developed several other important ideas, he kept true to his pledge, preferring to publish them in less impressive outlets.

But Fisher's reputation was growing, and he was not without professional options. He secured a position as Chief Statistician at the Rothamsted Experimental Station, the first laboratory devoted exclusively to researching agriculture. (Rothamsted is a city located a bit north of London.) While there, he researched problems in soil types and plant fertility. This job fit with his interests in farming and in genetics.

All his life, Fisher suffered from extremely poor eyesight, which was debilitating, but when young, he turned it to academic advantage. His weak eyesight made him ineligible for military service. Britain needed conscripts during years surrounding several of its conflicts, including its colonial campaigns, the Irish Troubles, and tensions across Europe leading to WWI. Because he couldn't serve, he had the opportunity to pursue academics—but, even more pointedly, because he could not easily read his mathematics texts, he was forced to visualize the subject, including memorizing formulas and entire tables. This unusual approach to studying gave him a perspective on the topic that few others had. It is reasonable to surmise, too, that it may have spurred on his imagination to conceive his innovative approaches to statistics and research designs. One never knows for certain from where creativity flows, but certainly Fisher had it.

While at Rothamsted, Fisher let his creative spirit fly, for he pursued all manner of studies on snails, poultry, and mice, and, seemingly with each one, he produced yet another modernization in statistical approaches and methodologies. He introduced the concept of randomized trials for experiments, and he designed experiments that had levels for each experimental factor, allowing for much more efficient testing. With such factorial experiments, several hypotheses could be tested simultaneously. If that were not enough, he devised his analysis of variance procedure to assess the viability of multiple hypotheses simultaneously. Each of these accomplishments is significant in itself. To think that all came from the creativity of one man—and in such quick succession—is testament to an astonishing level of productivity.

In 1922, he published a seminal paper titled "The Goodness of Fit of Regression Formulae and the Distribution of Regression Coefficients" (Fisher 1922b) in which he initially explored a distribution called the chi-square (χ^2) distribution and regression coefficients. But this exploration led him to offer a wholly new perspective on analyzing data from samples: a ratio of variances. This ratio uses a variance statistic that is (inferentially, at least) accounted by the experimental condition compared to all of the variance in the experiment. As some readers may immediately recognize, this is the famous analysis of variance (ANOVA) ratio. The statistic later came to be called the "F ratio." Before Fisher, no method had been devised to statistically account for differences between groups while simultaneously considering the amount of error. ANOVA's full approach gives veracity to interpreting research findings. As most readers know, too, ANOVA is foundational to experimental methods used in today's research. We look at this famous procedure momentarily.

Fisher continued innovating procedures and methods at an astounding rate. To cite one important article, "On the Mathematical Foundations of Theoretical Statistics" (Fisher 1922a), he lays out statistical vocabulary and concepts, including "efficiency," "location," "scaling," "maximum likelihood," and "parameter." The quantity and significance of his many contributions is awesome.

A few years later, in 1925, he documented all these procedures in a remarkable book titled *Statistical Methods for Research Workers* (Fisher 1925). This book quickly became the standard text for research methods for many years. Fisher updated it in numerous editions throughout his life. In the introduction to a later edition, Fisher wrote about the success of this first tome, saying,

> It is 25 years since *Statistical Methods for Research Workers* first appeared. Its importance to research workers in all branches of science is clear from the fact that it is now in its eleventh edition, and that translations have been made into French, German, Spanish, and Italian. (Fisher 1950, vi)

Today, more than ninety years later, its influence remains. Of course, the work has been overtaken by hundreds of complementary texts, yet virtually all of them include Fisher's experimental designs and his analysis of variance. It is no wonder that some hold fast to their attribution of Fisher as the father of modern statistics.

Ten years on at Rothamsted Experimental Station, Fisher had performed hundreds of agriculture experiments. He was focused particularly on methodology, and he developed an entire structure for many experimental designs. In particular, he was interested in controlling error between groups, which he called "batch-to-batch variation," as a refinement on Gossett's work at Guinness with variation in resins in hops. He set out to develop procedures for experiments that would control for this variation. He was enormously successful in his quest.

While Fisher did not invent the concept of random numbers, he did introduce randomization to experimental design, making it an operational feature in experiments. This development was momentous at the time, and it survives today as an essential feature of valid experiments. In a genuinely random experiment, say, as in a "control group vs. experimental group" design, all of the subjects for a sample are selected at random from a population. Next, the sample's subjects are randomly assigned to one of the two groups (treatment group or control group). And, finally, the treatment (called the "independent variable") is then assigned at random to one group but not the other. This practice was Fisher's devising, and it is the standard in use today.

When true randomization cannot be achieved but only an approximation of randomization is obtained (a common circumstance in human factors experiments), the experiment is labeled as following a "quasi-experimental" design.

Fisher had another important first in designing experiments: introducing the "null hypothesis" into research designs. This important feature is where a positive statement is not proved; instead, the unlikely occurrence of the opposite is *rejected*. This experimental design is called "hypothesis testing" or "inferential testing." To give an example, suppose a careful scientist wants to prove that the sun will rise in the morning. The scientist proffers a hypothesis to this effect: "H_1: The sun will rise in the a.m." In Aristotelian logic, as well as in research contexts, a hypothesis like this cannot be proved directly. So, the researcher revises the original hypothesis into a second, opposite hypothesis, a *null hypothesis*: "H_0: The sun will not rise in the a.m." The experimenter then gathers data by observing that, for many days, each morning, indeed, the bright yellow-orange orb shows up over the eastern horizon. Using some statistical criterion to analyze the data, the researcher finds that the null hypothesis is unlikely. The conclusion is reached to reject the null hypothesis and infer that the original hypothesis (the sun will rise in the morning) should be accepted as true. Significantly, the research has not *proven* the hypothesis but has reached a conclusion by inference.

Fisher continued on his road of firsts by devising a statistical test to reject a null hypothesis. In it, he compared variances (this is the standard deviation squared) in a systematic manner. He analyzed the variances of an experiment, calling his procedure the "analysis of variance," (referred to by the shorthand ANOVA). Likely, many readers have some familiarity with ANOVA, so my explanation here is brief. ANOVA is a ratio of two variances: the variance attributed to the experimental condition, and all of the variance in the experiment. The values for these variances are expressed in a "sum of squares," which is itself divided by the experiment's degrees of freedom to yield an average (the "mean square" for the sums) used in the ratio.

Mathematically, ANOVA can be written as an assumption wherein

$$X_{ij} = \mu_i + e_{ij}$$

This simple equation means that the sample observed value (for a random subject) equals the population mean plus some random amount of error. (Note that this is the ANOVA generalized form as an assumption and not the formula for the F ratio, which is its implementation with actual data.) The general formula is incredibly powerful. It is one of the most-used expressions all of research for comparing group differences. It is efficient, in that it can test several hypotheses simultaneously.

Because the ANOVA procedure is so common to research, in Figure 14.3 I present sample tables of the output procedure leading to the F ratio. Don't worry about interpreting the tables, or even making sense of its columns—the purpose here is merely to illustrate tables that are universally known in experimental research.

As you may have guessed, the "F" statistic is so named to honor Fisher. Well earned, Sir Ronald Fisher!

In modern statistics, the ANOVA has been extended into a more comprehensive tool in which differing contexts for the variable can also be considered. For example, suppose one is interested in investigating differences in teaching styles (e.g., supportive vs. authoritarian) on effectiveness in reading comprehension for late elementary school-age students (late primary level). Although the ANOVA can readily accommodate this type of between-group

Descriptive Statistics

Year in school	N	Mean	Std. Deviation	Variance
Freshman	52	309.10	65.424	4280.245
Sophomore	248	306.79	59.178	3502.053
Junior	287	296.96	60.685	3682.729
Senior	115	283.23	60.887	3707.234

Tests of Between-Subjects Effects

Dependent Variable: Subject-Math

Source	Sum of Squares	df	Mean Square	F	Sig.
Corrected model	50121.186[a]	3	16707.062	4.557	.004
Intercept	40363806.05	1	40363806.05	11008.949	.000
Year	50121.186	3	16707.062	4.557	.004
Error	2559184.855	698	3666.454		
Total	65404095.00	702			
Corrected total	2609306.041	701			

[a] R^2 Squared = .019 (adjusted R^2 Squared = .015).

Figure 14.3 Illustrative output showing descriptive statistics and an ANOVA table

comparison, it does not capture much of the complexity of a real-world scenario. Suppose further, then, that one is also interested in considering contextual variables, such as the locale of the school (neighborhood socioeconomics: high vs. medium vs. low) or a student's motivation (high vs. medium vs. low). These additional contrasts can be incorporated into Fisher's analysis of variance in the advanced procedure "hierarchical linear modeling" (or HLM). This form of variance analyses is very popularly employed in sophisticated research problems.

Just to show you how HLM looks, the ANOVA comparison is now rewritten to focus on an individual's outcome within the contexts mentioned (a linear regression model). The formula now is

$$Y_{ij} = y_{00} + y_{01}W_j + y_{10}\left(X_{ij} + \bar{X}\right) + y_{11}W_j\left(X_{ij} - \bar{X}_j\right) + u_{0j} + u_{1j}\left(X_{ij} - \bar{X}_j\right) + r_{ij}$$

You may recognize this as a special expansion of the binomial discussed earlier. Those early ideas do indeed carry forward! Don't worry about working through it. (I suppose it is possible to work it by hand, but that would be ridiculously tedious—even silly—with today's computers at hand.). As always in this book, my purpose with formulas is to just illustrate them so that you may see how they look, not to explain them didactically.

As mentioned earlier, Fisher organized his experiments according to their conditions for treatments, which he labeled "factors." His factorial approach was instrumental to most of his research. There are many ways in which factors can be arranged in an experiment. One common arrangement is called a "Latin squares" design. Fisher did not invent the notion of Latin squares (that happened much earlier, probably with Euler), but he devised an entire typology of applying the concept to research designs. This is a clever arrangement of variables and factors that controls for unwanted variance (e.g., things that interfere with learning the true effects of an experiment). At Cambridge University, a stained-glass window illustrating a 7×7 Latin squares design honors Fisher.

Incidentally, those who enjoy working on Sudoku puzzles may be interested to learn that any solution to a Sudoku puzzle is a Latin square of Fisher's devising!

****

In storybook fashion, Fisher summarized his work, including introducing many research principles, like his Latin squares design, in a foundational book on research methods titled *The Design of Experiments*, which was first published in 1935. As with so much of Fisher's work, this book is remarkable in many respects. For one, he presents a classic description of inferential testing, as we discussed above. He does this in Chapter 2 of the book, which he titled "The Principles of Experimentation, Illustrated by a Psycho-Physical Experiment." This is the chapter that contains the famous story of the lady tasting tea.

It is most fun to see the story in Fisher's own words. He describes the problem as follows.

> A lady declares that by tasting a cup of tea made with milk, she can discriminate whether the milk or the tea infusion was first added to the cup. We will consider the problem of designing an experiment by means of which this assertion can be tested. For this purpose, let us first lay down a simple form of experiment with a view of studying its limitations and its characteristics, both those which appear to be essential to the experimental method, when well developed, and those which are not essential but auxiliary. (Fisher 1956, 11; also Fisher 1971, 1512)

And here is the experiment:

> [It] consists in mixing eight cups of tea, four in one way and four in the other, and presenting them to the subject for judgment in a random order. The subject has been told in advance of that the test will consist, namely, that she will be asked to taste eight cups, that these shall be four of each kind. (Fisher 1956, 11; also Fisher 1971, 1512)

Milk poured first (4 cups)   Tea poured first (4 cups)

Figure 14.4 Illustration of Fisher's research problem "The Lady Tasting Tea."
(*Source*: http://commons.wikimedia.org/wiki/Category:Public_domain)

Table 14.1 *Order of experiments for Fisher's "the lady tasting tea" problem*

		True Order		
		Tea First	Milk First	Total
Lady's guesses	Tea first	a = 3	b = 1	a + b = 4
	Milk first	c = 1	d = 3	c + d = 4
Total		a + c = 4	b + d = 4	n = 8

By this, there are $\frac{8!}{4!4!} = 70$ distinct possible orderings of the cups of tea. This is computed from the fact that there are eight cups of tea: four had milk poured first, and four had tea poured first (see Figure 14.4).

The possibilities can be presented in a contingency table as in Table 14.1. This table shows the number of choices made for the lady's guesses by the true order (tea first vs. milk first).

In looking at the problem, one realizes that there is more than one solution, each with its associated assumptions and varying implications—but it is outside of our purpose to explain them fully. For those interested, there are numerous websites and other publications devoted to "the lady tasting tea" problem. I presume most are useful and even fun to explore, but I offer a caution, because some explanations are inaccurate and others seem to be unnecessarily complex. It is a relatively simple problem in statistics—its importance lies in Fisher's presenting it as a probability problem with a statistical solution, a first at the time.

In *The Design of Experiments*, Fisher presented a solution that is referred to as "Fisher's exact method." By his approach, the exact probability can be readily calculated for each of the seventy possible outcomes, although it would be a bit silly to figure them all (since most are near zero): three or four possibilities are common. The formula for Fisher's exact probability is given as

$$P_{Fisher} = \frac{(a+b)!(c+d)!(a+c)!(b+d!)}{N!a!b!c!d!}$$

Using values shown in the contingency table, the answer is $\frac{16}{70} = 0.22857$, meaning that the lady has about a 23 percent chance of selecting the four cups correctly. However, there is a bit more to it than that. Fisher's research hypothesis is a null hypothesis, which states the "empty" case: that is, "How likely is it that the lady cannot discriminate?" Or, rather, that she cannot correctly guess all eight cups, as one-in-seventy circumstances.

Here, the values in our contingency table change slightly to $a = 4$, $b = 1$, $c = 1$, and $d = 4$. By Fisher's exact probability, we see $\frac{1}{70} = 0.0142$, or more than one in a thousand of all correct choices. By summing these possibilities, the probability is $\frac{16+1}{70} = 0.24285$, or about a 24 percent likelihood of observing Fisher's hypothesis.

The "lady tasting tea" remains one of the most enduring stories for teaching probability and hypothesis testing.

CHAPTER 15

Related to Relativity

Two iconic structures, the Eiffel Tower and the Brooklyn Bridge (each built in the closing years of the nineteenth century), symbolize the inventions, discoveries, and innovations of the newly begun twentieth century. Both structures impose themselves on the landscape, bold in appearance, and stand as feats of engineering. Many other twentieth-century occurrences followed in the same spirit. In 1903, the Wright brothers made their maiden powered flight, and Henry Ford established the Ford Motor Company, soon to mass-produce the Model T. That same year, Madame Curie won the Nobel Prize in Physics, becoming the first woman so honored. A few years later she was awarded a second prize, this time for chemistry. And, pertinent to our story of quantification as a general mindset, Albert Einstein published his theory of special relativity in 1905. The world memorized the equation $E = mc^2$, although almost no one could explain it. Due to its universal recognition, however, Einstein plays an important role in cementing numeracy as a worldview of ordinary people. In this chapter, we explore some of his remarkable accomplishments to see how they advanced this perspective.

To set the context, however—and following Pearson's admonition that "it is impossible to understand a man's work unless you understand something of the environment"—I describe some features of the era, if only briefly.

To begin, this period was the age of the explorers. These daredevils (as they were called then) sojourned to every untouched corner of the earth, from the Himalayas, to the South Pole, to South American jungles, and elsewhere. We saw already that the half-cousins Darwin and Galton traveled (separately) to the Galapagos Islands and the Kalahari Desert in the middle of the nineteenth century. That spirit continued, propelled by many others.

The Error of Truth. Steven J. Osterlind. Oxford University Press (2019). © Steven J. Osterlind 2019.
DOI: 10.1093/oso/9780198831600.001.0001

The exploits of these explorers captured the public's attention on both sides of the Atlantic. Like waiting for the next installment of a Dickens novel, people looked for the next missive from Cook, Perry, Shackleton, or Teddy Roosevelt to be published on the front pages of newspapers from London to New York. The treks and adventures were commonly described as scientific explorations. Many were sponsored by the Royal Geographical Society. We know already that this society was a group of learned men (although now women can also join) who also had the bug for adventure. Founded in 1830, the prestigious society continues today and now declares itself with the statement, "We are the UK's learned society and professional body for geography, supporting geography and geographers across the world" (Royal Geographical Society 2018).

One early explorer was a man on a British government mission. Sir George Everest was the surveyor general of India and was the first to map the meridian arc, an area from the southernmost point of India north to Nepal, providing the first real maps of the area. He began his surveys at the turn of the nineteenth century and continued them for nearly his entire career, up to the middle of the twentieth century. He mapped the general region of the Himalayan mountains and was later honored by having its highest peak named in his honor.

Credit for being the first to climb Mount Everest (in 1953) originally went to New Zealander Sir Edmund Hillary and his Sherpa guide Tenzing Norgay (who was hailed as a hero in Nepal and throughout India). But, almost immediately, questions arose about whether two other climbers, George Mallory and Sandy Irvine, had actually conquered it much earlier, in 1923. For many years, the dispute was unresolved and held as legend among adventurers, because both Mallory and Irvine died on the climb and their bodies had not been found. Only relatively recently (in 1999) were their bodies finally discovered, trapped on a side glacier. Still, the questions remained: had they reached the mountain's summit, making them truly the first to climb to the top? Did they die while summiting or descending the mountain? Forensics could not determine their direction of travel, but on the basis of other evidence, such as the track of rips in their clothing and wear on the hemp ropes, experienced climbers now believe they died on the descent, making them truly the first to have summited Mount Everest.

Many of the explorers of this era were mariners. Their nautical adventures were genuinely dangerous, because weather forecasts were unavailable for most of the oceans and what forecasts were available were not long range. Still, oceanographic maps were relatively accurate and sextants (still in use at the time) could be synchronized with chronometers, making direction of travel generally known.

One maritime event in particular evoked massive public interest, but it was not an expedition by explorers at all. It is the curious case of a double-masted schooner named the *Mary Celeste*. On a clear day in the seas east of the Azores, a thousand miles east of Portugal, a merchant ship came upon the *Mary Celeste*. It was adrift and abandoned. There was no visible damage to the vessel, and a six-month supply of food was found intact. Most ships of the day carried a yawl (a rowboat that could double as a lifeboat) lashed to the main hull, but if the *Mary Celeste* had one at the start, it was now gone. Where was the crew? And why had they apparently abandoned a sound vessel? Public interest was rampant, and speculation ran wild. An inquiry came up empty. Stories of all sorts arose: pirates, mutiny, sea monsters...you can imagine that it was great fun at the time to offer a theory. Sir Arthur Conan Doyle wrote a short story about the mystery of the *Mary Celeste*, and a movie featured its explanation of a deranged madman who killed everyone and then rowed off in the missing lifeboat, committing a watery suicide. The truth is, to this day, no one knows what happened.

The most famous explorer of the day was David Livingstone, a young Scottish missionary and doctor. He was the first to traverse the African continent, and the first European to see Victoria Falls, one of the true natural wonders of the world. Indigenous Africans led him there in canoes, calling it "the smoke that thunders" because of the heavy mist and deafening sound. In one of his adventures, he sought to find the headwaters of the Nile River, a trek that took more than four years. He was out of contact for so long that the *New York Herald* newspaper (which had sponsored much of his journey) sent Henry Stanley to look for him. In 1871, Stanley finally found the missing missionary and approached him in a village with the immortal question, "Dr. Livingstone, I presume?" The quote ran in banner headlines, and people loved it.

Many of these adventurers were interested in advancing science, something we saw earlier in the adventures of Darwin and Galton. The Royal Geographical Society sponsored Sir Richard Burton, a famous intellectual–explorer of the time, although he is less well known today. Handsome, muscular, and rugged, he was also very intelligent. He read voraciously and spoke several languages fluently, including Arabic, which was quite unusual for the period. He seems to have been a true Renaissance man. He has been described as an explorer, a geographer, a translator, a writer, a soldier, an orientalist, a cartographer, an ethnologist, a spy, a linguist, a poet, a fencer, and a diplomat. Whew! Burton traveled extensively throughout India and eastern Africa; he often went to the Middle East, disguising himself as an Arab to gain entry to several cities forbidden to

Westerners. His writings were very popular in England and across the Continent. He was genuinely scientific in approach and was one of the first to conduct valid ethnographic studies. A prolific author, he penned forty-three volumes about his trips. Later in life, while an attaché in Trieste, Burton translated all sixteen volumes of *The Tales of the Arabian Nights*. These stories enchanted children across the globe. I remember reading many of them as a child (probably in Burton's translation). Burton was knighted by Queen Victoria in 1886.

The most popularly anticipated adventure, however, was the race to the South Pole. This race to be the first was the last great explorer's prize, and it captured the public's interest beyond anything else. People followed the explorers in dramatic newspaper accounts, and when the explorers were back home, their public lectures were very popular. The South Pole prize was ultimately captured in December 1911 by Norwegian Roland Amundsen, and word went out across the globe. He was an instant celebrity. But news at the pole itself was (obviously) nonexistent, and, only thirty-four days later, in January 1912, a rival British team led by Robert Falcon Scott arrived, thinking they were the first— that is, until the last mile when Scott looked across the ice and spotted Amundsen's flag at the pole. Scott's spirit immediately fell flat for he suddenly realized he was too late. It was a chastening defeat. To make the defeat more cruel, vicious storms arose, and Scott and five others died on the return trip to their ship.

But possibly the greatest adventure story of all time is the true tale of the Anglo-Irish explorer Sir Ernest Shackleton with his trans-Antarctica expedition of 1914 to 1917 on the ship *Endurance*.

I suspect many readers know the story already—if not, it is wonderfully entertaining and humbling to realize the bravery and determination Shackleton exhibited to save his men.

Prior to this famous trek, Shackleton had led three expeditions to Antarctica and was well respected as a leader and scientific explorer. Soon after Amundsen conquered the pole, Shackleton sought a slightly different prize: to make the first sea journey across the icy continent. It is widely reported that he recruited a crew by placing the following advertisement in the local newspaper, although no archival copy has been found to verify its authenticity.

> *MEN WANTED: FOR HAZARDOUS JOURNEY. SMALL WAGES, BITTER COLD, LONG MONTHS OF COMPLETE DARKNESS, CONSTANT DANGER, SAFE RETURN DOUBTFUL. HONOUR AND RECOGNITION IN CASE OF SUCCESS.*
>
> *— SIR ERNEST SHACKLETON*

What is known, however, is that Shackleton sailed on the day before WWI broke out, after receiving a telegram of encouragement from Winston Churchill. He headed to the Weddell Sea and began to push forward. At first, all appeared normal, although the ice pack was unusually heavy. Over the next six weeks, they slogged forward through the ice, progressing a thousand miles. But, on January 18, 1915, at 76° 34′ S, *Endurance* got stuck. One crew member is said to have described the *Endurance's* plight as being stuck "like an almond in toffee" (PBS 2002).

The journey's official photographer took a picture of *Endurance* trapped in the ice (Figure 15.1). At first, Shackleton and his crew thought it would be no problem, because they had provisions and so they believed they could just wait

Figure 15.1 *Endurance* trapped in ice while on the 1914 Imperial Trans-Antarctic Expedition

(*Source:* http://commons.wikimedia.org/wiki/Category:Public_domain)

until the spring thaw. Twenty-eight men were trapped on the drifting ice pack. To pass the time, they performed skits and told one another tall tales. Remember, at this time, there was no communication to any mainland, so no one knew of their predicament. But, after a while, things took a turn for the worse. Slowly and relentlessly, the ice closed in on the ship, breaking her apart one board at a time. The crew's mood changed, as shown in their journals. After eleven months, in November, the ice took the last bits of the *Endurance* under.

Describing their situation on New Year's Eve 1915, Shackleton wrote:

> Thus, after a year's incessant battle with the ice, we had returned to almost the same latitude we had left with such high hopes and aspirations 12 months previously; but under what different conditions now! Our ship crushed and lost and we ourselves drifting on a piece of ice at the mercy of the winds. (Shackleton 1919, 95)

By this time, they were desperate. Their supplies had run out, including oil for lamps and cooking, and they were forced to eat their sled dogs. Then, as a last-ditch effort, using the three lifeboats they were able to recover from *Endurance*, the party struck out to sea and made it to Elephant Island, where they waited for a miracle rescue. But that was unlikely, given that no one knew where they were and they didn't know whether anyone was looking for them. They waited and waited, sleeping under the lifeboats and clubbing seals and birds for food.

After nearly another half-year, in a desperate move, Shackleton and two of the crew set out to sea in search of dry land. After an incredible lifeboat journey, they finally reached South Georgia. They landed on the uninhabited side of the island, coming upon steep cliffs. It took unbelievable strength and determination to reach the small whaling station on the other side, but they finally made it, collapsing on the doorstep of the small shacks. Shackleton spent only days recuperating before he started his efforts to get back to Elephant Island to rescue the remaining crew members. Obstacles abounded, but he never quit. It took him many months and four attempts, but he finally made it back to his stranded shipmates, in another adventure story all its own. Yet, he finally got to his crew—he had rescued them! In the end, everyone on the expedition survived.

In a 1956 address about the ill-fated expedition, another explorer and contemporary of Shackleton, Sir Raymond Priestley, made a telling statement that cemented Shackleton's reputation as a man of courage and endurance:

> As a scientific leader give me Scott; for swift and efficient polar travel, Amundsen; but when things are hopeless and there seems no way out, get down on your knees and pray for Shackleton. (Wordie 1957, 25)

(The quote is possibly a paraphrase of a more lengthy *mot* from Cherry-Garrard's preface to the second English edition of *The Worst Journey in the World* (see Croome 1958)). I am omitting many important twists and turns in the incredible story—it is one of the most astonishing journeys of all time—a tale well told by Alfred Lansing in *Endurance: Shackleton's Incredible Voyage* (2014).

Interestingly, another, more recent explorer, the American astronaut Scott Kelly, wrote a personal memoir about his adventures in space after having spent a year on board the International Space Station that he pointedly titled *Endurance* (Kelly 2017).

In the opening years of the twentieth century, people were fascinated by these explorers and their scientific adventures. They are a significant step to quantification, as we near our own journey's end. First, they represent that science in the broadest sense is pervading society and culture—new exploration, new experiences, and answers to unknowns are sought. Second, these explorations are a common interest experienced across all social classes and lands. Regardless of wealth and irrespective of locale, from Asia, to the Continent, to America, there was widespread public interest in these science-based adventures. Quantification had been broadly accepted and embraced.

America experienced a parallel public interest in the 1960s with excitement over the US space program and the quest to put a man on the moon, another scientific adventure, albeit massively more sophisticated. As we all know, this astonishing feat was finally accomplished on July 20, 1969, when Neil Armstrong lightly jumped off the last step of the lunar module of *Apollo 11* and onto the gray, dusty lunar surface. It was shared quantitative experience. Ask anyone over a certain age where they were on that date, and nearly all can recall exactly—many will even remember seeing that step on a black-and-white TV, as it was broadcast worldwide. It is science in the extreme, and there was broad interest, evidence that science and numeracy was by then a part of normal life—as they still are.

****

A single remarkable lecture, delivered in the summer of 1900 at the Sorbonne in Paris by a bright young German named David Hilbert, ushered quantification into the twentieth century. At the time, Hilbert was a promising mathematician (he received his education at the famous Gauss institution, the University of Göttingen), and his famous lecture catapulted him into mathematical history. In the lecture, he presented twenty-three mathematical problems that set the agenda for theoretical exploration from then to now. "Hilbert's problems," as they have come to be called, have dominated theoretical mathematics for more than a hundred years and set mathematicians across the globe on a quest to solve them.

Prior to presenting his problems, however, Hilbert defined a modern philosophy for mathematics, stressing especially the relationship between mathematics and science. He broadened the discipline to place in perspective the role of proofs with axioms and formulas. Until then, Euclidian geometry and Newtonian calculus stressed axioms and formulas with strict expectations for an accompanying mathematical proof. Hilbert allowed that less-tangible concerns with symbolic algebra and abstract concepts were also legitimate areas for interest. To give a flavor of his argument, some of the celebrated problems are shown in Table 15.1.

To date (two decades into the twenty-first century), of the original twenty-three problems, ten have been solved, seven more have partial solutions, two (Problems 8 and 12) remain unsolved, and four others are abstractions that have no definitive proof. For the mathematically curious, Hilbert's original lecture is widely available in both its original German and the English translation (Hilbert 1900b, Newson 1902). Also available is a shortened version of the problems and description of current progress on their solutions (Hilbert 1900a).

Table 15.1 *Selected problems from Hilbert's set of unsolved problems in mathematics*

Problem No.	Problem Description
1	Cantor's problem of the cardinal number of the continuum
2	The compatibility of the arithmetical axioms
3	The equality of volumes of two tetrahedra of equal bases and altitudes
4	Problem of the straight line as the shortest distance between two points
6	Mathematical treatment of the axioms of physics
8	Problems of prime numbers (the Riemann hypothesis)
9	Proof of the most general law of reciprocity in any number field
11	Quadratic forms with any algebraic numerical coefficients
12	Kroneker's theorem on abelian fields to any algebraic realm of rationality
16	Problem of the topology of algebraic curves and surfaces
17	Expression of definite forms by squares

Hilbert's problems helped to set the stage for the remarkable accomplishments of another famous German who is also known for his influence on the philosophy of science: Albert Einstein.

****

Albert Einstein's life and accomplishments make up the final milestone on our road to quantification. As we know, he was a beloved public figure for most of his life, and one of the most recognized individuals on the global stage. In figurative terms, his very presence evokes scientific achievement, and his wild white hair and rumpled clothing hanging on a relatively thin frame are physical icons of professorship.

Einstein's life events are chronicled in more than fifty full biographies and thousands of shorter pieces. His theories and intellectual accomplishments are described—with varying degrees of sophistication and accuracy—in innumerable publications and other sources. Among the best books about Einstein's life are Abraham Pais's *Subtle Is the Lord: The Science and the Life of Albert Einstein* and its sequel, *Einstein Lived Here* (Pais 1982, 1994).

He left no autobiography in the conventional sense of recounting the facts of his life, but he did write something about himself that is perhaps even more valuable: his *Autobiographical Notes*, shown in Figure 15.2. It is a very personal reflection on the thoughts and emotions he had while developing his theories (Einstein 1949). With warmth and insight, he begins his "scientific self-portrait" (Einstein's words) with this reflection:

> Here I sit in order to write, at the age of 67, something like my own obituary. I am doing this not merely because Dr. Schlipp [his friend and translator] has persuaded me to do it, but because I do, in fact, believe that it is a good thing to show those who are striving alongside of us how our own striving and searching appears in retrospect. (Einstein 1949, 9)

Due to this surfeit of Einstein materials, then, I will be brief in my description of his life and accomplishments, mentioning just a few things that highlight his contribution to quantification as a worldview for ordinary people.

Certainly, Einstein's fame, both during his life and after, underscores the very notion of quantification to the general populace. Despite the fact that few understand his theories even superficially, and even fewer are those who can follow their logic and substance, any thought one has about Einstein virtually demands reference to quantification. As illustration, think of the $E = mc^2$. We immediately recognize this notation, and while we cannot explain it, we nonetheless associate it with the frazzled hair and endearing look of Albert Einstein.

Figure 15.2 Cover of the English-language edition of Einstein's *Autobiographical Notes*

More importantly, we realize that it is mathematically significant. It imbues public consciousness with an impression parallel to that of Newton's *Principia* in the late seventeenth century. These quantifying events suffuse the daily thoughts and impressions of everyday folks.

A lesser-known fact of Einstein's life is that, in 1952, he was offered the presidency of the then-young state of Israel, following the death of its first president, Chaim Weizmann. Some considered the offer a publicity stunt, as Einstein had no previous political or administrative experience and the offer specified that he could still focus on his scientific work. The offer came from the Israeli ambassador to America, Abba Eban, who wrote:

> Professor Einstein:
> Prime Minister Ben Gurion asked me to convey to you, namely, whether you would accept the Presidency of Israel if it were offered you by a vote of the Knesset. Acceptance would entail moving to Israel and taking its citizenship. The Prime Minister assures me that in such circumstances complete facility and freedom to pursue your great scientific work would be afforded by a government and people who are fully conscious of the supreme significance of your labors. (American-Israeli Cooperative Enterprise 2018)

Einstein, it seems, was not only brilliant about the stars "out there" but was also aware of his own inner nature. He declined, saying,

> All my life I have dealt with objective matters, hence I lack both the natural aptitude and the experience to deal properly with people and to exercise official functions. For these reasons alone I should be unsuited to fulfill the duties of that high office. (American-Israeli Cooperative Enterprise 2018)

Certainly, he was a celebrity, both among the general public and within the scientific community. The world's most notable physicists and chemists meet quadrennially at the International Conference on Electrons and Photons to discuss recent advances. Perhaps the most famous conference was the fifth conference of the Conseil du Physique Solvay (located in Brussels), held in October 1927; there they discussed the newly formulated quantum theory, mostly advanced by Werner Heisenberg. Marie Curie (in Figure 15.3, first row, third from the left) alone among them had won Nobel Prizes in two separate scientific disciplines, although seventeen of the twenty-nine attendees had been honored once, including Einstein, Niels Bohr, Werner Heisenberg, and Max Plank. Possibly at no other time in history has such a group of intellectuals been gathered together. Even today, this famous photograph has a following.

As a young man in Switzerland, Einstein applied for a job at the Bern Patent Office. Evidently, there was some gap before the job began, so to earn money he offered tutoring in mathematics and physics. He recruited students through an advertisement he placed in the miscellaneous section of the local Bern newspaper. This is shown in Figure 15.4. Imagine being tutored by Albert Einstein. But, of course, that was before he was… Einstein!

Einstein's time in Bern was an especially productive time in his life, for, during his employ at the patent office, he worked evenings on his theory of special relativity, which he published in 1905 along with two other important papers. Einstein's thinking and production that year represents a touchstone in modern astronomy and astrophysics.

Figure 15.3 Madame Curie, Albert Einstein, and other participants at the fifth conference of the Conseil du Physique Solvay, October 1927

(Source: http://commons.wikimedia.org/wiki/Category:Public_domain)

> Private lessons in
> **mathematics and physics**
> for students and pupils gives
> thoroughly
> **Albert Einstein**, owner of the
> Swiss polyt. subject teacher diploma,
> **Gerechtigkeitsgasse 32, 1. floor.**
> Trial lessons for free.

Figure 15.4 Einstein advertising for students in 1901 in the *Anzeiger der Stadt Bern* (*Newspaper of the City of Bern*)

(Source: http://www.einstein-website.de/z_biography/print/p_olympia-e.html)

Also during this period, he made the acquaintance of two men, Conrad Habicht and Maurice Solovine, who remained his lifelong friends and colleagues. As we saw earlier, together they founded the *Akademie Olympia* to discuss books and scientific ideas, starting with Pearson's delightful and informative *Grammar of Science*. Biographers of Einstein extol his time in Bern as a seminal period for both his personal growth and his intellectual development.

Although Einstein worked on several facets of quantum mechanics and thermodynamics, he is most often associated with the theory of relativity. We examine this only superficially because anything more is outside our scope. Even an elementary understanding of the theory, however, requires a bit of background, so we begin by going back to Newton's *Principia*. In it, Newton outlines his three laws of motion, including:

Law No. 1: Objects in motion (or at rest) remain in motion (or at rest) unless an external force imposes change.
Law No. 2: Force is equal to the change in momentum per change of time. For a constant mass, force equals mass times acceleration.
Law No. 3: For every action, there is an equal and opposite reaction.

For nearly 200 years, astronomers held these laws to be immutable. But, in 1865, they came into question when a Scottish physicist named James Clerk Maxwell established the speed of light as 186,000 miles per second. This led two others—Austrian physicist Ernst Mach and French mathematician Henri Poincaré—to question how light travels. Their discoveries did not conform to the required physics of Newton's laws of motion. In fact, Mach and Poincaré discovered that the laws only hold under certain circumstances. Their work engendered more questions than answers, often the sign of solid scholarship.

The problem of how light travels fascinated Einstein from his earliest days. While yet a teenager, he imagined that he was an explorer—akin to the explorers racing to the South Pole, but with a twist: standing on a beam of light while observing another light beam traveling next to him. Newtonian laws of physics specified that since light travels uniformly, the two beams of light would be parallel in all respects, and hence the beam he was observing would have a relative speed of zero. But this was not supported by Maxwell's work. Einstein knew that something was amiss, and he sought to find out what. He came to the realization that, by the laws of electromagnetism, the relative speed depended upon your vantage point.

He developed a simple but profound scenario to describe this point. Einstein imagined two people, one inside a moving train and the other some distance off, observing the train. The train was equidistant between two trees, with one behind it and the other in front of it. Suppose, Einstein said, that lightning struck both trees at exactly the same instant. The person inside the train would see the lightning strikes at different times, due to his being in motion, while the person outside would observe the lightning strikes simultaneously. Einstein concluded from this illustration that time is relative to one's state of motion.

This led him to several profound realizations, one being his special relativity equation, $E = mc^2$: energy equals mass times the speed of light squared. By it, energy and mass are one and the same, albeit in different states. He realized, too, that if the mass can be exploded, the amount of energy released is enormous. Inadvertently, Einstein had discovered a theoretical basis for the atomic bomb. Later, he shared this information with Wernher von Braun and others, who put it to use in making the first atomic bomb. Einstein, however, famously held pacifist views regarding the bomb throughout his entire life.

Also, from his train scenario, Einstein concluded that time is relative to the observer. Because the person inside the train saw the lightning strikes separately whereas the outside observer did not, time has a "dilation effect"—it moves more slowly for the person at motion than the one at rest. It has been noted that we have a real-life illustration of this effect in the US space program. Astronaut Scott Kelly and his twin brother Mark aged at different rates when Scott was in orbit for nearly a year (from 2015 to 2016) and hence traveling much slower than Mark. When Scott returned to Earth, he greeted his now slightly older brother. By some mathematics, it can be figured that if a fifteen-year-old traveled near the speed of light for five years and then returned to Earth, the individual would be about twenty years old whereas his earthbound contemporaries would be over sixty-five!

Another practical implication of this effect discovered by Einstein is that the clocks on GPS satellites move 38 microseconds faster than do clocks on the ground. Thanks to Einstein and his theory of relativity, we know about this and can make accommodations for our GPS systems to preserve their accuracy.

Later, in 1915, Einstein expanded this "special theory" into his "general theory," the more useful of the two today. This general theory is widely described as the most beautiful of all existing physical theories. It gives a geometric perspective on gravity, relating a "space–time curvature" to the energy and momentum for some mass. This means that a coordinate system, like that shown in Figure 15.5, is a continuum of space and time—the space–time continuum—with varying effects at different distances from an object and relative to its mass. Einstein's general theory of relativity provides the mathematics that explain the phenomenon.

As a way to describe the space–time continuum, consider the following scenario given by NASA. This space–time stretch means that a heavy celestial body like Earth (although a relatively small planet, it's nonetheless heavy enough) tends to push down on space and time, causing it to bend downward— but not equally everywhere. The curvature is larger when it is closer to an object

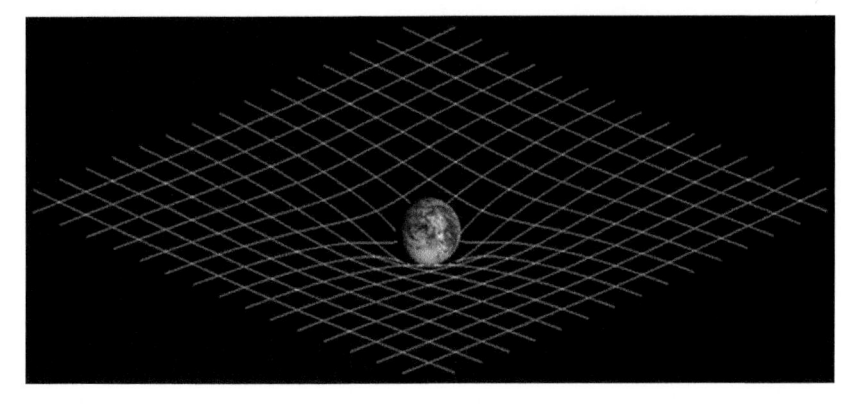

Figure 15.5 Illustration of Einstein's theory of relativity, showing geometric curvature of the space–time continuum
(*Source*: from CC BY-SA 3.0, http://commons.wikimedia.org/w/index.php?curid=86682)

and progressively less as the distance expands. This is a kind of distortion from our linear envisaging of space and time. Imagine dropping a baseball onto a trampoline. It causes the fabric to curve inward around it, even if only slightly. Space and time have changed, as illustrated by the depression. Now, imagine dropping a bowling ball, with its much greater mass, onto the trampoline. It would cause an identically shaped but much larger depression in the fabric. The space–time continuum with the larger mass is greater.

As humans, we cannot see this depression, and even our most sensitive and sophisticated instruments cannot directly detect it. Yet, we know it exists because we see its effects when measuring space and time. The space–time warp explains many phenomena in our world, both here on Earth and "out there" in infinite space. Einstein describes this in a set of equations known as the "Einstein field equations," a full system of differential equations. While his ideas for space and time are theoretical physics, they are also our practical reality.

The standard scholarly work on explaining gravitation and Einstein's space–time continuum is the textbook *Gravitation* by Charles Misner, Kip Thorne, and John Wheeler (Misner, Thorne, and Wheeler 1973). We will look more closely at this important book momentarily. For now, however, imaginably, reading it is demanding and requires some background in theoretical physics and calculus; nonetheless, it is a masterpiece of explanation and thoroughness. Two other excellent books that explain Einstein's theories and what they mean for us are by Stephen Hawking (formerly, Lucasian [Newton's]

Chair in Physics at Cambridge University): *The Future of Space Time*, which includes an excellent introduction by Richard Price (Hawking et al. 2003); and the long-popular *A Brief History of Time* (Hawking 1988).

Most compelling of all, however, is a book about relativity written for the general reader by Einstein himself. The book is entitled *Relativity: The Special and the General Theory* and has recently been reprinted in a hundredth-anniversary edition (Einstein 2015). In it, Einstein said that the book was "intended, as far as possible, to give an exact insight into the theory of Relativity to those readers who, from a general scientific and philosophical point of view, are interested in the theory, but who are not conversant with the mathematical apparatus of theoretical physics" (Einstein 2015, 10). When originally published in 1905, it was immediately popular and has been translated into at least sixteen languages since. This is an excellent read for people with a worldview of quantification and who are eager for new information, particularly because it may reinforce numeracy as the route for exploring the unknown.

Even a genius like Einstein makes mistakes. In what Einstein called the "biggest blunder of his life," he proposed a cosmological constant in his universal equations to make the force of gravity null (i.e., equal across the universe). He labeled it with the Greek lambda (Λ), and it is often referred to as "Einstein's lambda." Einstein's lambda represented a static and unchanging universe, although, within its sphere, there were orbital movements for the planets, stars, and other celestial bodies. When working on his general theory of relativity, he employed his lambda constant in his calculations. To Einstein and other astronomers of his day, the unchanging force of gravity across the universe was a given fact of astrophysics.

But Edwin Hubble, working at the same time on similar astronomy problems, realized that the stars furthest away were actually traveling faster than those closer to the sun. By this reasoning, gravity's force was not equal everywhere but relative to a source, which for the planets in our galaxy is the sun. The force of gravity is relative to the distance from it. This meant that the universe was not static after all, but was, in fact, expanding. Hubble reasoned, too, that Einstein's constant could not be correct. Hubble's discovery of an expanding universe is considered one of the most momentous discoveries in all astronomy, right up there with Newton's realization of gravity.

Einstein accepted Hubble's idea and made his admission of having "blundered." He was also frustrated with himself because he had stuck so long with including his constant in his calculations that he had missed the chance to be first to discover that the universe was expanding. Had he been more flexible in his thinking, he felt he would have been first to discover the expanding universe. This disappointment haunted Einstein for the rest of his life.

But, more recent thinking about the cosmos has come to include something that no one can describe except to say it exists—the so-called dark matter of the universe. Contemporary astronomers now believe that dark matter (whatever it is—its atoms move so fast and either explode or disappear before any measurements can be obtained) exists everywhere in the cosmos. In fact, it is thought to be the most common of all substances in the universe, despite our having no notion of what dark matter is. Now, for the best part: in modern astrophysics, dark matter is thought to be a static substance, and, in formulas, it can best be represented by Einstein's lambda! Thus, it seems Einstein was right all along to include a constant in his equations, but he just didn't know it. His blunder may not have been one, after all—he was right, even if for the wrong reason.

****

Probably the foremost scholar today advancing theories of gravitation is Kip Thorne, the Richard P. Feynman Professor of Theoretical Physics (emeritus) at Caltech (the California Institute of Technology), one of the most selective universities in the world (fewer than 2 percent of applicants are accepted) and a center for theoretical physics. Thorne is known for his contributions to gravitational physics and astrophysics. He is equally renowned for mentoring future physicists and for making his difficult topic comprehensible to a lay public. In 2017, he was awarded the Nobel Prize in Physics along with Rainer Weiss and Barry C. Barish for "decisive contributions to the LIGO (Laser Interferometer Gravitational-Wave Observatory) detector and the observation of gravitational waves," according to the Nobel Committee statement (Nobel Media AB 2018). Imagine the privilege of having Dr. Thorne as your doctoral adviser!

Thorne's seminal book is the aforementioned *Gravitation* (written with Charles Misner and John Wheeler)—often just called *MTW* after the initials of its three authors. Widely acknowledged as a landmark text, it is not light reading in any sense. More than 1,200 pages long, the preface promises that students can use this book to "grasp the laws of physics in flat spacetime, predict orders of magnitude...understand Einstein's geometric framework for physics, and explore applications, including pulsars" (Misner, Kip, and Wheeler 1973, v). Learn these, and you're all set!

The cover itself is creative. As seen in Figure 15.6, it is a line drawing by the artist Kenneth Gwin to show geodesics in general relativity. (Recall we saw

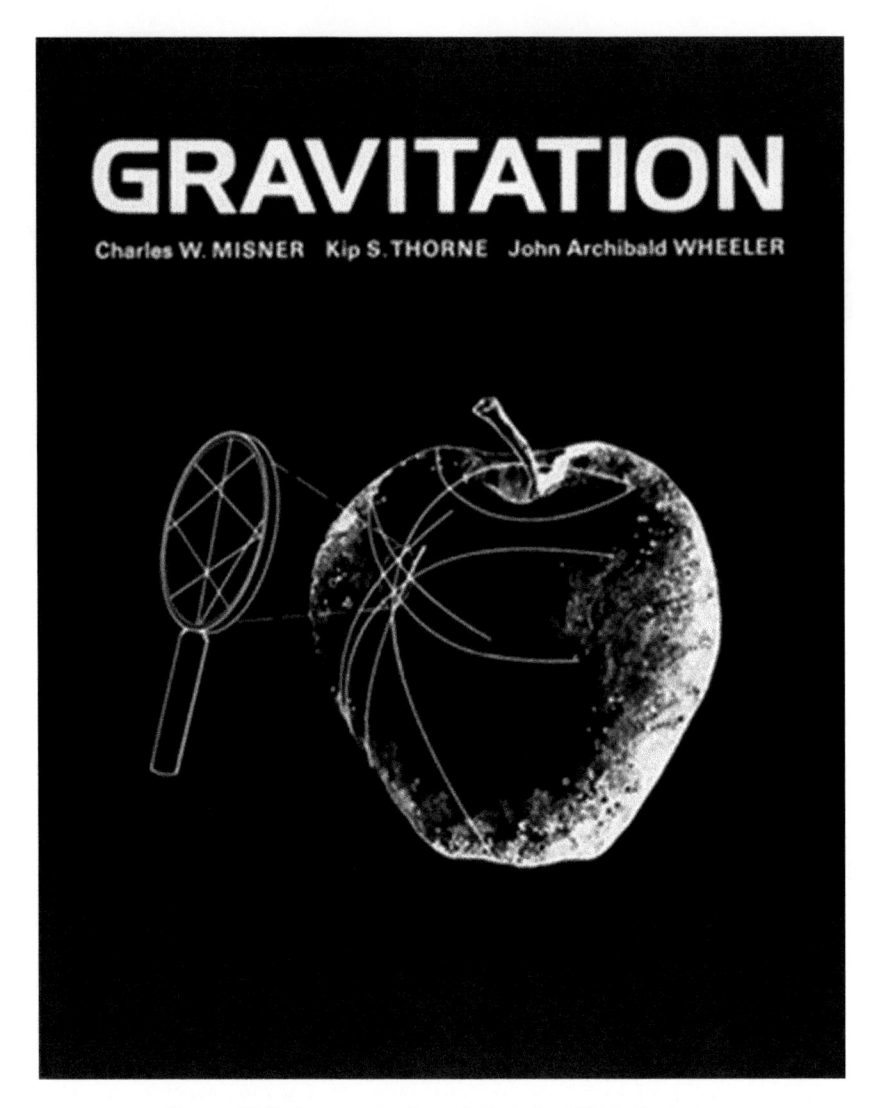

Figure 15.6 Cover art by Kenneth Gwin for MTW's *Gravitation*

this—a straight line bending to conform to a curved surface—earlier in Figure 8.1 in the projection for the orthodrome problem.) Unexpectedly, however, the curved surface in the book's artwork is not Earth or another celestial body; rather, it is an apple, which itself recalls the tale of Newton's apple in his "discovering gravity." The drawing shows a simple magnifying glass divided into ten parts reflecting onto forty-four sections of the apple, conforming to the organization of *Gravitation*, which was written in ten parts and forty-four chapters.

Following Einstein's lead of writing a popular book, Thorne offers an excellent read in *Black Holes and Time Warps: Einstein's Outrageous Legacy* with an engaging foreword by Stephen Hawking (Thorne 1994); Thorn writes,

> Gravitational-wave detectors will soon bring us observational maps of black holes, and the symphonic sounds of black holes colliding…Supercomputer simulations will attempt to replicate the symphonies and tell us what they mean, and black holes thereby will become objects of detailed experimental scrutiny. What will that scrutiny teach us? There will be surprises. (Thorne 1994, 524)

****

While all the central characters in our story of quantification were very bright, a few of them stand out, such as Bayes, Quetelet, and Laplace, and some others. Among these elites, some possessed an astonishing intellect, notably Newton and Gauss. Nearly all these individuals were men, owing to the lack of opportunity for women to participate in scholarly pursuits during much of the time of our story. Notwithstanding the obstacles, some women prevailed over their circumstances to exhibit their own exceptional intelligence and accomplish the remarkable. One such woman was Madame Marie Curie.

About 1891, a young girl named Maria Skłodowska was experiencing a troubled and difficult childhood in the Prussian section of Warsaw, in part because of the harsh circumstances of the foreign occupation. Yet, she was a very bright child and possessed a determined personality. While still young, she and her sister took the chance to move to Paris, although it meant leaving the rest of their family behind. In Paris, she enrolled in the University of Paris (then, "the Sorbonne") to study medicine and chemistry; there, she was recognized as a very able student with a strong work ethic.

Her life difficulties remained, but she single-mindedly continued her studies in the new field of radiology and eventually earned her doctoral degree, the first woman to do so in that field. Soon thereafter, she met Pierre Curie, and they married—per custom, she assumed his surname: Curie. Pierre called her Marie, and she thereafter used Marie as her Christian name rather than Maria. Hence, she became Madame Marie Curie. Pierre asked Marie to marry him, and she agreed to the marriage, believing it would help her keep her laboratory position. But, it seems that Pierre was an agreeable sort, and she eventually grew to love and cherish him as not only as a fellow scientist but also as a good husband. They had two children, both girls. Marie had at last realized a degree of happiness in her life. But later, Pierre was killed in a tragic accident. Marie remained devoted to her work and succeeded her husband as head of the Physics

Laboratory at the Sorbonne campus. She was promoted to full professor, the first woman to hold that prestigious position.

Her pioneering research was on radioactivity (she herself coined the term after discovering the element radium) and isolating radioactive isotopes. Eventually, her work led to a Nobel Prize (for physics, in 1903). The prize was first proposed for her male collaborator, but a Swedish committee member insisted that Marie be included since she had done most of the work. Again, this was a first for a woman. Later, she was awarded a second Nobel Prize (for chemistry, in 1911)—yet another first. By then, she was famous as Madame Curie. She went on make several other momentous discoveries, including a new element that she named *polonium* after her native Poland. Polonium is highly radioactive, and, at the time, its danger was not fully recognized. With overexposure to radiation, she contracted radiation poisoning. She died of it at a sanatorium in 1934.

The list of Madame Curie's firsts, both as a woman and as a pioneering scholar, is quite astonishing. Throughout her life, she received numerous awards and recognitions. Holding two Nobel Prizes in different fields is testament to both the depth and the breadth of her work. As noted earlier, she was also a member of the prestigious Conseil du Physique Solvay (see Figure 15.3).

Another remarkable woman working in more recent times was Maryam Mirzakhani, an Iranian mathematician and a professor of mathematics at Stanford University. Her research focused on pure mathematics, sometimes called "theoretical mathematics," as opposed to "applied mathematics." In pure mathematics, the focus is on the basic concepts and structures that underlie the entire mathematical domain. The goal of pure mathematics is to seek a deeper understanding of these structures so that mathematics may be more broadly applied. This is the level of exploration in which earlier giants such as Euler and Newton operated. Often, new theories and formulas arise from pure mathematics. Pure mathematicians study such difficult concepts as Teichmüller and ergodic theories and hyperbolic and symplectic geometry.

Mirzakhani was the first woman to win the prestigious Fields Medal for Mathematics (the highest prize for mathematics, awarded only once every four years—there is no mathematics category for the annual Nobel Prize). Clearly, she was woman of remarkable intellect and one who had much to offer. In a misfortune for us all, she died of breast cancer in 2017 at just forty-seven years old while still in her prime productive years—a true loss for humankind.

CHAPTER 16

Psychometrics and Psychological Tests

Earlier, in Chapters 1 and 14, I stated that while my organization of this book's content is approximately chronological by individuals in our story, there is a specialized area of mathematics and statistics—that of psychometrics and psychological testing—that seems more suited to being placed together, under one roof as it were. Some years ago (but still relevant), the journal *Science* cited the mental test as one of twenty discoveries that have shaped our lives forever (Miller 1984); and, hence, it needs attention all its own. The persons responsible for the invention and development of psychometrics and psychological testing were more focused on *applying* mathematical principles in new and important ways than on inventing original formulas and general processes. It is akin to ascribing your scholarly work to either pure (theoretical) mathematics or applied mathematics. The notables of psychometrics and psychological testing are decidedly in the latter group, dealing mostly with application. This chapter is devoted to describing these individuals and how their efforts contributed to quantification.

****

Quantification is both tangible and personal. Perhaps more than anywhere else, this is seen in the science of educational and psychological testing. It is decidedly personal, because we have all taken a test, and more likely, many of them. We know what they are like and realize that the test-taking experience is different from anything else described in this book. The events and inventions we have relived in these pages so far are, of course, amazing, but none of them engages our emotions directly like a psychological test.

With the test, we are vulnerable. It is a deep dive into our very nature and identity. An educational or psychological test looks into our brains—our achievements, aptitudes, beliefs, and opinions—and lays them bare for others to see. Making this very personal part of us public is scary because, before taking a test, we do not know how it will turn out. Upon learning the results, we can brag about them, hide them, or simply dismiss them. But, in all instances, we are directly engaging our emotions in the outcome. It is quantification up close and personal.

Our response is generally dependent upon the consequences of the assessment. When the stakes are high—such as with college admissions or when used in licensing and certification—our anxiety is correspondingly raised. When the stakes are low—such as when a marketer queries us about some opinion, we generally forget the experience almost immediately.

Although mental appraisal is most often associated with educational and psychological testing, it covers a broader range of issues than that, incorporating a concern for any kind of cognitive or psychomotor assessment. Actually, the types of tests that have psychometric properties are extensive, as shown in the list in Table 16.1.

Regardless of the type of assessment, we expect it to be accurate and thoughtfully done. Technically, these are concerns of validity and reliability. *Psychometrics* is the science of addressing such concerns of validity and reliability

Table 16.1 *Illustrative kinds of mental assessments*

Latent Traits	Analytical Skills	Mental Disorders and Abnormal Psychology	Psychomotor Skills
Achievement	Creative thinking	Depression	Talents
Ability	Deductive reasoning	Fear	Physical abilities
Aptitude	Problem solving	Panic	
Intelligence	Writing composition	Anxiety	
Opinions	Verbal communication	Stress	
Beliefs	Nonverbal communication	Paranoia	
Attitudes		Neurosis	
Preferences			

in cognitive measurements. Procedurally, it is the technical application of statistics and probability theory to address test development, test administration, and—most important—proper interpretation and use of a test's results.

Significant, and often misunderstood, is that validity is not a feature of a test instrument; rather, it is a concern for the appropriate, meaningful, and useful application of the scores yielded by a test for a given decision. In high-stakes testing, such as when the test's score is used for licensing or certifying an individual as competent in a given field (e.g., doctor, aircraft pilot, lawyer, certified public accountant, electrician), the validity of the decision is of paramount concern.

Further, establishing validity is complex. It is decidedly not a simple yes or no declaration. And, there is no single criterion or score for validity. Instead, *validity* is a matter of accumulating evidence to justify the decision made based on the yielded scores. Establishing validity is parallel to a lawyer making the case for the guilt or innocence of a defendant in a court case. It is justified by a preponderance of evidence, established anew with each circumstance.

Reliability, on the other hand, is a more tangible concept. It is an index calculated through mathematical determination of the consistency of a measure's score, generally from one testing occasion to the next, although there are some mathematical shortcuts wherein an index of reliability may be determined from a single administration. Most often, reliability is seen in either of two types of evidence: (1) temporal stability, as seen by consistency across multiple testing occasions, or (2) internal consistency, wherein the individual elements of a test (e.g., the questions) are arranged mathematically into a matrix that shows their consistency. Further, evidence of reliability is a necessary, but insufficient by itself, requirement of validity.

From even this brief description, you can see that the discipline of psychometrics has a utilitarian focus. But this does not mean it is a narrow or limited field. Actually, just the opposite: psychometrics is multifaceted, encompassing many topics, including developing a test, concerns for test administration, accommodations for examinees with disabilities, guidelines for fair use, and especially stipulations to ensure that a test's yielded scores are interpreted correctly and used properly.

Many of these issues are addressed in an industry-standard publication titled *Standards for Educational and Psychological Testing* (American Educational Research Association, American Psychological Association, and National Council on Measurement in Education 2014). In addition to the main topics of validity and reliability just mentioned, this publication gives direction to psychometricians for a host of related concerns, such as fairness to all test takers, how best to accommodate examinees with disabilities, and issues for test security. The standards are

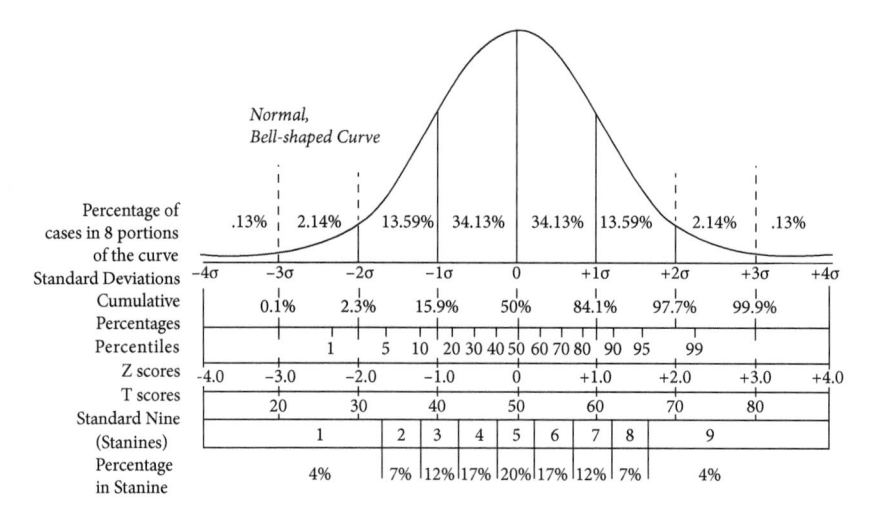

Percentage of cases in 8 portions of the curve	.13%	2.14%	13.59%	34.13%	34.13%	13.59%	2.14%	.13%
Standard Deviations	−4σ	−3σ	−2σ	−1σ	0	+1σ	+2σ	+3σ	+4σ
Cumulative Percentages		0.1%	2.3%	15.9%	50%	84.1%	97.7%	99.9%	
Percentiles				1 5 10 20 30 40 50 60 70 80 90 95 99					
Z scores	−4.0	−3.0	−2.0	−1.0	0	+1.0	+2.0	+3.0	+4.0
T scores		20	30	40	50	60	70	80	
Standard Nine (Stanines)		1	2	3	4	5	6	7	8	9		
Percentage in Stanine		4%	7%	12%	17%	20%	17%	12%	7%	4%	

Figure 16.1 Educational service bulletin to parents, showing a standard normal distribution with many derived scores

(Source: http://commons.wikimedia.org/wiki/Category:Public_domain)

reviewed and updated every few years, and examining older editions gives a history of testing, especially regarding evolving conceptions of validity.

Of course, when reporting the results of most tests, especially school-based tests, the standard normal curve is used for interpretation. A very famous depiction of the normal curve is shown in Figure 16.1. About 1950, this picture was distributed as an *Educational Service Bulletin* (this one is in *Bulletin No. 48* from The Psychological Corporation, the educational testing company founded by James Cattell, whom we'll meet momentarily). These testing bulletins were sent to parents of schoolchildren to help in their understanding of various kinds of test scores. At the time, such informational bulletins were common.

Below the normal curve, many types of derived test scores are given. Each scale has meaning for a different context. Among the scores presented are the standard deviations and the percentiles, quite familiar to us. Also shown are the *z*-scores and *t*-scores from the work of Gosset and Fisher. "Stanines" are yet another way to express test scores. For them, the full range of the scale is divided into nine standardized parts. The nine units were chosen because, at the time, only the values 0 to 9 could fit onto a punch card used for computer input.

****

The earliest documented use of testing on a large scale is with the Chinese Civil Service, which began its testing somewhere around 2000 BCE. These assessments were written tests to gauge language competence for prospective civil servants. The testing continued for several hundred years, although substantive changes in focus and administration were made along the way. After several variations of the assessment program, these early exams apparently ended at about 1370 BCE, when a new set of assessments, called the Imperial Examinations, were instituted. To take these imperial exams, candidates would sit in a small, isolated booth for at least 24 hours to prevent cheating. There, they would respond in writing to several questions of varying types, including asking them to compose a poem. Only a very few individuals passed, whereupon these achievers would progress to another set of tests. The imperial exams continued into the twentieth century, although numerous content and functional changes were made. Now, an entirely different set of examinations are used, but the practice of large-scale testing continues unabated.

Of course, large-scale testing is not confined to China. School-based achievement testing is used today in nearly every country. The current public school national assessment in the United States is the National Assessment of Educational Progress (NAEP; see National Center for Educational Statistics 2018). It is interesting, however, that NAEP assessment employs a complex administration procedure called "matrix sampling," wherein no individual student takes the whole exam; rather, each student takes only a part of it. The parts are then combined to yield a score for a group of students, such as within a school district or a state. No individual student receives a score. Thus, states' achievements may be compared, but those of individuals may not.

In Great Britain, there is compulsory testing in English, mathematics, and science at several ages along the educational track. Most exiting students also take the GCSE (General Certificate of Secondary Education) and its advanced levels to qualify for third-level education, and many take vocational qualifications (see British Government 2018).

While in Germany, five types of secondary schools are offered, each of which has a final examination. The most common of these is the *Abitur* (after *Gymnasium* (secondary schooling)) and the *Mittlere Reife* (after *Realschule*, or, in some states, *Oberschulen* or *Sekundarschulen*). These German school examinations are used by several other countries, too.

And many countries (more than sixty, at last count) participate in a cross-national testing with the TIMSS ("Trends in International Mathematics and Science Study") and PIRLS ("Progress in International Reading Literacy Study") assessment programs. The TIMMS, administered once every four years, measures how effective countries are in teaching mathematics and science.

And PIRLS documents worldwide trends in reading for students in the fourth grade, as well as school and teacher practices related to instruction (see IEA International Study Center 2018). These assessment programs do allow for some kinds of international comparisons.

****

Four persons were instrumental in founding modern testing: Francis Galton, Wilhelm Wundt, James McKeen Cattell, and Alfred Binet. Readers will recall that, earlier, we learned of Galton's work in his Anthropometric Laboratory, where he tested physical attributes of adults and children and made his discovery of regression to the mean, which led to the invention of statistical correlation. In addition, owing to his wide-ranging application of statistics and measurement theory to cognitive assessment (particularly with his early attention to issues of validity and reliability), he is credited with founding the field of psychometrics. Further, he formed the notion that intelligence can be viewed as a psychological construct, thereby making it a variable for research. As readers likely know, intelligence is one of the most researched—and illusory—variables in all of science. Quite a list of accomplishments—and that is from just one of the four!

These four individuals worked mostly in the mid- to late nineteenth century, with the later years of their careers spanning into the twentieth century. All were interested in early explorations into mental illness in children, particularly with its accurate and meaningful assessment (i.e., validity and reliability), which gave a common focus to their efforts. Accordingly, psychometrics was an important feature in their work.

Now, on to describing the accomplishments of each of these founders of modern testing individually—if only briefly. In doing so, I will keep in mind our focus: namely, how these individuals contributed to advancing quantification as a worldview, particularly in bringing it to the lives of ordinary folks. Each of these individuals made numerous and extraordinary accomplishments in psychology and psychometrics, and I must necessarily leave out much of their work. Of course, entire biographies exist for each of them, and other sources offer a thorough account of their accomplishments.

Wundt was a German physician and philosopher who worked to identify children with mental illnesses. He was interested exclusively in exploring psychological issues, and he called himself a "psychologist," possibly the first to do so. More than anyone else, he marked psychology as a distinct discipline, apart from biology and other fields of scholarship. Notably, he postulated a uniform theory of the science of psychology. Because Wundt was the first researcher to completely devote his career to psychological issues, he is often referred to as the "founder of modern psychology."

And, for his innovative experiments on appraising psychological characteristics in children, he is also called the "father of experimental psychology." To highlight his reputation, in 1980, the journal *American Psychologist* conducted a survey among historians of psychology to determine the most influential psychologists of all time. Wundt was the winner by a wide margin, even ahead of William James and Sigmund Freud.

Wundt was not subdued in his approach to the psychological research. Rather, he was a vociferous advocate for his new discipline of psychology, often encouraging students to the new field and giving frequent public lectures to educate the public at large. And he was a prolific writer. In 1874, he published *Principles of Physiological Psychology*, the first textbook on psychology, and he trained hundreds of students in psychology.

He believed that psychology should be studied with the same rigor as was brought to chemistry or the other hard sciences. He introduced the scientific method to the study of mental constructs and psychological issues with this direction: "the aid of the experimental method becomes indispensable whenever the problem set is the analysis of transient and impermanent phenomena, and not merely the observation of persistent and relatively constant objects" (Wundt 1904, vol. 1, 4). With this quote, he was the first to give scientific application to the study of psychology. Nearly all modern psychologists, whether in clinical practice or conducting research, have been influenced by Wundt.

Wundt was a professor at the University of Leipzig, which is one of the oldest third-level institutions in the world, having been founded in 1409. There, he created what is held to be the first formal laboratory devoted to studying psychological issues. His students went on to establish such centers for psychological research at many other universities, including Stanford University and Yale University. Today, they are commonplace on campuses throughout the world.

Much of his work focused on measuring mental chronometry, or more familiarly, reaction time—quite literally, the time it takes one to respond to stimuli. It is held that reaction time is related to general intelligence. With insight, Wundt said,

> For each person there must be a certain speed of thinking, which he can never exceed with his given mental constitution. But just as one steam engine can go faster than another, so this speed of thought will probably not be the same in all persons. (Wundt, 1862, 264, as translated in Rieber and Robinson, 2001, 35)

Most reaction-time tests measure in seconds and even milliseconds. To make his measurements, Wundt called upon the English physicist and instrument maker Charles Wheatstone to make a machine for measuring reaction time. Wundt called it a "thought meter." It is a clock with a pendulum device that has rods protruding down its sides. He set a calibrated scale behind the rods to measure how far they moved in a given time period (he used tenths of a second for his time increments). Wundt set the rods to begin at zero on his scale, and his reaction-time test began. A bell sounded. The subject's task was to anticipate the bell, and when it rang, the subject would stop the rods by pressing a button as quickly as his or her reactions allowed. By seeing where the rods stopped, Wundt could read on his calibrated scale how many tenths-of-a-second tick marks they had passed. This was his innovative measure of reaction time, and although rudimentary, it was fairly reliable in its measurements. Over time, Wundt employed a variety of similar devices. One instrument he used is shown in Figure 16.2.

Building from his belief that reaction time was correlated to intelligence, his work was instrumental in maturing the view that intelligence is itself a psychological construct that can be explored by cognitive means. He thought it could be assessed using both paper-and-pencil tests and, even more directly, experiments with reaction time. Prior to this view, researchers focused on simple sensory experiences, such as in Galton's anthropometric investigations.

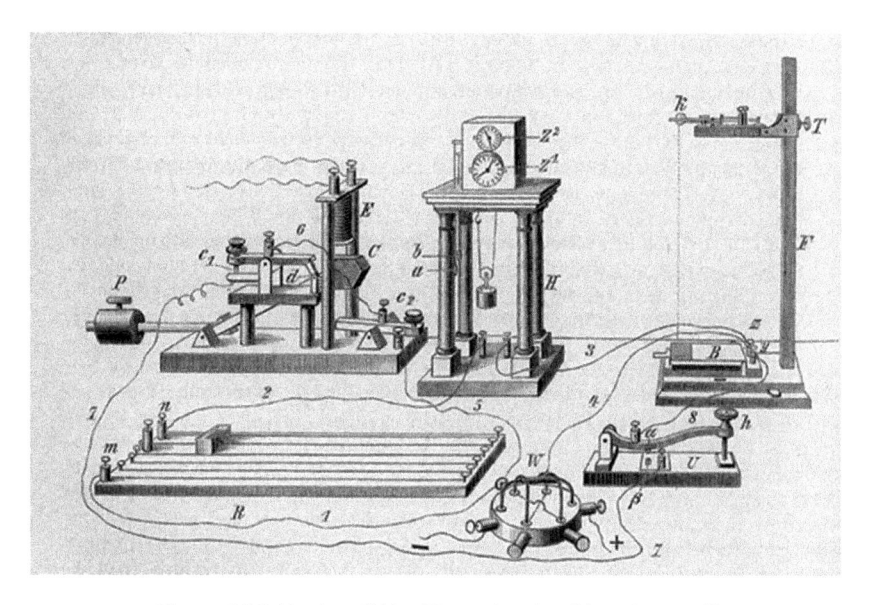

Figure 16.2 Version of Wundt's reaction time "thought meter"
(*Source:* http://commons.wikimedia.org/wiki/Category:Public_domain)

As a serendipitous consequence of his reaction-time experiments, Wundt also shed light on a classic problem in astronomy about measuring the density of the earth by noting differences in its position over time.

Interest in reaction time as a variable for assessment waned after Wundt (and slightly later, Cattell), but renewed attention to chronometric analysis arose in the late twentieth century. Modern scholars believe this kind of analysis has been fruitful in determining a general intelligence construct (the so-called Spearman's g), as well as in its application in psychomotor and other physical tasks. For example, assessing reaction time in athletes with procedures much like Wundt's thought meter is now common, although, of course, it is done for assessing physical abilities rather than mental constructs.

A leader in the revival of chronometric analysis was Arthur R. Jensen, a brilliant scholar at the University of California, Berkeley. Jensen worked in the mid- to late twentieth century (he passed away in 2012) and was renowned for his work in measuring individual differences. In his chronometric laboratory, he measured thousands of subjects from diverse backgrounds, noted relevant characteristics (like their age and other experiences), and presented his findings in hundreds of lectures, more than 200 articles, and six books. He summarized a lifetime of findings and drew several important conclusions in a book written near the end of his life, *Clocking the Mind: Mental Chronometry and Individual Differences* (2006). He was especially interested in exploring individual differences for school achievement, and he was an early advocate for extra funding and attention targeted to those with lower achievement and slower reaction time.

Following Wundt was his one-time assistant and advocate, James Cattell, who markedly advanced the discipline of psychology and psychological testing, especially in America. Cattell was born in Pennsylvania to a Presbyterian minister, who was himself an educated person and served as the president of Lafayette College, a small postsecondary institution in the town of Easton. Cattell spent his undergraduate years at that same college, where he studied English literature but also showed an ability and peripheral interest in mathematics. After graduating, but while still an undecided young man, he traveled to Europe, where he enrolled at the University of Leipzig. There he met Wilhelm Wundt, and they formed a friendship. Wundt convinced Cattell to study psychology and psychometrics, and, in doing so, Cattell found his calling. He excelled at his studies, and as a nascent researcher under the advising of Wundt, he wrote his doctoral dissertation (in German) on psychometrics. It was a first on the topic, and it set him apart from those studying more usual subjects.

The distinction of his dissertation led to a short-term lecturing appointment at Cambridge University. Soon, however, Cattell went back to Leipzig, where he worked at Wundt's center for psychological research. He continued learning about psychology from Wundt. There, the two of them conducted further research and had joint publications. During this time, Cattell gained knowledge in psychology and psychometrics, as well as valuable clinical experience. After a year as Wundt's assistant, Cattell returned to the United States, first working at the University of Pennsylvania and then moving to Columbia University, where his influence as a psychologist really took off. At Columbia, he founded the Department of Psychology, the first at an American university.

In 1890, he published an important paper entitled "Mental Tests and Measurements," in which he used the term "mental test," probably for the first time (Cattell 1890, Cattell and Poffenberger 1947). In the paper, he both developed a theory of intelligence and proposed a mental test for it that was intended for use with the general population. His theory of intelligence comprises two parts: (1) agile intellect and (2) crystallized intelligence. These parts encompass as many as 100 abilities. He postulated that while everyone has all the abilities, they exist in different amounts and combinations for each person. Thus, everyone has a unique intelligence. Cattell thus set one of the first theories of intelligence.

Following in the tradition of Galton, Cattell's tests of intelligence were a set of mental and sensory tasks, such as squeezing a hand, or the time needed for moving one's hand 50 centimeters or judging ten seconds of time. The combination of tasks of both types (i.e., sensory and judgment questions) illustrates the development of mental testing, moving beyond the purely sensory tasks of Galton but not yet reaching the full intelligence tasks of Binet.

This is an important touchstone on the road to modern intelligence testing. In one important first, the Cattell tasks are intended for the general population and not for a given subset, such as persons with abnormalities. Significantly, Cattell was one of the first psychologists to apply psychometric methods to test development and to make a meaningful score and scale for them.

It seems Cattell was not only an imaginative academic but also a person of strong political opinions. He was opposed to not only the United States' soon-to-be entry into WWI but to Great Britain's involvement, and to the fighting in Europe as a whole. He expressed his opposition vociferously and often, to colleagues, students, and, seemingly, anyone else who would listen. When he was called to duty in the military, he refused. His refusal was not based on religious beliefs (for which he would have likely been excused) but on the more general political notion that no war is justified.

Administrators at Columbia University felt Cattell had gone beyond his scope of employ, and he was dismissed. However, Cattell fought back and sued the university, claiming it was unfair to dismiss him solely due to his political opinions. After an extended trial period, he won his case in court and was awarded a substantial sum for damages. The case became important in eventually establishing tenure as a job protection, thereby protecting teachers at all levels from being dismissed merely for having contrary political opinions.

With the awarded money, Cattell and two others founded a company to develop educational and psychological testing. They called it The Psychological Corporation. The company still exists today and sells standardized tests to schools, some institutions, and to licensed psychologists. It is worth noting that such tests (from this company and most others) are not sold to the public at large for a number of obvious reasons, including security, the training required to administer them, and the knowledge needed to properly interpret the results.

****

Alfred Binet was another individual who promoted the idea that extra educational attention should be targeted to low achievers, although he worked much earlier than Jensen, and, of course, he was in France. Binet was a French psychologist who invented the first practical IQ test, the "Binet–Simon Scale." (Théodore Simon was a student of Binet.) In 1904, the French government was interested in codifying its education laws and wished to bring resources to children who were slow in their school progress. But the French ministry of education did not know a reliable way to measure cognitive abilities uniformly across the country. A governmental commission was formed to study how this might be done, and, in 1904, Binet was asked to become involved by devising a suitable test.

Binet was a well-known scholar in France, particularly for his early work in abnormal psychology. He studied physiology at the Sorbonne campus in Paris, where he developed in interest in mental infirmities. Upon graduation, he accepted a position at Salpêtrière Hospital, a neurological clinic in Paris. There he studied hysterical paralysis, and, following the teachings of his mentor, Jean-Martin Charcot, he began using magnetism (by reversing polarity) to treat patients. For Binet, however, the results from these magnetism experiments were not positive, since he was not actually helping those in his care. He was forced to admit that he was not and that he had made a mistake by following Charcot. It was a setback for his career, but, with his new charge to develop a mental test, he had a chance to regain his reputation.

Table 16.2 *A selection of tasks from Binet's 1905 test (out of thirty in the original)*

Task No.	Task
1	Follow a moving object with the eyes
2	Grasp a small object which is touched
4	Recognize the difference between a square of chocolate and a square of wood
8	Point to objects represented in pictures (e.g., "Put your finger on the window")
10	Compare two lines of markedly unequal length
15	Repeat a sentence of fifteen words
16	Tell how two common objects are different (e.g., "paper and cardboard")
21	Compare two lines of slightly unequal length
24	Produce rhymes (e.g., "What rhymes with 'school'?")
28	Reverse the hands of a clock
30	Define abstract words by designating the difference (e.g., "boredom and weariness")

About the same time, Sigmund Freud also studied hysteria under Charcot, and that experience is credited with influencing Freud to pursue psychoanalysis, which he did independently.

In 1905, Binet designed thirty tasks in pursuit of this aim, thereby developing the first formal intelligence test useful for assessing children. In a departure from the early sensory-based activities of Galton and Cattell (e.g., hearing, depth perception), his tasks were entirely focused on higher mental skills. As illustration, a few of these tasks are listed in Table 16.2.

He began his trial testing by first administering the tasks to his own two daughters! But once he had administered them, Binet realized that there was no prior scale for scoring his tasks—so he developed the first. Subsequently, Binet, now working with Simon, continued to develop his test. In 1908, Binet and Simon published an expanded test that had fifty-eight items and, in a significant change, introduced the notion of mental levels. In 1911, they published yet another revision that had tests for various age ranges, some of which extended clear into adulthood. These Binet–Simon scales constituted the first intelligence test, although the term *IQ* had not yet been introduced. Figure 16.3 shows the levels of his scale—they have been updated many times since his original invention.

Binet Scale of Human Intelligence		
IQ Score	Original Name	Modern Term
Over 140	Genius or Near-Genius	
120 - 139	Very Superior	
110 - 119	Superior	
90 - 109	Average or Normal	
80 - 89	Dull	Dull Normal
70 - 79	Borderline Deficiency	Mild
50 - 69	Moron	Moderate
20 - 49	Imbecile	Severe
Below 20	Idiot	Profound

Figure 16.3 Binet's early intelligence scale with levels

(*Source:* Public domain at http://www.freerepublic.com/focus/fbloggers/3267659/replies?c=1&q=1&page=37)

Then, Lewis Terman, a psychologist working at Stanford University in California in the early years of the twentieth century, revised the Binet–Simon scales into the Stanford–Binet intelligence test. Over the years, it has undergone a number of revisions and is still in popular use today.

Initially, Terman broadened the use of the test and began to explore its utility for examining giftedness in schoolchildren. He began a longitudinal study of characteristics of very bright schoolchildren: the "Genetic Studies of Genius." Working in California public schools, he identified about 1,400 such children and began to track them throughout their lives. The study continues to today; about two hundred of the subjects survive. There are plans to continue following them for the remainder of their lives but not to add new subjects. The study has been criticized for a variety of technical deficiencies, but it remains the largest and longest-running study of intelligence and ability testing ever conducted.

Terman was a central figure in another IQ testing experience, now commonly referred to as the Army Alpha and Beta tests. During WWI, he was asked to administer an intelligence test to Army recruits, with the intention that the information would help identify those who could be trained as officers. The test was administered to more than 1.7 million recruits, making it the first large-scale IQ testing ever done. The test was in two forms: Alpha, which required examinees to read a series of questions, and Beta, which had only pictures and could be administered to those with low reading proficiency or nonreaders. The nature of the assessment demonstrated that IQ testing could be used broadly with a large population.

From this, IQ testing took off in schools, where it was widely employed for the next fifty years. Today, IQ tests are seldom administered across the board to a large group; instead, they are given individually, to one person at a time, and typically after some preliminary evaluation of an individual. Even with these attentions, many thousands of such tests are administered to schoolchildren and adults every year.

While there are several other persons important to the history of mental assessment, I will mention but one more: L. L. Thurstone. The son of Swedish immigrants, Thurstone showed a high intellect at a very early age. He studied engineering at Cornell University, but, after graduating, he developed an interest in the neurosciences. He developed a theoretical underpinning for making choices, whether you are selecting among alternatives on an IQ test or in daily activities. He called it the "law of comparative judgment," and he derived a neurological rationale for it in psychophysics (Thurstone 1927). This work provides a theoretical foundation for modern methods of scoring a test.

In a related advance, Thurstone (working with Charles Spearman) developed a statistical methodology for finding overlap among a set of variables. The overlapping construct is termed a "factor," and the procedure is called "factor analysis" (see Thurstone 1932). As many readers will recognize, factor analysis is commonly employed in psychometric analyses as well as in many areas of psychological research.

****

Thurstone was interested in developing scales for tests and exploring new ways of scoring them. Rather than looking at just the number of test questions an examinee answered correctly, he posited that the *pattern* of responses might yield a more accurate score. Many diverse paths for analyses arise from this perspective, and I will mention one of them (the most popular today) momentarily. He developed his ideas in a classic work entitled *A Method of Scaling Psychological and Educational Tests* (Thurstone 1925). This work is more than simply classic research: Thurstone's ideas have really taken hold in the past half-century, and they are growing in influence in modern testing.

Only a few years later, a Danish mathematician and Fisher protégé named Georg Rasch developed a statistical modeling approach that came to be called the "Rasch item response model." This scaling model focused exclusively on the pattern of an examinee's responses to a set of test questions. Rasch's approach has attracted a group of ardent devotees. By the efforts of many others, the item-pattern response idea has evolved into a more general approach called "item response theory," generally referred to as IRT. Because Rasch worked in the 1960s—a time later than our storyline—I will not elaborate on his other accomplishments. In truth, IRT was not original to Rasch, although he is credited with developing it into a testing model. (As with nearly all other mathematical developments, it has important antecedents, but they do not concern us here.)

Nonetheless, in modern educational and psychological testing, IRT is one of the more important psychometric developments of the previous century, and, therefore, I will describe it briefly. Also, note that, worldwide, most large-scale tests today employ IRT scoring. Some examples include NAEP, TIMSS, college admission tests such as the SAT and ACT assessments, and college general education tests such as College Basic Subjects Examination (CBASE). Many second-level school tests are also scaled with IRT.

To describe IRT (again, this is the briefest of explanations and omits many important details), it is first necessary to understand a bit about test theory. Most educational and psychological tests are classed as following "classical test theory," also sometimes called "true-score theory." In this concept, there is an exactly accurate test score for each examinee, called the "true score." But realize, however, that a true score is only a theoretical possibility, because it is based on the assumption that there is no error in the testing (full validity and reliability). Errors in testing may emanate from many sources, including (1) an imperfect test instrument (e.g., imprecise wording of questions), (2) testing conditions (e.g., distractions during a test administration), or (3) the examinee (e.g., not fully engaged or inattentive during the test's administration). All of these real-world circumstances lead to error.

We have seen error many times through the history of probability theory and the developing quantification mindset, such as when Galileo noted variations in his observations, or with the "true outcome" in Bernoulli's law of large numbers. In all cases, these true scores or true outcomes exist only when the measurement is without any error from any source. By a simple formula, the true score is the observed score (here, termed X) minus any error, as seen in this simple formula:

$$T = X - e$$

In this assessment model, test questions (also called "items," given that not all of them are interrogatives) are typically scored as right or wrong, meaning that every item is scored equally as 1 point. These points sum to the total test score. The error is estimated for the test by a reliability index.

IRT, however, envisions a wholly different perspective on the assessment process. In IRT, the focus is not on capturing the true score (despite the sources of error just mentioned) but on determining an examinee's locale for a "latent trait" on a scale that (theoretically) extends from no ability in the trait to its complete mastery, expressed as minus to plus infinity. For example, for the latent trait of reading ability, the scale would stretch from wholly illiterate on the low end to a degree of complete mastery of reading on the high end. Now, the score is not the total of correct responses by an examinee but is shown in the pattern of responses. For instance, if any examinee responds inconsistently on a test by correctly answering some difficult items but missing easier ones, the score is less than if a consistent pattern of response is observed.

This idea can be seen in Figure 16.4, which is the trace line for a single test item. Note the horizontal and vertical axes. The horizontal axis (Trait level (θ)) shows the trait (say, reading comprehension) ranging from very low at the left to very high on the right. The vertical axis ($P(\theta)$) is the probability (likelihood) of the examinee getting the item correct; it ranges from 0 to 1, as in 0 to 100 percent probability. As shown in the item trace line (its technical form is an ogive—recall we saw earlier that Galton originated this idea and invented the term), the higher the trait level, the more probable it is the examinee will respond correctly. Note that, for very able examinees (high on the trait level), the ogive comes close to 1.0 (a perfect probability), but it never actually reaches it, since, theoretically, no one ever has a 100 percent chance of getting the item correct.

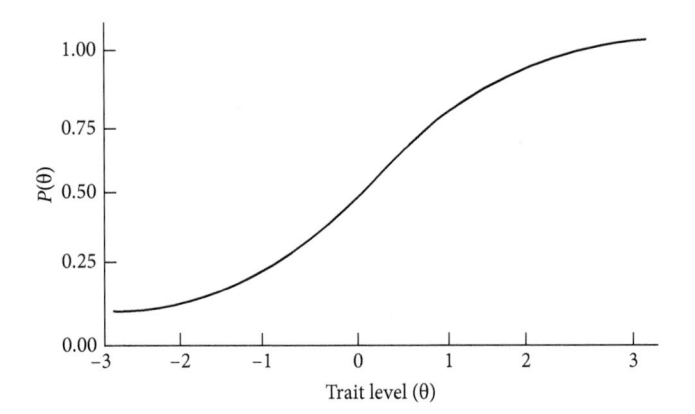

Figure 16.4 Item characteristic curve representing the relationship between an examinee's ability and the probability of a correct response

Also, this illustration makes the name—item response theory—readily apparent. The formula for IRT is illustrated here, just so you can see what it looks like. However, when solved, the formula shows the probability of responding correctly to a test item, with the probability being dependent upon an examinee's ability. To many psychometricians, it is all in a day's work:

$$P(X_{is}=1|\theta_s,\beta_i,\alpha_i,\gamma_i) = \gamma_i + (1+\gamma_i)\frac{\exp[\alpha_i(\theta_s-\beta_i)]}{1+\exp[\alpha_i(\theta_s-\beta_i)]}$$

Psychometrics, as manifest in the science of educational and psychological testing, brings quantification to a truly personal level.

CHAPTER 17

The Arts and the Age of the Chip

This chapter provides the capstone to this book's argument that humankind has adopted quantification as a worldview. Up to now, I have advanced a reasoned (and coherent, I hope) argument of how this came to be; namely, it is primarily explained by the intersection of the circumstances of history with mathematics. Here, we take a look at quantification in two directions not yet addressed. First, we explore quantification across various disciplines such as art and music; and second, I briefly mention quantification in our current twenty-first century, an era I call the "Age of the Chip."

Both areas of exploration are replete with elements of mathematics—and quantification explains the why and the how for each of them. Philosophers use terms like "ontology" and "aesthetics" to describe this framework. I do not advance my argument using the language and conventions of that discipline because I am not a philosopher. Regardless, we will explore how several domains not yet addressed are part of quantification.

Of course, art, music, and virtually all other disciplines are influenced by mathematics, some directly. My point, though, is more fundamental and subtle—maybe even profound—than simply calling attention to something so obvious. By mentioning the connection between mathematics and other disciplines, I hope to illustrate how quantification is our current reality. We perceive things quantitatively without consciously evoking our powers of perception; rather, quantification is part of our mental, automatic response system.

Admittedly, my selection of topics and examples here is arbitrary, but it is decidedly not capricious: all have something to do with advancing the quantification argument. There are many other examples you could choose. The point remains.

****

To start, let us look at quantification in art. Of course, the connection between mathematics and art has been recognized since antiquity. As far back as Plato (and certainly this point was made even earlier by others), aesthetics was suggested in the plane geometry of line and angles. According to Plato, Socrates said:

> I do not mean by beauty of form such beauty as that of animals or pictures… but … straight lines and circles, and the plane or solid figures which are formed out of them by turning-lathes and rulers and measurers of angles; for these I affirm to be not only relatively beautiful, like other things, but they are eternally and absolutely beautiful. (Plato (360–347 bce) 2013)

This declaration reminds us of the elemental beauty in the symbols used to represent numbers and math operations. This is seen throughout all numeric representations, from the basic symbols of elementary arithmetic, moving through those for higher mathematics and into the dimensional spaces of geometry. A few of the many hundreds of such symbols are shown in this line:

$$\neq \quad \leq \quad \infty \quad \div \quad \approx \quad \int \quad \pm \quad \%$$

Here, reflect on their form rather their meaning in mathematics. They are like basic line drawings: simple and pleasing in form. A person with no mathematical knowledge—say, someone who just walked out of a cave and had never seen these symbols—would, nonetheless, likely perceive them as pleasing, and maybe even beautiful.

As an even more beautiful example, the Greek alphabet is the epitome of visual grace in its symbols. Consider these letters as a visual experience:

$$\sigma \quad \beta \quad \eta \quad \lambda \quad \xi \quad \mu \quad \zeta \quad \gamma$$

The aesthetic quality of mathematics formulas is considered by Clara Moskowitz in "Images: The World's Most Beautiful Equations." She writes, "Mathematical equations aren't just useful—many are quite beautiful. And many scientists admit they are often fond of particular formulas not just for their function, but for their form, and the simple, poetic truths they contain" (Moskowitz 2013).

She explains the beauty of Einstein's relativity equation and Lagrange's "standard model" equation for physics, the fundamental model of calculus (recall that, in Chapter 4, we explored one application of an integral calculus formula in the central limit theorem), and even the Pythagorean theorem, and the balanced beauty of a very simple equation which states that the quantity 0.999, followed by an infinite string of nines, is equivalent to 1: that is, $1 = 0.999999\ldots$.

The grace of mathematics is, of course, not only visual, but literary, too. Lord Byron, the eighteenth-century Romantic poet (and father of Ada Lovelace, whom we saw earlier was possibly the first computer programmer by coding actual machine language for Babbage's Difference Engine) completed the intersection between quantification and poetry in these well-known lines, as

> When Newton saw an apple fall, he found
> ....
> A mode of proving that the earth turn'd round
>   in a most natural whirl, called "gravitation,"
> And this is the sole mortal who could grapple
> Since Adam, with a fall or with an apple.
>   (Byron, "Don Juan," "Canto the Tenth," Part I)

The notion of numeric symbols as art is extended, too, by looking at an example of mathematics in art from a long time ago, around the fifteenth century. This one is from the work of Leonardo da Vinci. Of course, we all know his *Mona Lisa* (and, upon seeing the actual painting in its frame on a solitary wall in the Louvre Museum, most are surprised by its small size) and *The Last Supper*, the single most reproduced religious painting of all time.

Further, most also recognize his drawing *Vitruvian Man* (Figure 17.1). This pen and ink drawing shows the anatomical view of two slightly superimposed male figures. They are upright, muscular men, with arms outstretched. A circle surrounds one of them, and a square the other, meant to accentuate the body's proportions, which are mathematically defined by da Vinci in text below the image; for example, he says, "The length of a man's outspread arms is equal to his height...., from the bottom of the chin to the top of the head is one eighth of his height (Richter, 1970, vol. 1, 182)" According to da Vinci, there was a "medical equilibrium" to the parts of the body, reflecting God's symmetry to man's purpose.

Da Vinci was as much scientist as artist; he was especially interested in mathematics, as reflected by the words of one art critic who said, "*Vitruvian Man* is a study of the human form visually perfected through the application of

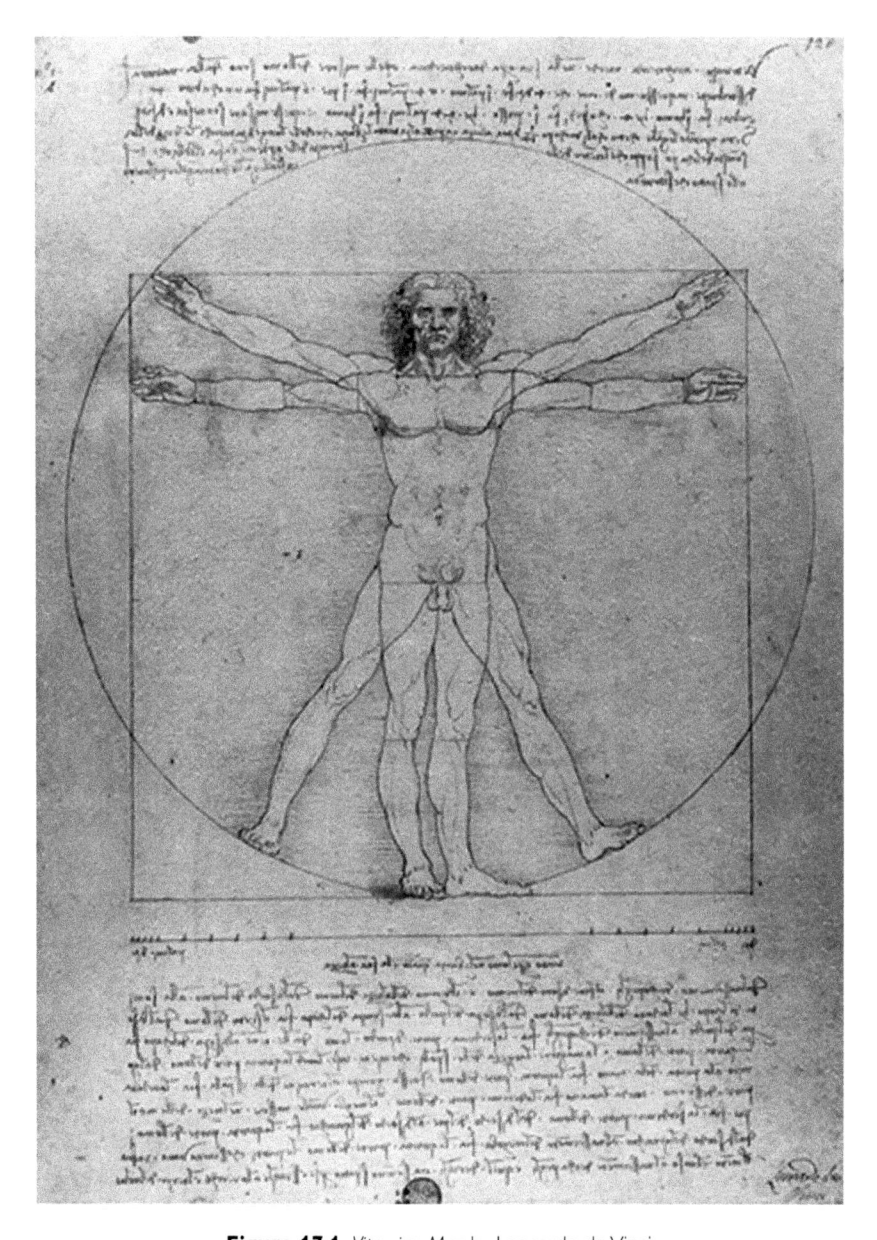

Figure 17.1 *Vitruvian Man* by Leonardo da Vinci
(*Source:* http://commons.wikimedia.org/wiki/Category:Public_domain)

mathematics" (Beyer, 2018). *Vitruvian Man* is the very quintessence of mathematics in art.

Less well known is the fact that da Vinci's illustration appeared in one of the best-known books on mathematics of the day: *De Devina Proportione* (*On the*

Divine Proportion), written by Luca Pacioli sometime around 1498. It is a book on proportions in mathematics, such as the golden ratio. One scholar explains the work as follows.

> In his study of *Divine Proportion*, Pacioli first dealt with current thinking on theology, philosophy, and music in the light of mathematics as expressed in the golden ratio (golden section or golden mean). He turned to a study of Euclid's *Elements* and then to a consideration of regular and dependent polyhedrons. (Codices Illustrates 2018)

Da Vinci provided more than sixty line drawings, each to illustrate a particular proportional view of some mathematics. Figure 17.2 displays two pages from this work, both of which show a three-dimensional object, called an "icosidodecahedron," that is mathematically defined. The book is replete with these interesting drawings.

Pacioli was not himself an artist but a Franciscan teacher of mathematics, especially interested in both architecture and how money was recorded in business. In fact, he introduced double-entry accounting, and because of this invention, he is sometimes called the "father of accounting." It is known that he taught da Vinci elementary geometry; da Vinci, in turn, informed Pacioli of the application of geometry to art and architecture. Although Pacioli is listed as the book's only author, the book is clearly the result of a close collaboration by these two Renaissance scholars.

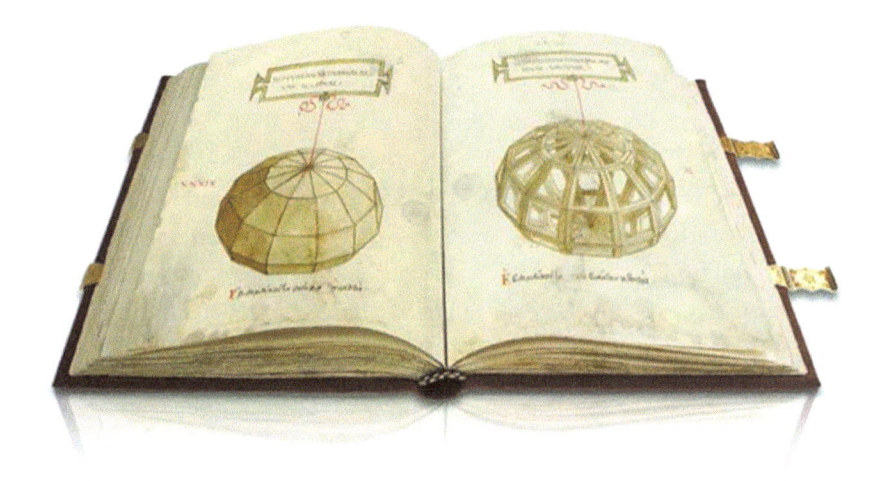

Figure 17.2 Dimensional drawing by da Vinci in Pacioli's *De Devina Proportione*
(*Source:* http://commons.wikimedia.org/wiki/Category:Public_domain)

As important as the work of Pacioli and da Vinci may be, the relationship between art and mathematics can be seen in many other art periods and styles, too, such as cubism, which Pablo Picasso (along with Georges Braque) introduced to the art world around 1905. This style emphasizes angles and applies a hard, one-dimensional structure to the human form. It was a wholly new perspective on art (rejecting the impressionism of predecessors) and brought a fresh worldview reflecting quantification in this discipline.

During his life, Picasso was the archetypal Parisian artist, always nonconformist and bohemian during the Gertrude Stein expatriate era. Like many other artists, he was fascinated by the world of mathematics. For him, though, mathematics was not a discipline for studying formulas and seeking solutions, but rather an art form unto itself—a kind of unexplored reality that became an outlet for his creativity. And, for Picasso, there was always ever more creative work. In what is an astonishing level of productivity, he produced a huge number of works—at least 20,000 pieces of art, both in painting and in ceramics. I have a friend who owns a small, simple line drawing by Picasso. My friend was disappointed to learn that his prized artwork was worth only a modest amount, due to the sheer volume of similar Picassos.

Picasso enjoyed a lifelong friendship with Henri Matisse, who had an opposite view of art and deliberately steered his work away from the lines and angles of cubism. They often dueled through their art, with the two geniuses of the art world working in contrapuntal effort. Together, though, they show an influence of mathematics in art. Unconsciously, quantification became a part of their art lives, and it leaves its impression for us to interpret.

A plethora of books and studies highlight the connection between mathematics and art. A fairly recent pertinent work is titled *Mathematics and Art* (Gamwell 2015). The book is both a coffee table piece and—more relevant to us—a scholarly treatise that explores associations between mathematics and art, looking at their relationship from antiquity to the Enlightenment during much the same period as our story. Concepts such as number form and infinity are explained as mathematics and then illustrated in one sumptuous image after another. The book's link to mathematics is further highlighted by having a noted astronomer, Neil deGrasse Tyson, contribute the foreword.

****

Almost instinctually, we realize there is a strong relationship between mathematics and music. Common to every genre—classical, blues, country, rock, and nearly all others—is the fact of an underlying mathematical structure. All of them

THE ARTS AND THE AGE OF THE CHIP | 309

can be quantitatively analyzed. Further, virtually any sound within the human range of hearing (and far beyond) can be reproduced in intensity, duration, frequency, and timbre with relatively simple electrical engineering, at first with analog circuits but now by digital signal processing. It is interesting (and we are thankful), however, that we can immediately detect whether a sound is produced by electricity or by an instrument—regardless of whether the instrument produces its sound by percussion (like a piano or a drum), with wind (such as blowing directly into a horn or over a reed, as in a clarinet), or by using strings (like a violin or a guitar).

For the interested musicologist or mathematician, music as engineered sounds governed by the laws of physics is explored in a two-volume scholarly work titled *Musimathics: The Mathematical Foundations of Music* (Loy 2007).

Music notation illustrates the mathematics of sound. Consider the single measure shown in Figure 17.3, from J. S. Bach's *The Well-Tempered Clavier*. This measure contains only six principal notes of rhythm (the others are complementary), yet it can be described mathematically in at least a dozen different ways. One of its descriptions is with simple addition, while another "solves" it with a set of complex linear equations. It is simultaneously simple and complex. It is beautiful to realize its sophistication, and even more luxurious to listen to it.

Following this systematic arrangement of tones on a musical scale is the more complete music theory, which is wholly mathematical. Much of music theory can be summarized in the "circle of fifths." One depiction of the circle of fifths is displayed in Figure 17.4.

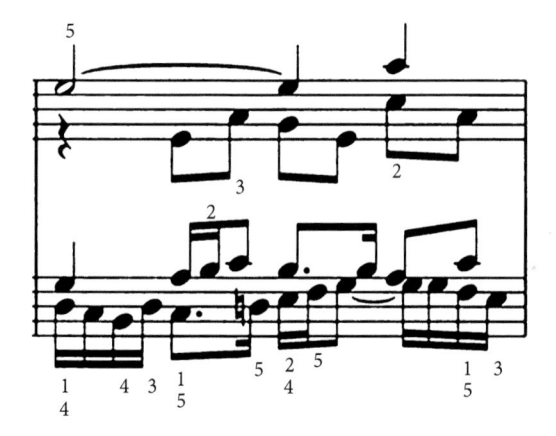

Figure 17.3 A measure from Bach's "Fugue no. 17 in A-flat major, BWV 862" from Book 1 of *The Well-Tempered Clavier*

(*Source:* http://commons.wikimedia.org/wiki/Category:Public_domain)

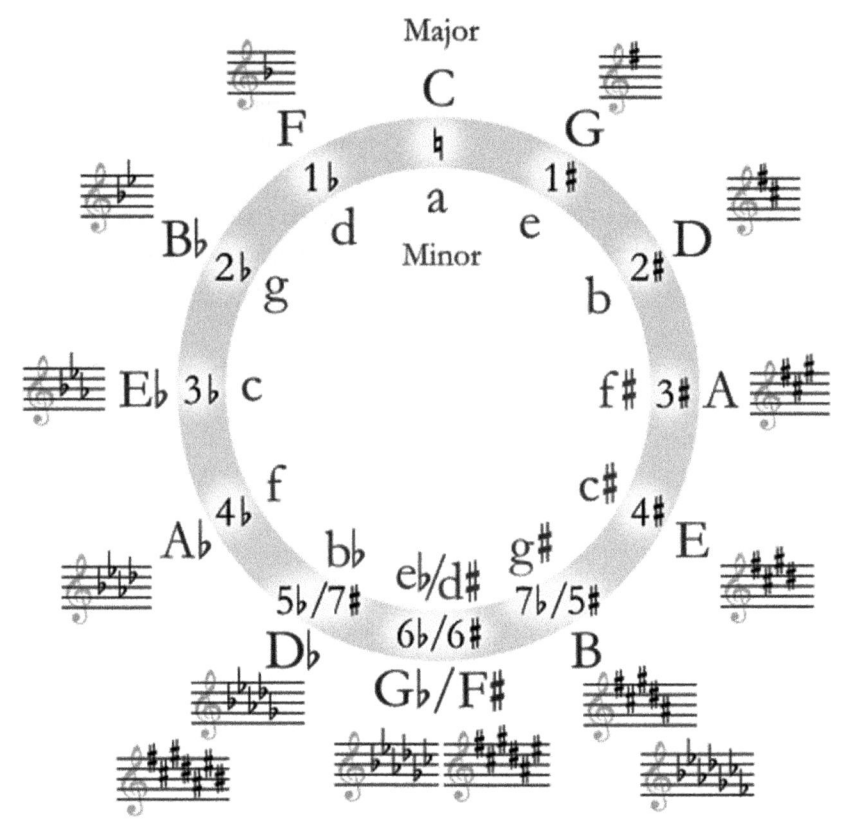

Figure 17.4 The circle of fifths in music theory
(*Source:* http://commons.wikimedia.org/wiki/Category:Public_domain)

Some of music's structure—but certainly not all of it—is organized into major and minor keys. The circle of fifths is a simple depiction of the relationship between them. When a major and minor key use the same sharps and flats, they are said to be in the same "signature." The circle of fifths shows this signature: a major key is on the outside, and its relative minor key is correspondingly on the inside. This musical representation can be read in either direction, clockwise or counterclockwise. Jazz musicians generally read it counterclockwise.

Most often, composers follow the pattern of the circle of fifths, but there are notable exceptions. Serial music, for example, deliberately violates the traditional pattern. Igor Stravinsky's ballet score *The Rite of Spring* is a jarring example of this. Appreciated for its sophistication, it is nonetheless not for the inexperienced listener. It brings to mind the quote about opera music by Mark Twain (in his *Autobiography*): "Wagner's music is better than it sounds!"

There is much more to music theory than is shown in the circle of fifths, and nearly all of it relates to mathematics. For example, you are probably aware of the Pythagorean theorem. The name of the ancient mathematician and philosopher Pythagoras occurs in music, too, in a special circumstance in music theory called the "Pythagorean comma." It is the interval (hence, "comma") between two enharmonically equivalent notes, as between C♯ and D♭. All of the black keys on a piano keyboard are enharmonically equivalent, defining that small space. Imagine (or actually look at) a piano keyboard. Here, we see the Pythagorean comma in all of the black keys.

Apparently, too, when tuning a musical instrument, the Pythagorean comma can be calculated to find an angle for harmonic tones and has a certain harmonic frequency ratio, and is defined as the ratio of twelve "just perfect" (theoretically perfectly tuned) fifths and the seven octaves:

$$\frac{twelve\ fifths}{seven\ octaves} = \frac{\left(\frac{3}{2}\right)^{12}}{2^7} = 1.013643264 \approx 23.46\ cents$$

The "cent" is a log value used in music to convey meaning for these very small values. Of course, cent is the root word for one hundred, as in centimeter, percentile, centurion, and centipede. For us, this mathematical underpinning of music is yet another idea that brings us ever closer to quantification, extending to and influencing every aspect of every person's worldview: *Everyman*.

Of course, literary descriptions of music and mathematics abound. I can think of none better than Leo Tolstoy describing its power in his story *The Kreutzer Sonata*. Here, he gives a literary description of music that captures an emotion felt by all of us:

> How shall I say it? Music makes me forget my real situation. It transports me into a state which is not my own. Under the influence of music, I really seem to feel what I do not feel, to understand what I do not understand, to have powers which I cannot have.... And music transports me immediately into the condition of soul in which he who wrote the music found himself at that time. I become confounded with his soul, and with him I pass from one condition to another.
>
> (Tolstoy and Katz 2008, 134)

I imagine we can each relate to Tolstoy's words, regardless of whether listening to Chopin, Eric Clapton, or an old man with a penny whistle.

****

Today, we see quantification most obviously in the technology of integrated circuitry in microchips. These little silicon and copper devices are so much a part of our daily activities that we do not even think of them as unusual or remarkable. Among just my kitchen appliances, I count more than a dozen that are controlled by microchips: even the toaster has a microchip built into it. (As a personal aside, I recall my own father who, drawing from his Swedish heritage, always referred to our refrigerator as an "ice box"—persons of his generation could not imagine this common kitchen necessity eventually having a microchip.) My car, too, probably has another dozen (at least), and individually some them are many times more sophisticated than the early programming language used by the Apollo Guidance Computer that Apollo astronauts relied on to carry them to the moon and back. Clearly, we live in the "Age of the Chip."

Further, unlike almost any other invention in history, the chip can be found worldwide. Even those living in countries cited in the latest United Nations' *Human Development Report* as the most repressive and poorest in the world—such as the communist nations of Cuba, Libya, North Korea, and Somalia, as well as Russia's Chechnya (United Nations Development Programme 2015)—have the microchip in their daily experiences. While most persons living in these countries have far fewer electronic devices than we consider common, their water treatment plants, electricity-producing sources, and essential aspects of transportation and medicine all rely on the chip.

If you view these facts as humdrum and promptly ask, "So, what else is new?"...well, that response is itself telling, because the microchip is a relatively recent invention, around just since the 1960s. The two pioneers usually credited with inventing the technology are Jack Kilby and Robert Noyce. In 1959, each of them was granted a patent for his remarkable invention: one to Kilby for a miniaturized electronic circuit, and another to Noyce for a silicon-based integrated circuit. (Noyce went on to found Intel, which is the world's largest producer of microchips, far ahead of Samsung, Taiwan Semiconductor, Qualcomm, and a host of smaller companies.) Some places where the chip was first used in a practical application (also in the 1960s) were the US Air Force's Minuteman II missile and NASA's Apollo project.

Recounting more examples of the technology quickly overplays the point, so I will not do so—you know them already. But I do call attention to one area where the technology of integrated circuits might become our new reality: in artificial intelligence, commonly known as "AI."

A precise definition of AI is hard to come by, doubtless due to its constantly changing status. We may think it as the capacity of a machine to perform a task that would require some degree of intelligent reasoning if it were performed by a human or an animal. This broad definition includes almost any activity, from simple mechanical grasping of an object to some level of logical reasoning. One feature of AI that makes it distinct from a robot is that AI employs a level of decision-making independent from human direction. A robot, on the other hand, is merely (!) able to carry out a mechanical activity through direct programming, although often even that requires a level of adaptation, such as when it spots a defective part on a manufacturing line (e.g., comparing each can of soup with a picture of what that can should look like).

Early efforts at AI had machines playing checkers, wherein a fairly small number of binary choices exist. But they quickly outgrew that. Today, they calculate conditional probabilities of Bayesian thinking so as to make decisions in many thousands of real-world circumstances, as in disease diagnosis; manufacturing processes, as well as in many applications in physics, engineering, and statistics. One example affects people more significantly than they may be aware: AI is at the very heart of control systems for the modern autopilot mode in airplanes.

Two early pioneers in the field of AI were Marvin Minsky and John McCarthy (who coined the term "artificial intelligence"). Both were professors at the Massachusetts Institute of Technology (MIT; McCarthy also taught at Stanford and Princeton). From the very beginning, they envisioned a machine that could perform tasks that humans would consider to be dependent on "common sense," such as deciding when to open a door or to not place your hand on a hot stove. They even wanted a machine that would appropriately say, "Good morning." Of course, today, even a smartphone can do that—but, *meaning it* is still, apparently, some ways off. As humans, we say, "Good morning," and imply a myriad of meanings and emotions. For the moment, then, humans are ahead.

Regardless, today AI is very advanced (relative to, say, the average home computer). These systems have "brains" with supercomputing power that can perform many thousands of calculations every second. The fastest computer today (at the time of this publication) is the Titan Supercomputer at the Oak Ridge National Laboratory of the US Department of Energy. However, a new version of the Titan Supercomputer is now being built that will have a theoretical peak performance of more than 20 petaflops: over 20,000 trillion calculations per second. Recall that, less than 200 years ago, Babbage's early Difference Engine could produce only a few calculations by each hand crank.

AI generally employs probabilistic reasoning for uncertain information, where things are known only incompletely, and the available information is often contradictory. It uses Bayesian networks, hidden Markov modeling, and obtuse technologies called Kalman filters and particle filters, as well as the more expected ideas of decision theory and utility theory.

Most significantly, AI systems learn progressively; that is, AI advances its knowledge on the basis of its own prior learning, independent of human intervention. We do not know what this means for its future capability or even how we should respond. But, given the progress made already, such infinite learning is already upon us—awaiting only faster processing units and mechanical advancements, such as moving electricity itself ever faster. In reality, developments in AI are coming along at a quicker pace than nearly anybody imagined. Within just the past five years, the progress has been astonishing. It is unsettling to think about what progress may take place in the next five years.

A Swedish futurist named Nick Bostrom leads a school of thought which holds that the near-fact of a computer having a "superintelligence" poses a supreme danger to humanity. He says, "The first ultraintelligent machine is the last invention that man need ever make, provided that the machine is docile enough to tell us how to keep it under control" (Bostrom 2014, 20). This is sobering to think—especially coming from Bostrom, a philosophy professor at Oxford University and founding director of the Future of Humanity Institute and of the Program on the Impacts of Future Technology.

Bostrom carefully elaborates his ideas with percipient insight in an engaging book titled *Superintelligence: Paths, Dangers, Strategies* (Bostrom 2014). The book is endorsed by Bill Gates and Elon Musk. I wonder how we will view his prescience in ten years.

Another brilliant Swedish intellect, Max Tegmark—an MIT physics professor and cosmologist—explores the same questions in his recent book *Life 3.0: Being Human in the Age of Artificial Intelligence*. He titles his first chapter "Welcome to the Most Important Conversation of Our Time" (Tegmark 2017). I believe Tegmark has posed the question correctly, and with the appropriate gravity implied.

Tegmark explores the outermost extension of quantification as a worldview in his role as cosmologist. In a physics-cum-philosophy argument, he posits a universe that is entirely mathematical. Arguing an extreme point of view, Tegmark says of his parallel universe, "There is only mathematics; that is all that exists" (Tegmark 2014, 124). Recall that, in Chapter 9, I described the numerical relationship between the nautilus shell and Bernoulli's logarithmic spiral and, similarly, for the mathematically based music of Mozart. The point made there is that many things in nature can be described mathematically because, in essence, they have a dimensional structure. Certainly, Tegmark would agree. Most of this current chapter is an elaboration of the same point, as we consider mathematical structures in art, music, and elsewhere.

In his thinking, Tegmark goes much further than to merely suggest the influence of mathematics. He postulates that there is another universe beyond our own, but not just one wherein all contents have a mathematical structure; rather, his is a universe that is itself mathematical, wholly and exclusively. Hence, his quote in the previous paragraph: the only reality is of math itself.

Certainly, Tegmark would embrace Galileo's opinion that "the universe...is written in the language of mathematics" (quoted in Drake 1957, 238). But I do not think Galileo meant it in the same literal sense as Tegmark. Rather, Tegmark theorizes that there are four levels to these universes, three of which have been explained by others. He says that he merely adds the fourth: the mathematical universe. Further, it is a level that subsumes all previous levels of universes. In this, he offers a taxonomy for levels of alternate universes beyond just this one universe that we can observe, at least in part.

He does not deny the existence of accepted physical laws or of an infinite ergodic universe. For instance, he believes in the Big Bang theory, which suggests a start for our universe of about fourteen billion years ago. But, to Tegmark, that explanation is not complete enough. There are other universes, as well. In fact, there exist an infinite number of them, something he calls a "multiverse." Accordingly, he surmises that there must also be another Earth out there somewhere:

> But if space goes on forever, then there must be other regions like ours—in fact, an infinite number of them. No matter how unlikely it is to have another planet just like Earth, we know that in an infinite universe it is bound to happen again.
>
> (Quoted in Frank 2008)

This means, too, that if he is correct, you have an exact doppelgänger (fully identical twin) in this parallel universe. This is a heady thought, indeed.

Apparently, cosmologists are split on whether there are parallel universes or not. Tegmark is the leading proponent for the multiverse side of the argument, and he has attracted a number of followers, both cosmologists and others, in a sort of cult following. Tegmark's "maths-is-a-reality" theory is his idea alone among astrophysicists. His ideas seem as much philosophical ontology as cosmology or physics because they focus on the concept of reality itself. For his "out-there" ideas and his engaging personality, some call Tegmark "Mad Max." I would say he is a twenty-first-century Allen Ginsberg (a San Francisco beat poet of the '60s; he also had a cult following, but his was for counterculture).

Tegmark does proffer some sophisticated physics arguments as support of his ideas. He explains them in a number of technical papers and a book for the lay audience titled *Our Mathematical Universe: My Quest for the Ultimate Nature of Reality* (Tegmark 2014). I can recommend it as an engaging read.

****

Through these various disciplines, we realize more fully how we view things and what we accept as true—the Truth described in the early chapters of this book. The pieces to this worldview with quantification at its core are fully drawn together in the next—and final—chapter.

CHAPTER 18

The Sum of It All

Now—after 150 years of history with concomitant mathematical accomplishments—quantification is firmly set, ensconced in our minds and thoughts. It is our very view of life, our *Weltanschauung* (worldview). Since my opening words in Chapter 1, we have traced how this worldview came about.

But, explaining the ineffable is difficult. The story of quantification is complex and occasioned by dynamic events, and only truly known thorough a carefully reasoned narrative that follows a coherent argument. No glib explanation or simple description can give it due. I hope that, through this telling, I have succeeded.

Most of the events of this story happened during a relatively short time frame in world history. With antecedents noted throughout the eighteenth century, our story begins (roughly) late in that period with events leading up to the Congress of Vienna (ending the drawn-out French Revolution and the era of Napoleon) and was not entirely complete until the outbreak of WWI in 1914: the so-called long century. This 130-year-long period brought forth a torrent of breathtaking advancements in mathematics and statistics, and, in particular, the ability to measure uncertainty through the developments of probability theory.

We have seen that the contemporaneous history provided an encouraging context for the intellectual advances. As I have said throughout the previous chapters, the astounding mathematical inventions happened because they *could* happen. History gave a kind of tacit assent to the quantitative developments. Thus, we can only understand the mathematical accomplishments by knowing their historical context: social, political, educational, and cultural. This is just what we have done in following our route to quantification.

The Error of Truth. Steven J. Osterlind. Oxford University Press (2019). © Steven J. Osterlind 2019.
DOI: 10.1093/oso/9780198831600.001.0001

Further, the developments were mostly attributed to a relatively few (about fifty or so) remarkable men and women working at the time. As we have seen in chronicling their achievements, our principals were individuals of almost unimaginably high intellect. It is much to our good fortune that they lived in the same era and worked on common problems in probability and statistics—a simply remarkable fact.

Quantification has come into our lives so elementally and indispensably that we have changed fundamentally. It is how we view things, and it is what we accept as true. Our Truth is explained by quantification. It defines who we are.

The consequence of this new perspective has been for us to change our behavior, in both interacting with our environment and in our relationships with other people. We have a worldview defined by our inclination to view natural and everyday phenomena through a lens of measurable events. It is a rethought ontology. We see things more expansively in both time and space, and, more boldly, in terms of impact.

As we have seen through this story, our worldview is now fully one of quantification, and it holds primacy in our lives, bringing us to an entirely new perspective on what we know about the world and how we know it. By it, we are changed in what we each think about ourselves.

****

Let us consider a "thought experiment." A thought experiment is an investigation into a scientific question that is carried out only in the imagination. Sometimes the question is even stated as a hypothesis, but there is no data or empirical verification. Nonetheless, it is more than mere conjecture, because a methodology is employed, even if only in the scientist's imagination. This is informed speculation. Thought experiments can be important and useful because they can carry ideas forward when it is not possible to actually perform a physical experiment. Such thought experiments are common in physics, theoretical mathematics, and philosophy. People who are known to have carried them out include historical luminaries like Galileo, Descartes, Newton, and Leibniz, as well as dozens of modern scholars.

One of the most famous scientific experiments in history may actually have been a thought experiment. In 1589, Galileo supposedly dropped cannonballs of different weights from the top of the Leaning Tower of Pisa to see whether they would fall at the same rate. But we have no record to verify that Galileo in fact performed the experiments other than as thought experiments.

The only written account is given by someone we saw earlier, Lagrange, in his *Mécanique Analytique*. There, he says Galileo described his actions as "rhetoric" in *De motu* (*Of Motion*), which is itself an unpublished text. Thus, Galileo's actually having climbed the steps of the tower and the rest of it is unverified. We know, too, from other contexts, that Galileo had conducted prior thought experiments. Therefore, from the information we have, Galileo's experiments with cannonballs on the nature of gravity—while important and certainly famous—may not have physically taken place at all, but instead were his "thought experiments."

Here, I'll carry out two thought experiments. The first is to imagine that quantification had not taken place and we had not transformed our worldview to it. In the second, I imagine our current quantified worldview but project it into the future, to speculate on how we might evolve in a next step.

Thought Experiment No. 1: "Quantification has not taken place and we have not transformed our worldview to it."

In this thought experiment, we are in present times but without quantification as a worldview. I am not suggesting that we imagine ourselves in the eighteenth century or at a time before the beginning of quantification. We live today, even in this thought experiment. Significantly, however, we do not have reliable predictions, regressions, conditional probabilities, Bayesian thinking, the method of least squares, or even a means to calculate correlational relationships.

As a consequence, nearly all of the dramatic inventions of the nineteenth and twentieth centuries would not exist or would be rudimentary versions of themselves, something like Pascal's wooden Pascaline calculating machine. As hard as it is to even imagine, we would not be living in the Age of the Chip. Cars, television, telephones, and many other modern inventions would still have been invented, but even those that require electricity (which we would have) would work only as mechanical devices; they would not perform with their current chip-enhanced capabilities. And, in this scenario, the Internet would not exist.

If this sounds like regressing to the Stone Age, remember that this world was *everyone's* reality only fifty or sixty years ago. An interesting side question, apart from our thought experiment, is to ask someone old enough to have lived in those days—say, most folks over fifty years of age—whether he or she is happier today than before these inventions. This would be a complex consideration, no doubt.

Clearly, humankind is greatly advantaged with these inventions. Consider developments in disease diagnosis and treatment, for example. But not all modern advances are ipso facto beneficial. Along with them come some deleterious effects. For instance, psychological stress is increased because of social media, wherein the anonymous sender of information can describe themselves as living the perfect life, leaving the receiver to worry that he or she is missing out on something important. And private information, everything from the personal details of your life to confidential financial records, can be covertly obtained. These concerns are indeed serious.

Without a quantified worldview, we would rely upon habits, incomplete (or unknown) information, superstition, and whims as the means to make our decisions, big or small. When looking back at something, we would be at a loss to understand how events might be interrelated in a meaningful way. Fate would still play a large role in our thinking. Independence of thought would be less. Actions taken would be more haphazard.

I conclude this thought experiment by deducing that quantification—the worldview that gives us a numerical perspective on all things—is, overall, a huge plus in our lives. It brings us advantages and conveniences that we would not have otherwise had. It is hardly an absolute good, but certainly more so than not.

Thus concludes Thought Experiment No. 1.

> **Thought Experiment No. 2:** "From our current quantified worldview, project into the future to speculate on how we might evolve in a next step."

In this thought experiment, our worldview is our present version: one in which we have quantification at its core. We view things from a perspective of knowing that correlations and probability are calculable events, although rare is the circumstance that we would actually have the data or the inclination to do so. Still, we base decisions on what we may believe to be a correlational relationship or on the probability of a perceived outcome. I am mindful that this thought experiment is distinct from making predictions about the future; hence, I will avoid sliding into the projections of a futurist. Serious futurists attempt to make predictions of what could happen someday, which could be either close at hand or far off. Of course, they have a lot to offer in their conversations, and we can learn much from them.

Here, I am less ambitious. I merely project quantification as our worldview into the future.

With quantification carried forward, then, I expect three effects. First among them is the continuance of developments in statistics and probability theory, which, although likely, is by itself hardly a remarkable observation. However, it is important because it naturally leads to the conclusion of their yielding more precise and reliable predictions. In other words, we will know the outcomes of quantification more exactly, and we will have greater confidence in them.

Possibly, maybe even likely, the future developments will allow us to move predictions into fields that are now only incompletely known. For instance, we may be able to reliably predict natural disasters, such as earthquakes, so that emergency measures can be started sooner, even ahead of the event. Or, possibly, our predictions into human behavior will be more accurate. For instance, maybe we will be able to better predict characteristics for harmful addiction and dependency of all kinds and bring useful, effective treatments to those at risk. Clearly, this would be good for humankind—but, more importantly, it would benefit each particular person, possibly in life-changing ways.

Second, with the procedures for quantifying events and behaviors becoming more advanced and moving into new areas, the entirety of quantification as our worldview will become more foundational to our thinking and decision-making—that is to say, we will not erode in quantification as an outlook.

This prediction leads to my third conclusion in this thought experiment: namely, our increased reliance upon probability and statistics as a part of our routine behaviors. We will exist ever more in the realm of quantification. In fact, for as far as I can anticipate, we will depend more and more upon our worldview being that of quantification.

Thus concludes Thought Experiment No. 2.

****

A quantified worldview allows for us to achieve a peace in our lives, following our notion of what we believe to be true—our Truth. In the end, we live with eternal hope, even optimism, for we know that forever things will always get better. I mean this statement in the best sense possible. We are hopeful beings—and we individually strive to be better.

Such a life-sustaining peace is seen in one of the most beautiful poems in English, given us by Shakespeare through his character King Lear (Figure 18.1 shows a page from the only surviving manuscript in Shakespeare's handwriting). At this point in the play *The Tragedy of King Lear*, the king is old, frail, and beaten. Knowing he will spend the rest of his life in prison, he has a vision of true happiness:

> And pray, and sing, and tell old tales, and laugh
> At gilded butterflies, and hear poor rogues
> Talk of court news; and we'll talk with them too,
> Who loses and who wins, who's in, who's out;
> And take upon's the mystery of things,
> As if we were God's spies: and we'll wear out
> In a walled prison, packs and sects of great ones
> That ebb and flow by the moon.
>
> (*King Lear*, act 5, scene 3)

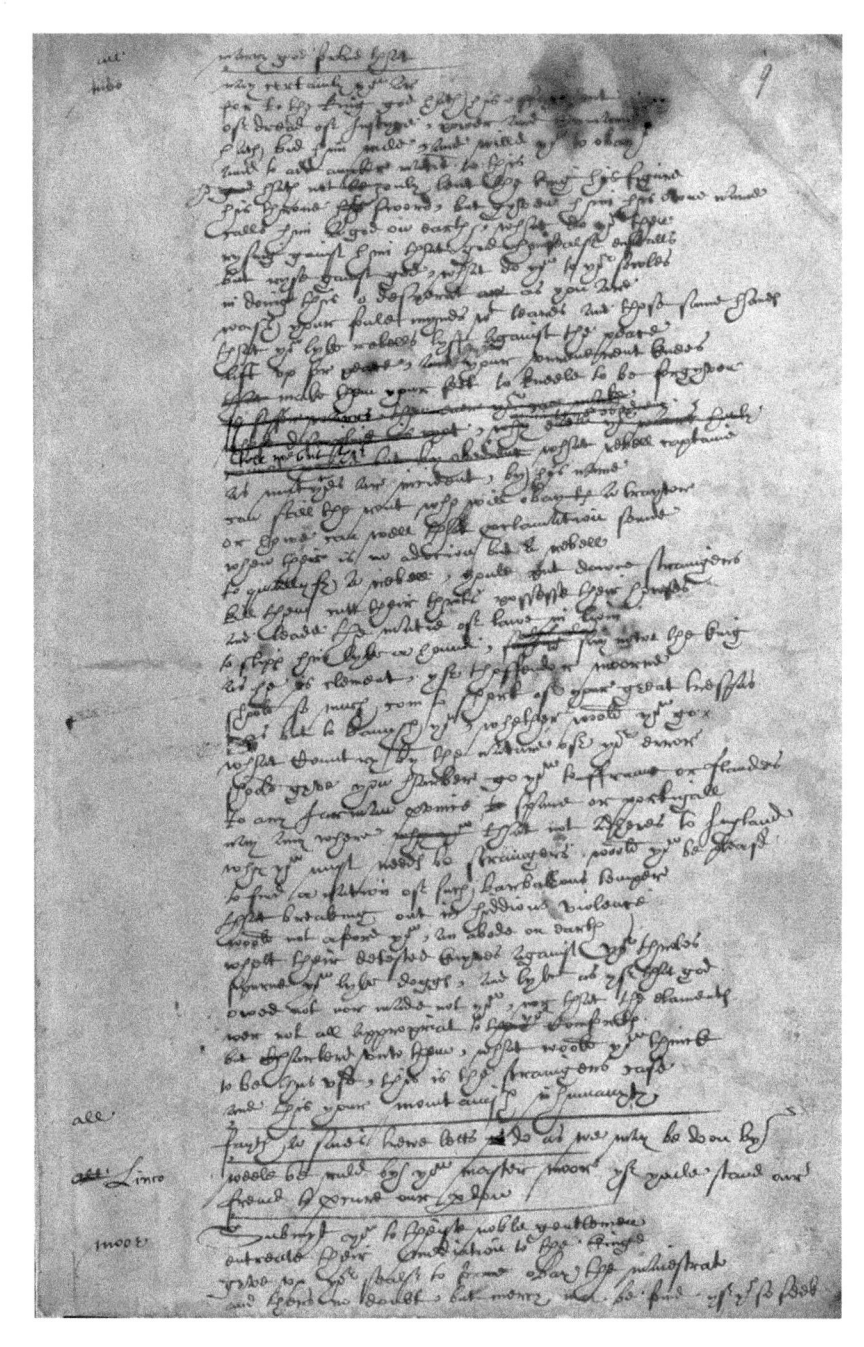

Figure 18.1 Page from only surviving manuscript in Shakespeare's handwriting, sometime between 1603 to 1606

(*Source:* http://commons.wikimedia.org/wiki/Category:Public_domain)

Dominus illuminatio mea ("The Lord is my light")
Motto of the University of Oxford, and opening words of Psalm 27

BIBLIOGRAPHY

American Educational Research Association, American Psychological Association, and National Council on Measurement in Education. 2014. *Standards for Educational and Psychological Testing*. Washington, DC: American Educational Research Association.

American-Israeli Cooperative Enterprise. 2018. "Israel Modern History: Offering Albert Einstein the Presidency of Israel to Albert Einstein (November 17, 1952)." http//www.jewishvirtuallibrary.org/offering-the-presidency-of-israel-to-albert-einstein, accessed September 3, 2018.

Andriesse, C. D. 2005. *Huygens: The Man Behind the Principle*. Cambridge: Cambridge University Press.

Babbage, Charles. 1834. "Letter to James Stewart, July 16, 1834." The Babbage Papers, vol. 7. London: British Museum.

Bayes, Thomas, and Richard Price. 1963. *Facsimiles of Two Papers by Bayes: An Essay Toward Solving a Problem in the Doctrine of Chances, with Richard Price's Forward and Discussion; and, A Letter on Asymptotic Series from Bayes to John Canton*. New York: Hafner Pub. Co. Original edition, 1763.

Bayley, Melanie. 2009. "Alice's Adventures in Algebra: Wonderland Solved." *New Scientist* December 16, 2009, http://www.newscientist.com/article/mg20427391-600-alices-adventures-in-algebra-wonderland-solved/, accessed September 3, 2018.

Bell, Eric Temple. 1956. "Gauss: The Prince of Mathematicians." In *The World of Mathematics; A Small Library of the Literature of Mathematics from A'h-mosé The Scribe to Albert Einstein*. Ed. James R. Newman, 295–339. New York: Simon and Schuster.

Bell, Eric Temple. 1986. *Men of Mathematics* (1st Touchstone edn). New York: Simon & Schuster.

Bergman, J. 2002. "Did Darwin Plagiarize His Evolution Theory?" *Journal of Creation* 16(3): 58–63.

Bernoulli, Daniel. 1954. "Exposition of a New Theory on the Measurement of Risk." *Econometrica* 22 (1):23–36.

Bernoulli, Jakob. 1968. *Ars conjectandi, opus posthumum*. Bruxelles: Culture et civilisation. Original edition, 1713.

Bernoulli Society. 1990. "Uppsala Abstracts." Abstracts of Papers Presented in Uppsala, Sweden, August 13–18 1990, Montréal.

Bernstein, P. L. 1998. *Against the Gods: The Remarkable Story of Risk*. New York: John Wiley & Sons.

Beyer, C. 2018. "Da Vinci's The Vitruvian Man: History & Golden Ratio." http://study.com/academy/lesson/da-vincis-the-vitruvian-man-history-golden-ratio-quiz.html, accessed September 4, 2018.

Bortkiewicz., Ladislaus von. 1898. *Das Gesetz der kleinen Zahlen*. Leipzig: B. G. Teubner.

Bostrom, Nick. 2014. *Superintelligence: Paths, Dangers, Strategies*. Oxford: Oxford University Press.

Box, Joan Fisher. 1981. "Gosset, Fisher, and the *t* Distribution." *The American Statistician* 35(2): 61–6.

British Government. 2018. "Project Britain: British Life and Culture." http://projectbritain.com/education/tests.html, accessed January 15, 2018.

Brunt, David. 1931. *The Combination of Observations* (2nd edn). Cambridge: University Press. Original edition, 1917.

Buffon, Georges Louis Leclerc. 1780. "Natural History General and Particular, by the Count de Buffon, translated into English with notes and observations." Edinburgh: William Creech. https://writersinspire.org/content/natural-history-general-particular-count-de-buffon-translated-english-illustrated-above-26-1, accessed September 6, 2018.

Burns, Ken, et al. 2004. *The Civil War*. [Video recording (in five disks)]. Burbank, CA: PBS Home Video: Paramount Home Entertainment.

Cain, Harel. 2017. "C. F. Gauss's Proof of the Fundamental Theorem of Algebra." http://math.huji.ac.il/~ehud/MH/Gauss-HarelCain.pdf, accessed September 4, 2017.

Carroll, Lewis, and John Tenniel. 1939. *The Complete Works of Lewis Carroll*. London: Nonesuch Press. Original edition, 1903; The Hunting of the Snark and Other Poems and Verses.

Cassirer, Ernst. 1951. *The Philosophy of the Enlightenment*. Trans. Fritz C. A. Koelln and James P. Pettegrove. Princeton: Princeton University Press.

Cattell, J. McKeen. 1890. "Mental Tests and Measurements." *Mind* 15:373–81.

Cattell, James McKeen, and Albert Theodor Poffenberger. 1947. *James McKeen Cattell, 1860–1944. Man of Science*. Lancaster, PA: Science Press.

Chesterton, G. K. 1994. *Orthodoxy, The Wheaton Literary Series*. Wheaton, IL: H. Shaw Publishers.

Codices Illustrates. 2018. "The Text of De Divina Proportione." http://www.codicesillustres.com/pdf/De_Divina_Proportione.pdf, accessed March 18, 2018.

Computer History Museum, and Microsoft Research. 2012. "The Babbage Difference Engine #2 at Computer History Museum." Computer History Museum, Menlo Park, CA. https://www.youtube.com/watch?v=be1EM3gQkAY, accessed February 1, 2018.

Croome, Angela. 1958. "Antarctic Paraphrase." *The New Scientist* 3 (59): 35.

Crosby, Alfred W. 1998. *The Measure of Reality: Quantification and Western Society, 1250–1600*. Cambridge: Cambridge University Press.

Crowley, Roger. 2005. *1453: The Holy War for Constantinople and the Clash of Islam and the West*. New York: Hyperion.

Curley, Robert. 2010. *The 100 Most Influential Inventors of All Time*. New York, NY: Britannica Educational Publication in association with Rosen Educational Services.

de Moivre, Abraham. 1725. *Annuities Upon Lives: or, The Valuation of Annuities Upon any Number of Lives; as also, of Reversions. To which is added, An Appendix concerning the Expectations of Life, and Probabilities of Survivorship*. London: Printed by W.P. and sold by F. Fayram; Benj. Motte; and W. Pearson.

de Moivre, Abraham (1738) 1967. *The Doctrine of Chances: or, A Method of Calculating the Probabilities of Events in Play* (2nd edn), Cass Library of Science Classics. London: Cass. Original edition, 1718.

Devlin, Keith. 2008. *The Unfinished Game: Pascal, Fermat, and the Seventeenth-Century Letter that Made the World Modern*. New York: Basic Books.

Devlin, Keith. 2010. "The Hidden Math behind *Alice in Wonderland*." *Devlin's Angle*, March 2010. http://www.maa.org/external_archive/devlin/devlin_03_10.html, accessed September 3, 2018.

Dodgson, Charles A. 1879. *Euclid and His Modern Rivals*. London: Macmillan and Co.

Doyle, Arthur Conan, Christopher Morley, and Eric Mottram. 1981. *The Penguin Complete Sherlock Holmes*. Harmondsworth: Penguin.

Drake, Stillman. 1957. *Discoveries and Opinions of Galileo*. New York: Doubleday & Company.

Edwards, Jonathan. 1741. *Sinners in the Hands of an Angry God* [microform]. Boston: S. Kneeland & T. Green.

Efron, Bradley. 1998. "R. A. Fisher in the 21st Century: Invited Paper Presented at the 1996: R. A. Fisher Lecture." *Statistical Science* 13(2): 95–122.

Einstein, Albert. 1949. *Autobiographical Notes*. Trans. Paul Arthur Schilpp. Peru, Illinois: Open Court Publishing.

Einstein, Albert. 2015. *Relativity: The Special and the General Theory, 100th Anniversary Edition*. Ed. Hanoch Gutfreund and Jürgen Renn. Princeton, NJ: Princeton University Press.

Euclid. 1491. *Elementa Geometria*. Vicenza: Leonardus (Achates) de Basilea and Gulielmus de Papia.

Euclid, Oliver Byrne, Werner Oechslin, and Petra Lamers-Schütze. 2010. *The First Six Books of the Elements of Euclid*. Köln: Taschen. Original edition, 1491.

Fermat, Pierre de, and Blaise Pascal. 1654. "Fermat and Pascal on Probability." http://www.york.ac.uk/depts/maths/histstat/pascal.pdf, accessed August 17, 2018.

Fisher, Ronald A. 1922a. "On the Mathematical Foundations of Theoretical Statistics." *Philosophical Transactions of the Royal Society of London* 222(594–604): 309-68.

Fisher, Ronald A. 1922b. "The Goodness of Fit of Regression Formulae and the Distribution of Regression Coefficients." *Journal of the Royal Statistical Society* 85(4): 597–612.

Fisher, Ronald A. 1925. *Statistical Methods for Research Workers*. Edinburgh: Oliver and Boyd.

Fisher, Ronald A. 1935. *The Design of Experiments*. Edinburgh: Oliver and Boyd.

Fisher, Ronald A. 1950. *Statistical Methods for Research Workers* (11th edn). Edinburgh: Oliver and Boyd.

Fisher, Ronald A. 1956. "Mathematics of a Lady Tasting Tea." In *The World of Mathematics; A Small Library of the Literature of Mathematics from A'h-mosé the Scribe to Albert Einstein*. Ed. James R. Newman, 1512–21. New York: Simon and Schuster.

Fisher, Ronald A. 1971. *The Design of Experiments* (9th edn). New York: Macmillan. Original edition, 1935.

Foote, Shelby. 1986. *The Civil War: a Narrative*. New York: Vintage Books.

Forbes, C., Evans, M., Hastings, N., & Peacock, B. 2011. *Statistical Distributions* (4th edn). Hoboken, NJ: John Wiley & Sons.

Frank, Adam. 2008. "Is the Universe Actually Made of Math?" *Discover Magazine*, June 16, 2008, http://discovermagazine.com/2008/jul/16-is-the-universe-actually-made-of-math, accessed September 4, 2018.

Franklin, Benjamin (1706–57) 2016. *The Autobiography of Benjamin Franklin*. Edited by Charles Eliot. Urbana, IL: Project Gutenberg. http://www.gutenberg.org/files/148/148-h/148-h.htm (accessed September 5, 2018).

Franklin, Benjamin. 1732. *Poor Richard's Almanack*. New York: Barnes & Noble Books.

Franklin, Benjamin, and Jared Sparks. 1836. *The Works of Benjamin Franklin: Containing Several Political and Historical Tracts Not Included in any Former Edition and Many Letters Official and Orivate, Not Hitherto Published: with Notes and a Life of the Author*. Boston: Hilliard, Gray and Company.

Furuti, Carlos A. 2012. "Useful Map Properties: The Geodesic." http://progonos.com/furuti/MapProj/Normal/CartProp/Geodesic/geodesic.html, accessed September 6, 2018.

Galton, F. 1908. *Memories of My Life*. London: Methuen.

Galton, Francis. 1855. *The Art of Travel: Or, Shifts and Contrivances Available in Wild Countries*. London: J. Murray.

Galton, Francis. 1869. *Hereditary Genius: An Inquiry into its Laws and Consequences*. London: Macmillan.

Galton, Francis. 1875. "Statistics by Intercomparison, with Remarks on the Law of Frequency of Error." *The London, Edinburgh, and Dublin Philosophical Magazine and Journal of Science* 49(322), 33–46.

Galton, Francis. 1877. "Typical Laws of Heredity." *Nature* 15, 492–5, 512–14, 532–3.

Galton, Francis. 1885. "Anthropometric Laboratory: For the Measurement in Various Ways of Human Form and Faculty." http://galton.org/images/anthro-lab-poster.jpg, accessed September 6, 2018.

Galton, Francis. 1886. "Family Likeness in Stature." *Proceedings of the Royal Society of London* 40 (242–5), 42–73.

Galton, Francis. 1889a. "Co-Relations and their Measurement, Chiefly from Anthropometric Data." *Proceedings of the Royal Society of London* 45(273–9), 135–45.

Galton, Francis. 1889b. *Natural Inheritance*. New York: Macmillan.

Galton, Francis. 1907. *Probability, the Foundation of Eugenics: The Herbert Spencer Lecture Delivered on June 5, 1907*. Oxford: Clarendon Press.

Galton, Francis. 2012. *Hereditary Genius: An Inquiry into its Laws and Consequences*. New York, NY: Barnes & Noble Digital Library. Original edition, 1869.

Galton Institute. 2017. "The Art of Travel (Murray, 1855)." Galton Institute. http://galton.org/books/art-of-travel/, accessed September 6, 2018.

Gamwell, Lynn 2015. *Mathematics and Art: A Cultural History*. Princeton: Princeton University Press.

Gauss, Carl Friedrich. 1801. "Disquisitiones Arithmeticae." Ann Arbor, MI: HathiTrust Digital Library. http://catalog.hathitrust.org/Record/012312864, accessed September 6, 2018.

Gauss, Carl Friedrich. 1871. *Theoria motus corporum coelestium in sectionibus conicis solem ambientium, His Werke*. Gotha: F.A. Perthes.

Gauss, Carl Friedrich, and G. W. Stewart. 1995. *Theory of the Combination of Observations Least Subject to Error: Part One, Part Two, Supplement, Classics in Applied Mathematics*. Philadelphia: Society for Industrial and Applied Mathematics.

Gauss, Carl Friedrich, and William C. Waterhouse. 1986. *Disquisitiones arithmeticae* (English edn). New York: Springer-Verlag.

Gauss Society. 2018. "Gauss-Gesellschaft Göttingen." Göttingen: University of Göttingen. http://www.hs.uni-hamburg.de/DE/GNT/w.htm, accessed March 18, 2018.

Gorroochurn, Prakash. 2016a. *Classic Topics on the History of Modern Mathematical Statistics: From Laplace to More Recent Times*. Hoboken, NJ: John Wiley & Sons, Inc.

Gorroochurn, Prakash. 2016b. "On Galton's Change from 'Reversion' to 'Regression'." *The American Statistician* 70(3): 227–31.

Gould, Stephen Jay. 2002. *The Structure of Evolutionary Theory*. Cambridge, MA: Belknap Press of Harvard University Press.

Gowers, Timothy, June Barrow-Green, and Imre Leader. 2008. *The Princeton Companion to Mathematics*. Princeton: Princeton University Press.

Halacy, Daniel Stephen. 1970. *Charles Babbage, Father of the Computer*. New York: Crowell-Collier Press.

Hawking, Stephen. 1988. *A Brief History of Time*. New York: Bantam.

Hawking, Stephen W., Kip Thorne, Igor Novikov, Timothy Ferris, and Alan Lightman. 2003. *The Future of Spacetime*. New York: W. W. Norton & Company.

Herrnstein, R., and C. Murray. 1994. *The Bell Curve*. New York: The Free Press.

Hilbert, David. 1900a. "Hilbert's Mathematical Problems: Table of Contents." http://mathcs.clarku.edu/~djoyce/hilbert/toc.html, accessed February 2, 2018.

Hilbert, David. 1900b. "Mathematical Problems." http://mathcs.clarku.edu/~djoyce/hilbert/problems.html#note1, accessed February 2, 2018.

Howe, Daniel Walker. 2007. *What Hath God Wrought: The Transformation of America, 1815-1848, Oxford History of the United States*. Oxford: Oxford University Press.

IEA International Study Center. 2018. "TIMMS & PIRLS." Boston: Lynch School of Education, Boston College. https://timssandpirls.bc.edu/index.html#, accessed February 4, 2018.

Isaac Newton Institute for Mathematical Sciences. 2018. "Books about Sir Isaac Newton." http://www.newton.ac.uk/facilities/library/books-about-newton, accessed February 20, 2018.

Jeffreys, Harold. 1973. *Scientific Inference* (3rd edn). Cambridge: Cambridge University Press.

Jensen, Arthur R. 2002. "Galton's Legacy to Research on Intelligence." *Journal of Biological Science* 34(2): 145–72.

Jensen, Arthur, R. 2006. *Clocking the Mind: Mental Chronometry and Individual Differences*. Boston: Elsevier.

Kant, Immanuel, and Lewis White Beck. 1995. *Foundations of the Metaphysics of Morals and What is Enlightenment* (2nd edn). The Library of Liberal Arts. Upper Saddle River, NJ: Prentice Hall. Original edition, 1797.

Kelly, Scott. 2017. *Endurance: A Year in Space, A Lifetime of Discovery*. New York: Knopf.

Kennan, George. 1891. *Siberia and the Exile System*. London: James R. Osgood, McIlvaine & Co.

Keynes, John Maynard. (1936) 1973. *The General Theory of Employment, Interest and Money. The Collected Writings of John Maynard Keynes, Volume 7*. London: Macmillan/St. Martin's Press.

Keynes, John Maynard. 1956. "The Application of Probability to Conduct." In *The World of Mathematics: A Small Library of the Literature of Mathematics from A'h-mosé the Scribe to Albert Einstein*. Ed. James R. Newman, 1360-73. New York: Simon and Schuster.

Lansing, Alfred. 2014. *Endurance: Shackleton's Incredible Voyage*. New York: Basic Books. Original edition, 1959.

Laplace, Pierre Simon, and Andrew I. Dale. 1995. *Philosophical Essay on Probabilities, Sources in the History of Mathematics and Physical Sciences*. New York: Springer-Verlag. Original edition, 1843–7.

Laplace, Simon Pierre. 1878–1912. *Oeuvres Complètes de Laplace*. Ed. Gauthier-Villars. Surviving Works of Laplace. Paris: Academy of Sciences of Paris.

Le Corvec, Veronique, Jeffrey Donatelli, and Jeffrey Hunt. 2007. "How Gauss Determined the Orbit of Ceres." http://math.berkeley.edu/~mgu/MA221/Ceres_Presentation.pdf, accessed August 17.

Legendre, A. M. 1805. "Nouvelles méthodes pour la détermination des orbites des comètes." Paris: F. Didot. HathiTrust Digital Library. http://catalog.hathitrust.org/Record/008630090, accessed September 6, 2018.

Legendre, A. M. 1810. *Mémoires sur la méthode des moindres quarrés et sur l'attraction des ellipsoïdes homogènes.* Paris.

Lewis, Meriwether, and William Clark. (1804–5) 2005. "The Journals of Lewis and Clark 1804–1806." Urbana, IL: Project Gutenberg. http://www.gutenberg.org/files/8419/8419-h/8419-h.htm, accessed September 5, 2018.

Linton, Marisa. 2006. "Robespierre and the Terror." http://www.historytoday.com/marisa-linton/robespierre-and-terror, accessed September 5, 2018.

Locke, John, and Peter H. Nidditch. 1987. *An Essay Concerning Human Understanding, The Clarendon Edition of the Works of John Locke.* Oxford: Clarendon Press.

Louisiana, Cammie G. Henry Research Center at Northwestern State University of. 2018. "Carl Fredrich Gauss Papers." http://www.nsula.edu/documentprovider/docs/library/Melrose%20Collection/gausspage2.pdf, accessed March 18, 2018.

Loy, Gareth. 2007. *Musimathics: The Mathematical Foundations of Music.* Vol. 1. Cambridge: MIT Press.

Mack, Maynard. 1985. *Alexander Pope: A Life.* New Haven, CT: Yale University Press.

McGrayne, Sharon Bertsch. 2011. *The Theory That Would Not Die: How Bayes' Rule Cracked the Enigma Code, Hunted Down Russian Submarines, and Emerged Triumphant from Two Centuries of Controversy.* New Haven, CT: Yale University Press.

Miller, G. A. 1984. "The Test." *Science* 84(5): 55–60.

Misner, Charles W., Kip S. Thorne, and John Archibald Wheeler. 1973. *Gravitation.* New York: W. H. Freeman & Company.

Moskowitz, Clara. 2013. "Images: The World's Most Beautiful Equations." http://www.livescience.com/26681-most-beautiful-mathematical-equations.html, accessed September 5, 2018.

National Cancer Institute. 2018. "Cancer Statistics." National Cancer Institute. https://www.cancer.gov/, accessed March 20, 2018.

National Center for Educational Statistics. 2018. "National Assessment of Educational Progress." http://nces.ed.gov/nationsreportcard/, accessed February 2, 2018.

Newman, James R. 1956. *The World of Mathematics: A Small Library of the Literature of Mathematics from A'h-mosé the Scribe to Albert Einstein.* New York: Simon and Schuster.

Newson, Maby Winton. 1902. "Mathematical Problems by David Hilbert." *Bulletin of the American Mathematical Society* 8(10): 437–79.

Newton, Isaac. (1675) 2017. "Isaac Newton Letter to Robert Hooke." http://digitallibrary.hsp.org/index.php/Detail/objects/9792, accessed September 5, 2018).

Newton, Isaac. 1687. *Philosophiæ Naturalis Principia Mathematica*. London: Jussu Societatis Regiæ ac Typis Joseph Streater.

Newton, Isaac, Andrew Motte, and N. W. Chittenden. 1846. *Newton's Principia. The Mathematical Principles of Natural Philosophy* (1st American edn). New York: D. Adee.

Nobel Media AB. 2018. "The Nobel Prize in Physics 2017." http://www.nobelprize.org/prizes/physics/2017/prize-announcement/>, accessed September 3, 2018.

Oates, J. C. T. 1986. *Cambridge University Library: A History from the Beginnings to the Copyright Act of Queen Anne*. Cambridge: Cambridge University Press.

Osterlind, S. J. 2010. *Modern Measurement: Theory, Principles, and Applications of Mental Appraisal* (2nd edn). Boston: Pearson.

Pais, Abraham. 1982. *Subtle is the Lord: The Science and the Life of Albert Einstein*. Oxford: Oxford University Press.

Pais, Abraham. 1994. *Einstein Lived Here*. Oxford: Clarendon Press.

Pasles, Paul. 2007. *Benjamin Franklin's Numbers: An Unsung Mathematical Odyssey*. Princeton, NJ: Princeton University Press.

PBS. 2002. "Tending Sir Ernest's Legacy: An Interview with Alexandra Shackleton." http://www.pbs.org/wgbh/nova/shackleton/1914/alexandra.html, accessed September 6, 2018.

Pearson, F. S. 1938. *Karl Pearson: An Appreciation of Some Aspects of His Life and Work*. Cambridge: Cambridge Press.

Pearson, Karl. (1892) 2004. *The Grammar of Science*. London: Dover Publications.

Pearson, Karl. 1897. *The Chances of Death and Other Studies in Evolution*, Vols I & II. London: E. Arnold.

Pearson, Karl. 1900a. "On the Criterion That a Given System of Deviations from the Probable in the Case of a Correlated System of Variables Is Such That It Can Be Reasonably Supposed To Have Arisen from Random Sampling." *Philosophical Magazine Series 5* 50(302): 157–75.

Pearson, Karl. 1900b. *The Grammar of Science* (2nd edn). London: Adam and Charles Black.

Pearson, Karl. 1905. *On the General Theory of Skew Correlation and Non-Linear Regression*. London: Dulau and Co.

Pearson, Karl. 1907. *On Further Methods of Determining Correlation*. London: Dulau and Co.

Pearson, Karl, and E. S. Pearson. 1978. *The History of Statistics in the 17th and 18th Centuries Against the Changing Background of Intellectual, Scientific, and Religious Thought: Lectures by Karl Pearson Given at University College, London, During the Academic Sessions, 1921–1933*. London: C. Griffin.

Plato. (360–347 BCE) 2013. *Philebus*. Trans. Benjamin Jowett. Urbana, IL: Project Gutenberg. http://www.gutenberg.org/files/1744/1744-h/1744-h.htm, accessed September 4, 2018.

Poisson, Siméon-Denis, and O. B. Sheĭnin. 2013. *Researches into the Probabilities of Judgements in Criminal and Civil Cases: "Recherches sur la probabilité des jugements en matière criminelle et en matière civile."* Berlin: NG-Verlag (Viatcheslav Demidov Inhaber).

Porter, Theodore M. 1986. *The Rise of Statistical Thinking, 1820–1900*. Princeton, NJ: Princeton University Press.

Quetelet, Adolphe. 1968. *A Treatise on Man and the Development of His Faculties, Burt Franklin Philosophy Monograph Series*. New York: B. Franklin.

Quetelet, Adolphe. 1996. *Instructions populaires sur le calcul des probabilités*. Roma: Istituto nazionale di statistica.

Quetelet, Adolphe, and Richard Beamish. 1839. *Popular Instructions on the Calculation of Probabilities*. London: J. Weale.

Richter, Jean Paul. 1970. *The Notebooks of Leonardo da Vinci*. New York: Dover Publications.

Rieber, Robert W., and David K. Robinson, eds. 2001. *Wilhelm Wundt in History: The Making of a Scientific Psychology*. New York: Kluwer Academic/Plenum Publishers.

Royal Geographical Society. 2018. *The Royal Geographical Society (with IBG)*. http://www.rgs.org/HomePage.htm, accessed September 3, 2018.

Schocken, Wolfgang Alexander. 2015. *Mathematics of the Jewish Calendar/Gauss' Formula for the Date of Pesach*. Wikibooks, The Free Textbook Project. http://en.wikibooks.org/w/index.php?title=Mathematics_of_the_Jewish_Calendar/Gauss%27_Formula_for_the_Date_of_Pesach&oldid=3016598, accessed September 6, 2018.

Shackleton, Ernest. *1919. South: The Story of Shackleton's Last Expedition 1914–1917*. New York: The Macmillan Company.

Shakespeare, William, and A. L. Rowse. 1978. *The Annotated Shakespeare*. New York: C. N. Potter.

Silver, Brian L. 1998. *The Ascent of Science*. New York: Oxford University Press.

Stein, Mark. 2008. *How the States Got Their Shape*. New York: Harper Collins.

Stigler, Stephen M. 1980. "Stigler's Law of Eponymy." *Transactions of the New York Academy of Sciences* 39(1): 147–58.

Stigler, Stephen M. 1986. *The History of Statistics: The Measurement of Uncertainty Before 1900*. Cambridge, MA: Belknap Press.

Stigler, Stephen M. 1989. "Francis Galton's Account of the Invention of Correlation." *Statistical Science* 4(2): 73–86.

Student (William S. Gosset). 1908. "The Probable Error of a Mean." *Biometrika* 6(1): 1–25.

Teets, D., and Whitehead, K. 1999. "The Discovery of Ceres: How Gauss Became Famous." *Mathematics Magazine* 72(2): 83–91.

Tegmark, Max. 2014. *Our Mathematical Universe: My Quest for the Ultimate Nature of Reality*. New York: Knopf.

Tegmark, Max. 2017. *Life 3.0: Being Human in the Age of Artificial Intelligence.* New York: Knopf.

TFE Times. 2017. "The General Theory of Employment, Interest and Money." *TFE Times.* http://tfetimes.com/10-greatest-economics-books-of-all-time/6/, accessed September 6, 2018.

The Newton Project. 2011. "Draft of the "Hypothesis concerning Light and Colors." http://www.newtonproject.ox.ac.uk/view/texts/normalized/NATP00121, accessed September 5, 2018.

Thompson, B. 1981. *The History of Evolutionary Thought.* Fort Worth: Star Bible & Tract Corp.

Thompson, Clive. 2017. "Stereographs Were the Original Virtual Reality." *Smithsonian Magazine,* October 2017. https://www.smithsonianmag.com/innovation/sterographs-original-virtual-reality-180964771/, accessed September 6, 2018.

Thorne, Kip. 1994. *Black Holes and Time Warps: Einstein's Outrageous Legacy.* New York: W. W. Norton & Company.

Thurstone, L. L. 1925. "A Method of Scaling Psychological and Educational Tests." *Journal of Educational Psychology* 16(7): 412–33.

Thurstone, L. L. 1927. "A Law of Comparative Judgment." *Psychological Review* 34(4): 273–386.

Thurstone, L. L. 1932. *The Theory of Multiple Factors.* Ann Arbor, MI: Edwards Brothers.

Todhunter, Issac. 1865. *A History of the Mathematical Theory of Probability: From the Time of Pascal to that of Laplace.* London: Macmillan.

Tolstoy, Leo, and Michael R. Katz. 2008. *Tolstoy's Short Fiction: Revised Translations, Backgrounds and Sources, Criticism* (2nd edn). New York: Norton & Co. Original edition, 1889. Reprint, Geneve edition.

Tomatis, Alfred. 1991. *Pourquoi Mozart?* Paris: Fixot, Diffusion, Hachette.

US Court. 1803. Marbury v. Madison. Washington, DC: United States Supreme Court.

UDHR. 1948. "The Universal Declaration of Human Rights (UDHR). Resolution 217. Paris: United Nations General Assembly.

UNICEF. 2016. *The State of the World's Children 2016: A Fair Chance for Every Child.* Washington, DC: United Nations.

United Nations Development Programme. 2015. *Human Development Report, 2015.* New York: United Nations.

University of Massachusetts. 2017. "The Birthday Game." http://www.umass.edu/wsp/resources/poisson/birthday.html, accessed September 1, 2018.

UNODC. 2014. *Global Report on Trafficking in Persons.* Washington, DC: United Nations, Office on Drugs and Crime.

Van der Post, Laurens. 1958. *The Lost World of the Kalahari.* London: Hogarth Press.

Wikisource. 2018. "1911 Encyclopædia Britannica/Poisson, Siméon Denis." http://en.wikisource.org/wiki/1911_Encyclop%C3%A6dia_Britannica/Poisson,_Sim%C3%A9on_Denis, accessed September 1, 2018.

Will, G. F. 1990. *Men at Work*. New York: Harper Perennial.

Wordie, Sir James. 1957. "Exploration's Elder Statesman." *The New Scientist*, December 3, 1957, 25.

Wundt, Wilhelm Max. 1862. "*Die Geschwindigkeit des Gedankens* [The Velocity of Thought]." *Gartenlaube* 17: 263–5.

Wundt, Wilhelm Max. 1904. *Principles of Physiological Psychology*. Trans. Edward Bradford Titchener. London: Swan Sonnenschein & Co., Lim. Original edition, 1874.

INDEX

A Method of Scaling Psychological and Educational Tests 298
Achenwall, Gottfried 36
AI. *See* artificial intelligence
Akademie Olympia 235, 276
algebra 1–12, 33–6, 40–8, 73, 75, 109–22, 143, 147, 151, 167, 245, 247
An Essay Toward Solving a Problem in the Doctrine of Chances 89
analysis of variance 176, 257–60
Annuities Upon Lives 74
ANOVA. *See* analysis of variance
Aristotle 24, 80, 174
Arithmetic Machine. *See* Pascaline
arithmetic mean 24–7, 52–60, 70–6, 87, 105, 113, 156–7, 170–6, 203, 213–21, 244, 250–2, 256, 259, 307
Army Alpha and Beta tests 297
Ars Conjectandi 55, 70
artificial intelligence 3, 312–14
Autobiographical Notes 273
average [statistical]. *See* arithmetic mean
average man 171–5, 205, 237

Babbage, Charles 188–92, 205
Bacon, Francis 18, 91
Bayes, Thomas 5, 77, 85–98, 119, 128, 136, 161, 216, 283
Bayesian 3, 17, 86–99, 128, 136, 140, 161, 313
bell-shaped curve 17, 34, 58–9, 72–80, 91, 150–60, 171, 193, 216–18, 251, 288, 292

Bernoulli, Jacob (or Jacques) 5, 53–8, 60–1, 70–80, 108, 119, 132, 136, 167
Binet, Alfred 250, 290, 294–7
binomial 36–7, 47–51, 60, 69, 71–6, 81, 89–91, 115, 193, 196–7, 200, 213, 216, 242, 260
[*Also* binomial distribution]
[*Also* binomial expansions]
[*Also* binomial theorem]
black holes 125–6, 129, 279, 283
BMI. *See* Body Mass Index
Body Mass Index 174
Bonaparte, Napoleon, *See* Napoleon
Bostrom, Nick 314
Buffon's needle problem 179

calculus 12–17, 27–9, 33–47, 54–7, 60–3, 70, 74–5, 85–9, 106, 109, 116–17, 122–31, 143–56, 158, 180, 194, 201–16, 246–7, 272, 279, 305
calculus of variations 122–3, 125
Cardano, Girolamo 38–9
Carroll, Lewis. *See* Dodgson
Cattell, James 250, 288, 290, 293–6
central limit theorem 47, 51, 58–60, 69–76, 87–90, 150, 170, 200–1, 213–16, 221, 251, 305
Chesterton, G. K. 24
chi-square 72, 241–4, 257
Circle of Fifths 309–11
combination of observations. *See* observation
conditional probability 92–6
Congress of Vienna 15, 102, 163, 317

Copernicus 9–10, 24, 33
correlational relationship. *See* correlation
correlation 2, 17, 92, 109–16, 147, 209, 220–5, 236–41, 290, 319–20
Curie, Marie 265, 275, 283–4

d'Alembert, Jean le Rond 131–3
da Vinci, Leonardo 37, 305–8
Darwin, Charles 28, 179, 190, 206–23, 240–1, 265–7
de Buffon, Comte 179–80
De Devina Proportione 37, 306
de Moivre, Abraham 69–89, 108, 132, 154, 167
Decembrist Revolution 184–6
De Ratiociniis in Ludo Aleae 39
degrees of freedom 255, 259
Delambre, Joseph 108
density function 17, 60, 73–80, 151–8, 205
Descartes, Rene 18, 61, 63, 77–8, 318
Difference Engine 189–93, 205, 305, 313
Disquisitiones arithmeticae 106, 147
Dodgson, Charles 35, 246–7, 249

Edwards, Jonathan 84–5, 88, 176
Einstein, Albert 5–6, 17, 29, 126–46, 235–47, 265, 273–83, 305
Elements of Geometry 36, 40, 147, 247
Encyclopedia 132
Enlightenment 4, 8, 15, 18–19, 45, 67, 84, 103, 119

error [statistical]. *See* measurement error
Essai philosophique sur les probabilités 137
Essay Concerning Human Understanding 19
Essay on Man 78
Euler, Leonhard 26–8, 40, 93, 122–3, 131, 143, 167, 261, 284
Everyman 119–21, 134, 169, 178, 203, 311

F ratio 156, 255–9
factor [statistical] 257, 298
Federalist Papers 185
Fermat, Pierre 38–9, 43, 96, 128
Fields Medal 148, 284
Fisher, Ronald 3, 5, 36, 51, 220–4, 252, 254–63, 288, 298
Fourier, Jean-Baptiste 127, 194–5
Franklin, Benjamin 5, 63–5, 67, 80, 131
French Revolution 4, 15, 18, 46, 102–3, 122, 133–5, 163, 317
frequentist 87

Galileo 18, 24, 38, 75, 124–5, 160, 224, 299, 315–19
Galton, Francis 5, 53, 177–9, 208–27, 236–41, 252–67, 290–6, 300
gambler's fallacy 57–8
Gates, Bill 90, 98, 191, 314
Gauss, Carl (or Fredrich) 5, 26–7, 60, 73, 88, 104–19, 130–6, 143–61, 167, 189, 193, 205, 213–16, 244–7, 271, 283
Gaussian quadrature. *See* Gauss
general linear model 12
General Principles for the Calculus of Probabilities 139
geometry 12, 31–40, 62, 89, 122–6, 143, 147, 237, 245–51, 272, 284, 304, 307
Goethe 19
Gosset, William 250–4, 258, 288
Grammar of Science 235–8, 276
Graunt, John 39
Gravitation 279, 281
Great Lisbon Earthquake 6–8, 18

Hawking, Stephen 29, 188, 279–80, 283
Helmert, Friedrich 244

Hereditary Genius 177, 209, 212–13
hierarchical linear modeling 161, 260
Hilbert, David 271–3
Histoire naturelle générale et particulière 179
HLM. *See* hierarchical linear modeling
Hubble, Edwin 207, 280
Hume, David 8, 67, 185

Industrial Revolution 4, 18, 21, 119, 203–4, 228, 232–5
Instructions populaires sur le calcul des probability 175
International Health Exhibition 217, 220
IQ testing 79, 213, 250, 295–8
IRT. *See* item response theory
item response theory 299–301

Jensen, Arthur 212–13, 219, 293, 295
Johnson, Andrew 184
Johnson, Samuel 5, 23–8, 31, 65, 80, 206
joint probability 92

Kant, Immanuel 18, 240
Kepler, Johannes 19

l'homme moyen. See average man
Laplace, Pierre-Simon 5, 13, 28, 54–60, 72–7, 87, 98, 121–41, 148–9, 154, 160, 169–73, 178, 182, 189, 193–6, 200–13, 237, 283
Latin square 3, 261
law of averages 57–8
law of large numbers 47, 51–8, 60, 69–70, 72, 87–90, 200–1, 242, 299
least squares [method of] 12, 17, 26–7, 101–22, 130, 145–58, 205, 244, 319
Legendre, Adrien-Marie 5, 104–9, 111, 115, 125, 130, 136, 147–52, 193–5, 205
Leibniz, Gottfried 19, 28, 36, 55, 104, 318
likelihood [statistical] 12, 80–2, 150, 257, 300
line of best fit. *See* regression

location [statistical]. *See* arithmetic mean
Locke, John 8, 19, 67, 84, 185, 240
long century 15–16, 19–20, 102, 317
Lovelace, Ada 189, 305
Lucasian Professor of Mathematics 29, 188–90

maximum likelihood estimation 82
Mayer, Tobias 24–7, 39, 54, 91, 105, 173, 208
mean. *See* arithmetic mean
measurement error 24–5, 39, 52–60, 72, 75–6, 170–6, 213–16, 242, 252–9, 299–300
methodology 12, 24, 26, 28, 44–6, 108, 161, 171–3, 193, 200, 207, 236–7, 298, 318
Microsoft. *See* Gates
Mirzakhani, Maryam 284
MLE. *See* maximum likelihood estimation
Monte Carlo fallacy 58
Monty Hall problem 96–7
Mozart, Wolfgang Amadeus 5, 65–6, 80, 145, 165, 315
multinomial distribution 72

Napoleon 6, 15, 18, 48, 68, 102, 134–5, 148–9, 163–88, 195, 229, 317
naturalistic inquiry 28, 207
Newton, Isaac 6, 17–18, 27–33, 40–55, 69–88, 104, 108, 122–39, 141–8, 167, 188, 205–6, 237, 247, 272–84, 305, 318
Newton's Chair. *See* Lucasian Professor of Mathematics
normal curve 73, 155, 157–8, 160, 170, 216, 288
Nouvelles méthodes pour la détermination des orbites des comètes 105
NSD. *See* statistical significance
number theory 27, 36, 143, 147, 247

observation 2, 23–43, 51–6, 65–91, 105–20, 151–3, 173–5, 199, 207, 210, 218–35, 281–3, 291, 299, 320

observational research.
    See observation
odds 1–4, 12–15, 39, 70, 77–82, 94–8, 149, 197, 199
OLS. See least squares
On the Origin of Species 28, 206
order of operations 27, 93
orthodrome problem 123–5, 245, 282

Pacioli, Luca 37, 307–8
Pascal, Blaise 28–9, 36–46, 50–1, 65, 70, 73, 104, 128, 145, 319
Pascaline 36–7, 319
Pasteur, Louis 10, 21
Pearson, Karl 5, 158, 213, 217, 223–7, 235–44, 252–6, 265, 276
Pensées 39
Peter the Great 6, 46
Philosophiæ Naturalis Principia Mathematica 17–18, 29, 31–2, 45–6, 274–7
Piazzi, Giuseppe 152
Poisson, Simon 72, 178, 193–201, 209
Poor Richard's Almanack 63, 65
Pope, Alexander 29, 78
population [statistical] 51–60, 70, 75–6, 82, 87, 152–7, 239–42, 258–9
Principia. See *Philosophiæ Naturalis Principia Mathematica*
Principles of Physiological Psychology 291
prior [statistical] 86–9, 91, 93–4, 161
probability theory. See probability
probability 2–17, 33–43, 59, 68–89, 102–7, 119–48, 163, 170–1, 181–8, 201, 207–25, 245, 250, 287, 299, 317–20
problem of points 37–8, 43
psychometrics 209, 220–7, 245, 285–303
Pythagorean theorem 89, 207, 305, 311

quantification 1–2, 4–5, 10–11, 22, 56–7, 80, 164, 201, 208
quantitative thinking.
    See quantification

Queen Victoria 6, 46, 217
Quetelet, Adolphe 169–79, 190, 193, 205–10, 213, 237, 283

random variable 87, 150, 155, 201
    [*Also* discrete random variable]
    [*Also* continuous random variable]
Recherches sur la probabilité 196, 199–200
Reformation 9, 46
regression 12, 17, 104, 110–17, 139, 151, 161, 205–25, 257–60, 319
    regression to the mean 53, 203–25, 290
    reversion to the mean 53, 209, 221
relativity [theory of] 29, 126, 237, 265, 275–81, 305
Renaissance 21, 39, 46, 307
research methodology.
    See methodology
Roget, Peter 168
Romanov 20, 46, 184, 186
Royal Geographical Society 211–12, 266–7

sample(ing) [statistical] 51–60, 70–6, 81, 87, 150–5, 173–6, 197, 212–21, 239–44, 251–9
sample space 150–1, 155
scale [statistical] 7–8, 20, 72, 86–90, 110–15, 132, 150–68, 176, 189, 198, 201, 238, 288–300, 309
Simon, Theodore 250, 295–7
spacetime continuum 281
St. Petersburg Paradox 57
standard deviation 60, 75–6, 87, 156–7, 171–6, 239–42, 251, 259, 288
Standards for Educational and Psychological Testing 287
Statistical Methods for Research Workers 257
statistical significance 156
Student. See Gosset
Sur l'homme et le développement de ses facultés 172
Sur le Calcul integral aux differences infiniment petites et aux differences finies 131

Tartaglia, Niccolo 38
t-distribution 251–2, 254–5
Tegmark, Max 314–16
Terman, Lewis 250, 297
The Bell Curve 79
The Chances of Death and Other Studies in Evolution 241, 243
The Descent of Man 28
The Design of Experiments 261–2
The Doctrine of Chances 73, 76, 89
The General Theory of Employment, Interest and Money 178
The Gleaners 44
The Grammar of Science 235–8
The History of Statistics 13
the lady tasting tea 255, 261–2
the unfinished game. See problem of points
The Vanity of Human Wishes 23
The Weimar Court of the Muses 18–19
Theoria motus corporum celestium 147
Théorie analytique des probabilités 136
theory of relativity 29, 126, 237, 277–80
theory of error. See measurement error
Theory of the Combination of Observations Least Subject to Error 152
Thorne, Kip 126, 279, 281–3
thought experiment 319–21
Thurstone, L. L. 250, 298
Traité de mécanique céleste 125–6, 135

Voltaire (*nom de plume* for François-Marie Arouet) 18, 65, 132, 179

Wechsler, David 250
Weltanschauung. See worldview
Westward the Course of Empire Takes Its Way 183–4
Will, George 231
worldview 1–33, 40–80, 90, 101–21, 141, 160–8, 181, 203–8, 227–37, 265–73, 280, 290, 303–21